Materials Physics in Thermoelectric Materials

Materials Physics in Thermoelectric Materials

Bao-Tian Wang
Peng-Fei Liu

Basel • Beijing • Wuhan • Barcelona • Belgrade • Novi Sad • Cluj • Manchester

Bao-Tian Wang
Institute of High Energy Physics
Chinese Academy of Science
Beijing
China

Peng-Fei Liu
Institute of High Energy Physics
Chinese Academy of Science
Beijing
China

Editorial Office
MDPI AG
Grosspeteranlage 5
4052 Basel, Switzerland

This is a reprint of articles from the Special Issue published online in the open access journal *Materials* (ISSN 1996-1944) (available at: www.mdpi.com/journal/materials/special_issues/materials_physics_thermoelectric).

For citation purposes, cite each article independently as indicated on the article page online and using the guide below:

Lastname, A.A.; Lastname, B.B. Article Title. *Journal Name* **Year**, *Volume Number*, Page Range.

ISBN 978-3-7258-2128-0 (Hbk)
ISBN 978-3-7258-2127-3 (PDF)
https://doi.org/10.3390/books978-3-7258-2127-3

© 2024 by the authors. Articles in this book are Open Access and distributed under the Creative Commons Attribution (CC BY) license. The book as a whole is distributed by MDPI under the terms and conditions of the Creative Commons Attribution-NonCommercial-NoDerivs (CC BY-NC-ND) license.

Contents

About the Editors vii

Preface ix

Ying Fang and Hezhu Shao
Wenzhou TE: A First-Principle-Calculated Thermoelectric Materials Database
Reprinted from: *Materials* **2024**, *17*, 2200, doi:10.3390/ma17102200 1

Junfeng Jin, Fang Lv, Wei Cao and Ziyu Wang
First-Principles Study of Doped $CdX(X = Te, Se)$ Compounds: Enhancing Thermoelectric Properties
Reprinted from: *Materials* **2024**, *17*, 1797, doi:10.3390/ma17081797 14

Ruipeng Zhang, Jianbiao Kong, Yangbo Hou, Linghao Zhao, Junliang Zhu and Changcun Li et al.
Enhanced Thermoelectric Properties of Nb-Doped Ti(FeCoNi)Sb Pseudo-Ternary Half-Heusler Alloys Prepared Using the Microwave Method
Reprinted from: *Materials* **2023**, *16*, 5528, doi:10.3390/ma16165528 26

Wenying Wang, Lin Bo, Junliang Zhu and Degang Zhao
Copper-Based Diamond-like Thermoelectric Compounds: Looking Back and Stepping Forward
Reprinted from: *Materials* **2023**, *16*, 3512, doi:10.3390/ma16093512 38

Dmitry Pshenay-Severin, Satya Narayan Guin, Petr Konstantinov, Sergey Novikov, Ekashmi Rathore and Kanishka Biswas et al.
Band Structure, Phonon Spectrum and Thermoelectric Properties of Ag_3CuS_2
Reprinted from: *Materials* **2023**, *16*, 1130, doi:10.3390/ma16031130 63

Sopheap Sam, Soma Odagawa, Hiroshi Nakatsugawa and Yoichi Okamoto
Effect of Ni Substitution on Thermoelectric Properties of Bulk β-$Fe_{1-x}Ni_xSi_2$ ($0 \leq x \leq 0.03$)
Reprinted from: *Materials* **2023**, *16*, 927, doi:10.3390/ma16030927 80

Rodrigo Coelho, Yassine De Abreu, Francisco Carvalho, Elsa Branco Lopes and António Pereira Gonçalves
An Electrical Contacts Study for Tetrahedrite-Based Thermoelectric Generators
Reprinted from: *Materials* **2022**, *15*, 6698, doi:10.3390/ma15196698 97

Wenyu Fang, Yue Chen, Kuan Kuang and Mingkai Li
Excellent Thermoelectric Performance of 2D $CuMN_2$ (M = Sb, Bi; N = S, Se) at Room Temperature
Reprinted from: *Materials* **2022**, *15*, 6700, doi:10.3390/ma15196700 117

Haishan Shen, In-Yea Kim, Jea-Hong Lim, Hong-Baek Cho and Yong-Ho Choa
Microstructure Evolution in Plastic Deformed Bismuth Telluride for the Enhancement of Thermoelectric Properties
Reprinted from: *Materials* **2022**, *15*, 4204, doi:10.3390/ma15124204 130

Xinying Ruan, Rui Xiong, Zhou Cui, Cuilian Wen, Jiang-Jiang Ma and Bao-Tian Wang et al.
Strain-Enhanced Thermoelectric Performance in GeS_2 Monolayer
Reprinted from: *Materials* **2022**, *15*, 4016, doi:10.3390/ma15114016 141

Qing-Yu Xie, Peng-Fei Liu, Jiang-Jiang Ma, Fang-Guang Kuang, Kai-Wang Zhang and Bao-Tian Wang
Monolayer SnI_2: An Excellent p-Type Thermoelectric Material with Ultralow Lattice Thermal Conductivity
Reprinted from: *Materials* **2022**, *15*, 3147, doi:10.3390/ma15093147 . . . . **154**

About the Editors

Bao-Tian Wang

Prof. Bao-Tian Wang received his Ph.D. degree in theoretical physics from the Institute of Theoretical Physics at Shanxi University in 2011 and worked as a postdoctoral researcher at Uppsala University of Sweden in 2011–2013 and also at Binghamton University-SUNY of the USA in 2014–2015. He joined the China Spallation Neutron Source (CSNS), Institute of High Energy Physics, CAS, since 2016 and was appointed as an associate professor. His research focuses on thermoelectric materials, actinides, superconductors, and topological systems using first-principles calculations and neutron scattering experiments. Prof. Wang has published more than 160 SCI papers with 4000+ citations, and his H-index is 36 (Google Scholar).

Peng-Fei Liu

Dr. Peng-Fei Liu received his Ph.D. degree in materials physics and chemistry from the Fujian Institute of Research on the Structure of Matter, Chinese Academy of Sciences (CAS), in 2017. He joined the China Spallation Neutron Source (CSNS), Institute of High Energy Physics, CAS, since 2017 and worked as an assistant professor. His research focuses on using first-principles calculations, neutron scattering, and synchrotron X-ray techniques to study the interactions of electrons, phonons, and orbitals in materials and the novel physical phenomena resulting from these interactions, as well as the impact of these effects on various properties of functional materials. He has published more than 160 SCI papers in top international journals such as *Nature Communications*, *National Science Review*, *Applied Physics Reviews*, and *Physical Review B*, with 3600+ citations and an H-index of 37 (Google Scholar).

Preface

Thermoelectric materials, which could directly convert a temperature gradient into electrical energy, provide a promising solution for sustainable energy harvesting. The development of thermoelectric materials has recently gained tremendous attention in the fields of solid-state physics, chemistry, materials science, and engineering. Many strategies have been implemented to achieve high-efficiency thermoelectric conversion efficiency, e.g., doping, defect, intercalation, band engineering, strain, nanostructures, and molecule junctions, which greatly promote further applications of thermoelectrics.

This Special Issue on "Materials Physics in Thermoelectric Materials" aims to provide a unique international forum for researchers working in thermoelectric materials to report their latest endeavors in advancing this field, including new pristine thermoelectric materials, strategies used to improve thermoelectric performance, theoretical understanding of thermoelectrics, physical insights into engineering high-performance thermoelectrics, computational discovery of new thermoelectric materials, and so on.

Bao-Tian Wang and Peng-Fei Liu
Editors

Article

Wenzhou TE: A First-Principle-Calculated Thermoelectric Materials Database

Ying Fang and Hezhu Shao *

School of Electrical and Electronic Engineering, Wenzhou University, Wenzhou 325035, China
* Correspondence: hzshao@wzu.edu.cn

Abstract: Since the implementation of the Materials Genome Project by the Obama administration in the United States, the development of various computational materials' databases has fundamentally expanded the choice of industries such as materials and energy. In the field of thermoelectric materials, the thermoelectric figure of merit (ZT) quantifies the performance of the material. From the viewpoint of calculations for vast materials, the ZT values are not easily obtained due to their computational complexity. Here, we show how to build a database of thermoelectric materials based on first-principle calculations for the electronic and heat transport of materials. Firstly, the initial structures are classified according to the values of bandgap and other basic properties using the clustering algorithm K-means in machine learning, and high-throughput first principle calculations are carried out for narrow-bandgap semiconductors which exhibit a potential thermoelectric application. The present framework of calculations mainly includes a deformation potential module, an electrical transport performance module, a mechanical and a thermodynamic properties module. We have also set up a search webpage for the calculated database of thermoelectric materials, providing search facilities and the ability to view the related physical properties of materials. Our work may inspire the construction of more computational databases of first-principle thermoelectric materials and accelerate research progress in the field of thermoelectrics.

Keywords: thermoelectric materials; material databases; high-throughput computing

Citation: Fang, Y.; Shao, H. Wenzhou TE: A First-Principle-Calculated Thermoelectric Materials Database. *Materials* **2024**, *17*, 2200. https://doi.org/10.3390/ma17102200

Academic Editors: Bao-Tian Wang and Peng-Fei Liu

Received: 3 April 2024
Revised: 2 May 2024
Accepted: 5 May 2024
Published: 8 May 2024

Copyright: © 2024 by the authors. Licensee MDPI, Basel, Switzerland. This article is an open access article distributed under the terms and conditions of the Creative Commons Attribution (CC BY) license (https://creativecommons.org/licenses/by/4.0/).

1. Introduction

In 2011, the Obama administration of the United States officially proposed the "Material Genome Project", which utilized high-throughput computing and experiments to obtain massive quantities of material data, combined with data analysis technology by artificial intelligence for new material development. The goal was to shorten the cycle of new material development and applications, as well as reduce the costs for material research and development, so that the United States could continue to maintain its leading position in manufacturing technology. In 2016, the US government released the "First Five Years of the Materials Genome Initiative: Accommodations and Technical Highlights" report, which pointed out that during the five years of the implementation of the Materials Genome Engineering program, federal research institutions such as the Department of Energy, the Department of Defense, the Natural Science Foundation, the National Bureau of Standards and Technology, and the National Aeronautics and Space Administration had invested over USD 500 million, establishing computational material research and development centers including the National Network for Virtual High throughput Preparation (NIST&NREL) and the Center for Cross Scale Material Design and Multi Scale Materials Research (NIST, ANL, ARL), forming three major computational material databases: the Materials Project (MP) [1], AFLOW [2], and OQMD [3,4]; several auxiliary databases such as the Materials Data Repository (MDR), the Materials Resource Registry, and the Energy Materials Network, as well as database-related analytical tools.

Shortly after the proposal of the Materials Genome Project by the United States, the European Science Foundation launched the Accelerated Metallurgy (ACCMET) program, which cost over EUR 2 billion, with the aim of keeping up with the pace of the United States. The European Commission funded the Horizon 2020 project NoMaD, led by the Max Planck Institute in Germany, for a period of three years in 2015. The project aimed to use the "centralized data warehouse" method to involve various research groups and provide data related to computational materials science, with the aim of building a "Encyclopedia of Materials" and a tool for analyzing big data on materials. In the UK, the government has also implemented an e-science program, with funding to carry out high-throughput material computing simulations and the construction of material computing basic databases, such as eMinerals and the "Material Grid" project. The Swiss EPFL University has led the development of the European materials database, AiiDA [5].

Nowadays, with the vigorous development of big data and artificial intelligence technology, the material genome project research characterized by high-throughput experiments, high-throughput computing, and artificial intelligence big data analysis is in full swing, and has shown astonishing advantages in many materials' fields. The paper "Machine-learning-assisted materials discovery using failed experiments", published in Nature in May 2016 [6], showed that based on years of accumulated experimental data, various catalytic new materials could be discovered using artificial intelligence (AI) technology. This work indicated that AI will profoundly transform the research methods in the field of materials. The centuries long history of human scientific development has formed three research paradigms: experimental, theoretical, and computational. However, in the fields of complex systems such as biology, astronomy, and materials, there are very complex interactions involved, coupled with a large number of variables, which greatly limits the effectiveness of theoretical and computational research models and requires the combination of big data and AI as the "fourth paradigm". In 2017, AlphaGo defeated the human Go master of the boardgame, but Google disbanded the DeepMind team responsible for developing the program, and then formed an AI research and development team engaged in material genome engineering. At present, American high-tech companies including Apple, Google, IBM, Tesla, etc., are all laying out the use of AI for the research and development of new materials based on material genomic methods. The fourth paradigm of materials science requires the ability to generate and process massive amounts of data, thus obtaining massive amounts of material data has become a key aspect of the Materials Genome Project. With the improvement of computing power, the accumulation of material data based on high-throughput computing is receiving more and more attention, and its application in the research and development of new thermoelectric materials is expected to greatly accelerate its application process.

The performance of thermoelectric materials is described by the figure of merit (ZT), which can be expressed as follows:

$$ZT = \frac{S^2 \sigma T}{\kappa_e + \kappa_l} \tag{1}$$

where S is the Seebeck coefficient; σ is the conductivity; T is the temperature, κ_e; and κ_l is the thermal conductivity contributed by carriers and phonons, respectively. These parameters of S, σ, and κ are coupled with each other, and it is difficult to independently regulate them. For example, for semiconductor materials, increasing doping concentration can increase conductivity, while at the same time reducing the Seebeck coefficient and increasing the carrier thermal conductivity. At present, the three major material databases, Materials Project, AFLOW, and OQMD, have data on several common physical quantities, including atomic and band structure, and other physical properties are also being added. However, the thermoelectric performance of materials, due to their particularity and the complexity in calculating electrical and thermal transport properties, generally require a large amount of computation.

Here, we have selected the Materials Project as the structural source for constructing a thermoelectric material database. Specifically, we employed the atomic structure files POSCAR and CIF (currently 19952 materials) in MP materials with an id number below

100,000 through the Materials Project API as the initial materials for building the present thermoelectric material database—**Wenzhou TE**. We built deformation potential modules, mechanical properties (elastic properties) modules, and electronic transport using BoltzTraP modules. And then, we collected data via Python scripts and displayed them on a web site, https://hezhu2024.github.io (accessed on 3 April 2024), for others to use.

2. Methodology

2.1. Clustering (K-Means)

At present, the excellent thermoelectric materials obtained in experiments are mainly semiconductors with narrow bandgaps; then, we chose the bandgap as a major feature for material screening. At the same time, we selected free energy, volume, density, and average atomic energy as the other features from the descriptors obtained from the MP database. They form the five featured variables for the K-means clustering algorithm.

Here is a brief introduction to the K-means principle [7]. K-means clustering is a non-hierarchical supervised pattern recognition method where a pre-defined numbers of clusters are formed. It divides the data into K classes. Firstly, K class random points are randomly generated, denoted as $O_1, O_2, \cdots O_l, \cdots O_K$. Assuming that the j-th feature of the i-th data is represented as x_{ij}, the distance from the i-th data sample to the l-th class random point is:

$$d_{il} = \sqrt{\sum_{j=0}^{j=J}(x_{ij} - O_{lj})^2} \quad (2)$$

among them, J represents a total of J features in the data. The random class point with the smallest distance represents the same class. After the first iteration, each data sample will be classified into a certain class. Then, we calculated the average value of each class of data as the new random class point. The new random class point can be represented as:

$$O_{lj} = \frac{1}{N}\sum_{i=0}^{i=N} x_{ij} \quad (3)$$

among them, $j \in [1, 2, \ldots, J], l \in [1, 2, \ldots, K]$.

Then, we re-calculated these distances, and reclassified them, and this process was repeated until convergence was achieved. And finally, the data were classified into K classes. In the present work, we also standardized the data before classification. In order to illustrate how many categories were most reasonable, we could assume that the formula for the total loss was as follows:

$$Loss = \sum_{i=0}^{i=n} d_{il} \quad (4)$$

where n represents the number of samples. This formula represented the sum of distances from all sample points to their random class points. When there was a significant inflection point on the line of loss with respect to class K, the value of K at the inflection point should be considered as a reasonable classification. Through the K-means method, we divided the initial materials from MP into five categories. Their quantities were 6602, 5425, 3770, 2800, and 1355, respectively.

2.2. Deformation Potential Theory (DPT)

The deformation potential theory was proposed by Bardeen and Shockley [8] in the 1950s to describe the charge transfer in non-polar semiconductors. The charge mobility can be expressed as $\mu_x = e\tau_x/m^*$, where the relaxation time for bulk materials could be written as follows [8,9]:

$$\tau_x = \frac{2\sqrt{2\pi}\hbar^4 C_x}{3(k_B T m*)^{\frac{3}{2}} E_{DPx}^2} \quad (5)$$

where $C_x = \partial^2 E / \left(\partial(\Delta a_x/a_x)^2 V_0\right)$ is the elastic constant; $E_{DPx} = \Delta V_i / (\Delta a_x/a_x)$, ΔV_i is the deformation potential energy, which is the difference between the energy level of the i-th energy band and the energy level of the deep nuclear state; and $m^* = \hbar^2 / \left(\partial^2 E/\partial k^2\right)$ is the effective mass.

2.3. Elastic and Thermal Properties

We can obtain the elastic properties, group velocity, Poisson's ratio, Debye temperature, Grüneisen coefficients, and lattice thermal conductivity, after calculating the elastic constants of materials [10], which could be easily achieved during the high-throughput calculations.

In the case of uniform deformation for a crystal, the generalized form of Hooke's law of stress–strain [11] is:

$$f_{ij} = C_{ijkl}\epsilon_{kl} \tag{6}$$

where f_{ij} and ϵ_{kl} is a homogeneous second-order stress tensor and a strain tensor, respectively [12]. C_{ijkl} represents the fourth order elastic stiffness tensor. Using matrix representation, we can abbreviate the stiffness tensor C_{ijkl} of four suffixes to the stiffness tensor C_{ij} of two suffixes, which can be represented as follows:

$$C_{ij} = \begin{bmatrix} C_{11} & C_{12} & C_{13} & C_{14} & C_{15} & C_{16} \\ C_{21} & C_{22} & C_{23} & C_{24} & C_{25} & C_{26} \\ C_{31} & C_{32} & C_{33} & C_{34} & C_{35} & C_{36} \\ C_{41} & C_{42} & C_{43} & C_{44} & C_{45} & C_{46} \\ C_{51} & C_{52} & C_{53} & C_{54} & C_{55} & C_{56} \\ C_{61} & C_{62} & C_{63} & C_{64} & C_{65} & C_{66} \end{bmatrix} \tag{7}$$

The elastic flexibility tensor $\left(s_{ij} = C_{ij}^{-1}\right)$ can be written as:

$$s_{ij} = \begin{bmatrix} s_{11} & s_{12} & s_{13} & s_{14} & s_{15} & s_{16} \\ s_{21} & s_{22} & s_{23} & s_{24} & s_{25} & s_{26} \\ s_{31} & s_{32} & s_{33} & s_{34} & s_{35} & s_{36} \\ s_{41} & s_{42} & s_{43} & s_{44} & s_{45} & s_{46} \\ s_{51} & s_{52} & s_{53} & s_{54} & s_{55} & s_{56} \\ s_{61} & s_{62} & s_{63} & s_{64} & s_{65} & s_{66} \end{bmatrix} \tag{8}$$

The Voigt [13] bulk modules can be calculated by:

$$B_v = \frac{1}{9}[(C_{11} + C_{22} + C_{33}) + 2(C_{12} + C_{23} + C_{31})] \tag{9}$$

The shear modulus can be obtained by:

$$G_v = \frac{1}{15}[(C_{11} + C_{22} + C_{33}) - (C_{12} + C_{23} + C_{31}) + 3(C_{44} + C_{55} + C_{66})] \tag{10}$$

The Reuss [14] bulk and shear modulus can be calculated by,

$$\frac{1}{B_r} = (s_{11} + s_{22} + s_{33}) + 2(s_{12} + s_{23} + s_{31}) \tag{11}$$

and

$$\frac{15}{G_r} = 4(s_{11} + s_{22} + s_{33}) - 4(s_{12} + s_{23} + s_{31}) + 3(s_{44} + s_{55} + s_{66}) \tag{12}$$

In the present work, we took the arithmetic mean of the boundaries between Voigt and Reuss, Voigt–Reuss–Hill (VRH) [14]:

$$B_h = \frac{B_r + B_v}{2} \tag{13}$$

$$G_h = \frac{G_r + G_v}{2} \tag{14}$$

The longitudinal (v_l), transverse (v_t), and average (v_a) elastic wave velocities can be calculated by:

$$v_l = \sqrt{\frac{3B_h + 4G_h}{3\rho}}, \tag{15}$$

$$v_t = \sqrt{\frac{G_h}{\rho}}, \tag{16}$$

$$v_a = \left[\frac{1}{3}\left(\frac{2}{v_t^3} + \frac{1}{v_l^3}\right)\right]^{-1/3} \tag{17}$$

The Debye temperature (θ_D) was obtained by:

$$\theta_D = \frac{h}{k_B}\left[\frac{3q}{4\pi}\frac{N\rho}{M}\right]^{1/3} v_a \tag{18}$$

And the Grüneisen coefficient was calculated by:

$$\gamma = \frac{3}{2}\left(\frac{1 + v_{poi}}{2 - 3v_{poi}}\right) \tag{19}$$

where $v_{poi} = (1 - 2\left(\frac{v_t}{v_l}\right)^2)/(2 - 2\left(\frac{v_t}{v_l}\right)^2)$ is the Poisson's ratio.

According to the Slack formula [15,16], the lattice thermal conductivity can be expressed as:

$$k_l = A\frac{\overline{M}\theta_D^3 \delta}{\gamma^2 n^{2/3} T} \tag{20}$$

where $\overline{M}$ is the average atomic mass; θ_D is the Debye temperature; δ is the volume of each atom; n is the number of atoms in the original cell; γ is the Grüneisen coefficient; A is a constant of 3.1×10^{-6}; and T is the temperature.

2.4. Methods for the First-Principles Calculations and Transport Properties

In the process of building a thermoelectric material database, first-principles calculations were performed using the Vienna Ab initio Simulation Package (VASP) 5.4.1 [17,18]. The generalized gradient approximation (GGA) with Perdew, Burke, and Ernzerhof functional (PBE) was employed [19]. The calculation of electronic transport required the use of the BoltzTraP2 program package [20]. In order to minimize computational costs while ensuring data reliability, during optimizing calculations, we set the plane-wave energy cutoff to be 1.4 times the maximum ENMAX of POTCAR of the composed elements, the electronic energy convergence to be 10^{-4} eV, the force convergence for ions to be 10^{-2} eV/Å, and the density k-mesh to be $0.04 \times 2\pi$ Å^{-1}.

All the processes were controlled through shell scripts. Data collection and calculation were implemented using Python 3 scripts. These codes were all home-made.

3. Capabilities and Workflow

3.1. The Application of K-Means on Datasets from MP

From Figure 1a, it can be seen that the number of points with an obvious inflection is six, which means that the initial structures can be divided into six categories. Considering the reasonable distribution of the average-bandgap values, we ultimately divided it into five categories. The featured distribution map and various information of K-means are shown in Figure 1c–g. The average value of bandgap for the first class is merely 0.025 eV, so this class of material contains many metals. The second class with an average bandgap value of 0.14 eV was mainly composed of semiconductors with narrow bandgaps. The third, fourth, and fifth categories were mainly composed of semiconductors and insulators with wide bandgaps. As a starting point, we focused on calculating the physical properties of candidate material sets for the first and second categories.

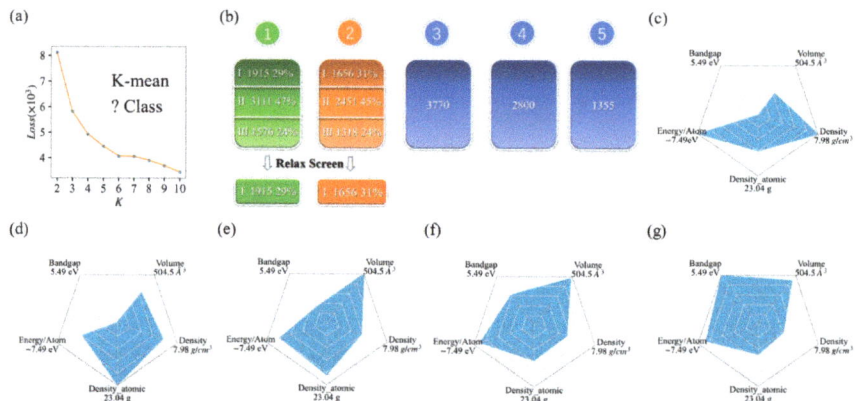

Figure 1. The application of K-means on MP databases: (**a**) the line chart of loss for K-class; (**b**) the classification data and relaxation screening results of the initial structures under K-means; (**c**–**g**) the average distribution of 5 features for each K-means class.

The material data collected from the MP database are finally divided into five categories using the K-means method. We identified the elements which are contained in the compounds of every class, and selected those larger than 5% of the total materials in number. For instance, in the first class, there are 6602 materials, if some element contained more than 330 (6602 × 5%) materials, we picked it out. As shown in Figure 2a, the first class of materials are mainly contributed by O, Si, P, Mn, Fe, Co, Ni, Ge, Rh, U, B, C, etc. In the second class, the materials are mainly composed of the III, IV, V, and VI main groups, with the alkali and alkaline-earth metals. In the sets of elements in the first and especially second class, the main group elements of III, IV, V, and VI are the habitats of many known semiconductors, suggesting the possible existence of excellent thermoelectric materials. Compared with the first and second class, many compounds in the third one are contributed by the halogen elements (such as F, Cl), and many of them take perovskite structures (such as $CsPbCl_3$, $CsGeCl_3$, etc.). The appearance of the H element in the fourth category indicates the presence of many hydrides and even organic compounds. And the fifth category has added Cs and Mg elements, most of which have larger bandgaps. By using the K-means method to classify the raw data for the first time, we can select the material dataset that needs to be calculated, which could help to reduce the calculation time.

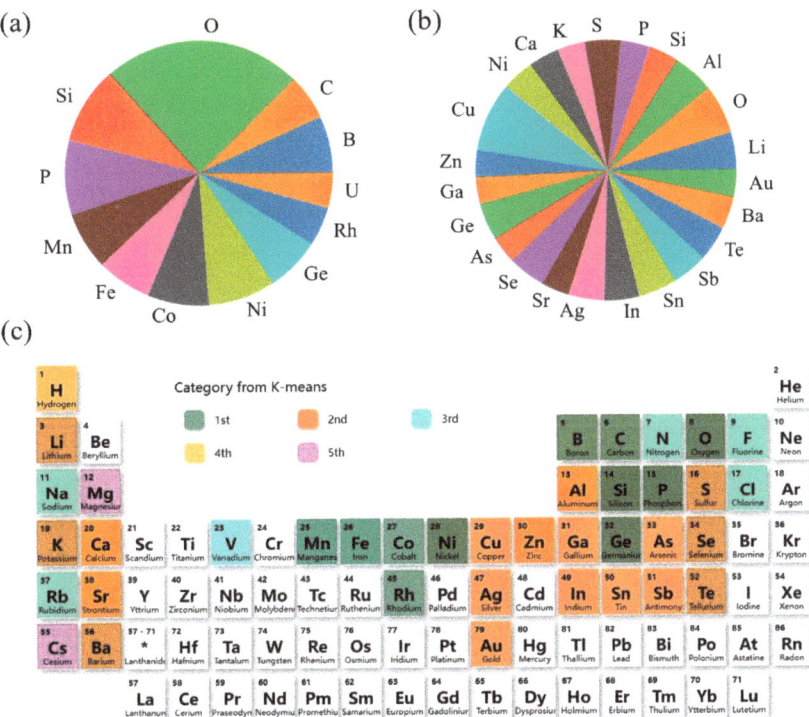

Figure 2. Element statistics for 5 classes of materials: (**a**) pie charts of elements for the first class; (**b**) pie charts of elements for the second class; (**c**) distribution charts of elements for 5 classes, and * represents for the rare-earth metal.

3.2. Computational Framework and Relaxation Process

After obtaining the structural file, we firstly performed structural relaxation and static calculation. Structural relaxation refers to the optimization process of atomic positions and lattice constants. We have employed VASP 5.4.1 for the first-principle calculations. Actually, several mainstream databases such as AFLOW, MP, OQMD, etc., were also calculated using VASP software.

For the first and second classes of materials obtained through K-means initial screening, there were more than 12,000 materials, many of which contained too many element classes and numbers of atoms in the primitive cell. In the present work, we firstly calculated the material system with a relatively simple structure. Therefore, a computational control process was employed during the structural relaxation to further screen them, resulting in a total of more than 3000 materials with relatively simple structures in the first and second classes. Nevertheless, conducting structural relaxation for so many materials was a computationally demanding task. In order to accelerate the calculation, we wrote several shell scripts to control the process of structural relaxation. The flowchart is shown in Figure 3.

After performing relaxation calculations on the data of the first and second classes of materials, we screened 1915 and 1656 materials, respectively, for further calculations, as shown in Figure 1b. In the first class, there are 3111 materials remaining with atomic numbers greater than 10 or element classes greater than 4, and the other 1576 materials are unrelaxed structures which are hard to obtain convergent relaxation for in our present setup calculations. In the second category, there are also 2451 materials with atomic numbers greater than 10 or element classes greater than 4, and 1318 materials that are difficult to be relaxed. After the relaxation calculation process, the convergent structures are saved for further calculations.

Then, we performed the calculations of the parameters of deformation potential theory. Firstly, we performed an anisotropic property judgment on the material, and then we performed static calculations with a density of $0.04 \times 2\pi$ Å^{-1} k-mesh set on the deformed structures in various directions.

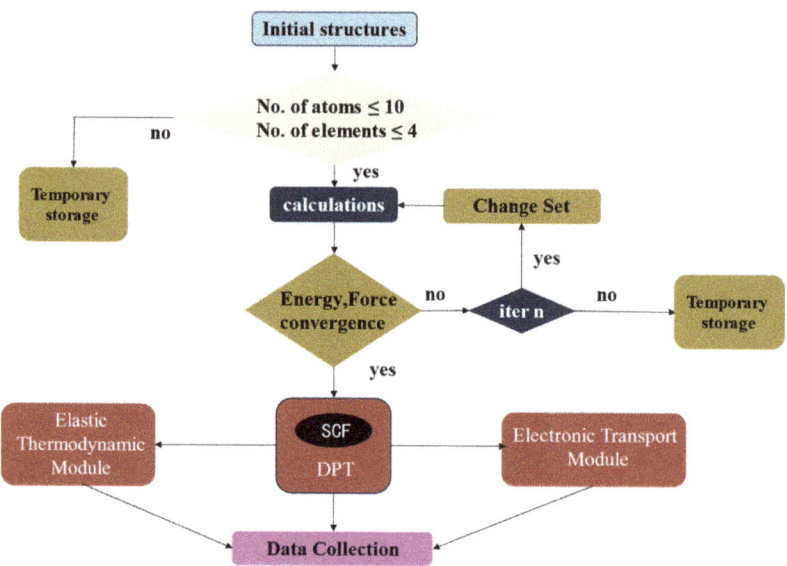

Figure 3. Process flowchart for constructing thermoelectric material database.

3.3. Analysis of Results of Deformation Potential Theory (Using Si as an Example)

The deformation potential method considered acoustic phonons as the main scattering sources for electrons. The relaxation time obtained by ignoring the contributions of optical phonon branches and other scattering mechanisms could be larger than the real one, but the calculation of deformation potential is relatively simple, easily employed in high-throughput calculations. The coefficients for applying deformation to the lattice vector are {0.98, 0.99, 1.00, 1.01, 1.02} of relaxed volumes, respectively. Such calculations could ensure the reliability of fitting with the second-order function for the elastic constant and the first-order function for the elastic potential energy. We took Si as an example, as shown in Figure 4.

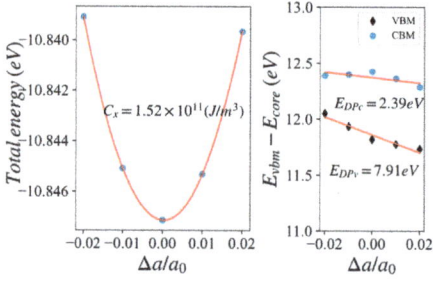

Figure 4. Schematic diagram of second-order fitting elastic constant C_x and first-order fitting of elastic potential energy E_{DP} for Si.

After calculating the deformation potential parameters, we could obtain the relaxation time of carriers by combing the effective masses.

3.4. Energy Band and Effective Mass Calculation

There are many methods to obtain the band structure of a material. Here, we compare three feasible schemes. The first scheme is calculating the energy band along the high symmetry point by VASP, the second one is using BoltzTraP2 [20] to fit the band structure by VASP, and the third one is using maximally localized Wannier function to interpolate the VASP results [21]. Considering the accuracy and efficiency, the second scheme was chosen in the present high-throughput calculations. The bandgap of Si in the MP database is 0.61 eV, which is consistent with VASP calculation. The relative error of the bandgap (0.59 eV) calculated by BoltzTraP is within 5%. The effective mass of Si calculated by BoltzTraP is similar to that of VASP calculation, and the relaxation time of electrons is around 1113 fs, as listed in the Table 1. The energy band of Si by three schemes is shown in Figure 5. Although set relatively coarse for the k-mesh set, the electronic properties' calculations could also converge to some reliable results, which will be discussed in followed text, due to the good fitting for the energy at the high symmetry point. We note here that the BoltzTraP calculation for the band structure is the fastest one; then, it is suitable for accelerating the high-throughput calculation.

Table 1. Calculated deformation potential parameters, effective mases, relaxation time of carriers, elastic coefficients C_{ij} (in GPa), bulk modulus B (in GPa), and shear modulus G (in GPa) for Si.

Carrier Type	E_{DPx} (eV)	C_x (10^{11} Jm^{-3})	m^*/m_0	τ_x (fs)
Electron (Hole)	2.39 (7.91)	1.5 (1.5)	0.89 (2.61)	1113 (20)
C_{11}	C_{12}	C_{44}	B	G
153.7 (165.8 *)	57.2 (64.0 *)	74.8 (79.6 *)	89.4 (97.9 *)	62.8 (66.5 *)

* is the experimental results [22].

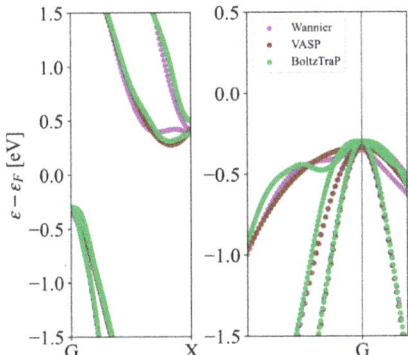

Figure 5. The band structure of Si by three schemes: VASP, BoltzTraP, and Wannier.

To facilitate the high-throughput calculation, we used the formula $m^* = \hbar^2 / \left(\partial^2 E / \partial k^2 \right)$ to calculate the effective mass. The effective masses of Si by the BoltzTraP scheme are shown in Figure 6. A series of effective masses of conduction and valence bands were obtained near the high symmetry points of Γ and X. We selected the maximum values of 0.89 m_0 and 2.61 m_0 as the effective masses for the conduction band and valence band, respectively. In addition, our program was designed to automatically determine whether the band is degenerate and calculate the effective mass for each degenerate band. We note here that the reason for selecting the maximum effective mass is that the deformation potential method could overestimate the relaxation time. By selecting the maximum effective mass, the relaxation time can be effectively reduced to compensate for the shortcomings of the deformation potential theory. In high-throughput calculations, the program also selected representative effective masses for other materials such as Si.

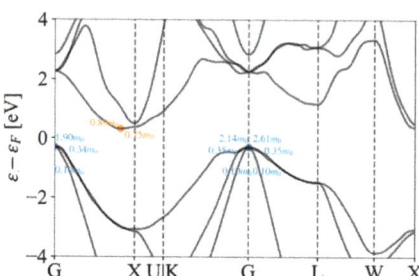

Figure 6. The effective mass of Si by the BoltzTraP scheme.

3.5. High-Throughput Electrical Transport Properties (BoltzTraP)

BoltzTraP is a program package calculating the semi-classic transport coefficients, based on a smoothed Fourier interpolation of the bands. The electronic transport properties such as Seebeck coefficient, electronic conductivity, and electronic thermal conductivity can be obtained at different temperatures and different doping concentrations. The BoltzTraP program has an input interface for VASP files, which could meet the needs of presenting high-throughput processes. After completing static calculations, the BoltzTraP module can be performed. Meanwhile, BoltzTraP based on Python can be properly embedded into our high-throughput Python data processing scripts, which are written for quickly obtaining the calculated quantities such as Seebeck coefficient, electronic conductivity, and electronic thermal conductivity. Combined with the lattice thermal conductivities estimated from the elastic property calculations, we could obtain the ZT values for the materials. For semiconductors, doping concentration is set within the range from 1.0×10^{18} cm^{-3} to 1.0×10^{21} cm^{-3}. During the statistic process, we left materials with a large relaxation time ($>10^{12}$ fs) for further testifying, and it remained for renewal in the next version of the database. For some materials, it is hard to obtain a reliable value for the deformation potential energy with the current calculation set, and then obtain a large relaxation time. When such relaxation times were fed into BoltzTraP, it would lead to unreliable conductivity and electronic thermal conductivity. The choice of n-type or p-type depends on where the maximum ZT value occurs. We list the top five semiconductor materials with ZT values in Table 2.

Table 2. Top 5 semiconductor materials sorted by ZT value at 300 K.

Id	Formula	N (cm^{-3})	κ_l (WK^{-1}m^{-1})	S (μV/K)	σ (kS/m)	κ_e (WK^{-1}m^{-1})	ZT	Type
mp-23231	AgBr	1.11×10^{19}	0.76	433.58	187.71	1.14	5.56	P
mp-22919	AgI	5.81×10^{18}	1.31	421.47	279.56	1.76	4.85	P
mp-27484	Tl$_4$O$_2$	2.69×10^{19}	0.67	361.13	196.98	0.92	4.83	P
mp-22922	AgCl	7.05×10^{19}	0.51	383.05	100.90	0.51	4.34	P
mp-32791	Ag$_4$S$_2$	1.78×10^{19}	0.39	310.31	144.72	0.80	3.52	N

To further discuss the effectiveness of the present calculations, we took Si as an example to discuss the performance of electronic transport properties under different k-mesh sets during static calculations. For Si, a density of $0.04 \times 2\pi$ Å^{-1} k-mesh set corresponds to a k-point grid of $8 \times 8 \times 8$, and that of $0.01 \times 2\pi$ Å^{-1} is for a $32 \times 32 \times 32$ k-mesh set. In BoltzTraP, the more k-mesh points provided, the more accuracy reached for the calculated electronic transport performance. As shown in Figure 7, a k-mesh density of $0.04 \times 2\pi$ Å^{-1}

could guarantee moderately accurate results and greatly reduces the calculation time. We also gave the HSE [23] results for comparison. The computational cost of HSE was demanding; we will provide the results using the HSE method in the next version of the database.

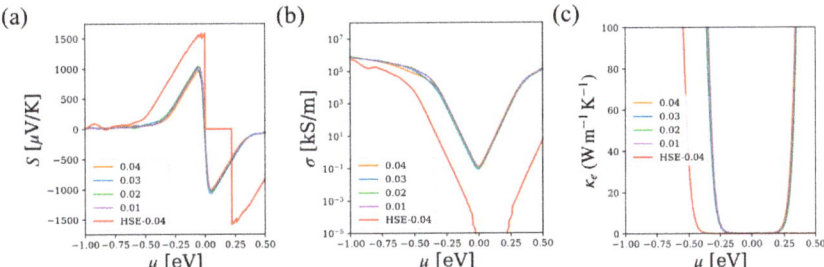

Figure 7. The performance of electronic transport under different k-mesh density set of PBE or HSE (0.04) at 300 K for Si: (**a**) Seebeck coefficient S; (**b**) conductivity σ; and (**c**) the electronic thermal conductivity κ_e with respect to different chemical potentials μ.

3.6. ZT Value and BE Value

As an example for the application of our database, we associated the thermoelectric ZT values with the electronic quality factor of $B_E T / \kappa_L$. By S and σ, the electronic quality factor B_E can be defined by [24]:

$$B_E = S^2 \sigma \left[\frac{S_r^2 exp(2 - S_r)}{1 + exp[5 - 5S_r]} + \frac{S_r \pi^2 / 3}{1 + exp[5(S_r - 1)]} \right] \quad (21)$$

where $S_r = |S|e/k_B$. As shown in Figure 8, the ZT values of most materials are positively correlated to the electronic quality factor of $B_E T / \kappa_L$, so the $B_E T / \kappa_L$ values could also serve as another criterion for judging the performance of a thermoelectric material.

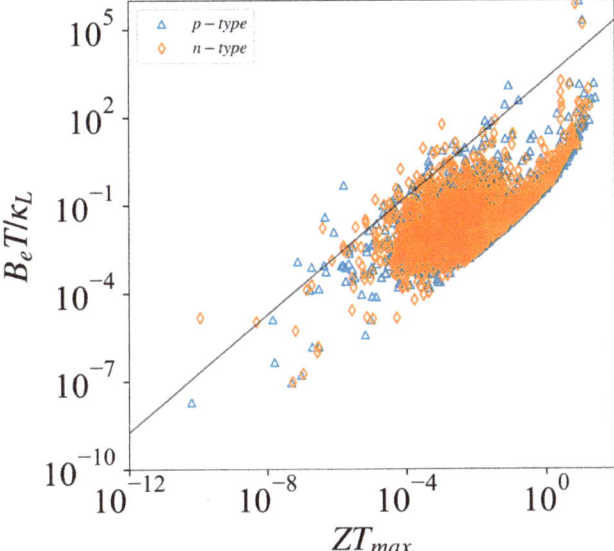

Figure 8. Thermal power quality factor $B_E T / \kappa_L$ and maximum ZT_{max} at 300 K. The grey line is a linear fitting for all the ZT and $B_E T / \kappa_L$ values, which used to show a trend of linear positive correlation.

3.7. Validation and Future Development of the Database

We built a database of thermoelectric materials. We note here that the errors in our data come from the following aspects: (1) the PBE method was employed to calculate the band structures, which generally results into the underestimation for the band gap; (2) the rigid band model was used in BoltzTraP, which may fail to give the change for the band structures during doping; (3) the deformation potential model was used to obtain the electron lifetimes, which could be much higher than the experimental results, for such a model generally underestimates the effect of electron–phonon scattering; and (4) the lattice thermal conductivities were estimated from the calculated elastic coefficients, which hardly described the phonon–phonon scattering in materials.

In the next versions of the Wenzhou TE database, we will gradually improve the reliability of the data, starting with a high-throughput calculation of the thermal conductivity by accounting the three-phonon scattering. Secondly, we will use HSE method to perform high-throughput band structure calculations. We will also use some other models to improve the prediction of electronic lifetimes. And lastly in the calculation, we will add the convergence judgment.

4. Conclusions

In this work, we built a thermoelectric material database—Wenzhou TE. We designed several modules to obtain the electronic and heat transport parameters for materials, including structural screening, deformation potential, elastic constant, and BoltzTraP electronic transport performance calculations module. In addition, we wrote several Python scripts to collect data and process results. Furthermore, we also built a webpage for Wenzhou TE: a first-principle-calculated thermoelectric materials database (https://hezhu2024.github.io (accessed on 3 April 2024)), which could be used for searching and viewing the physical properties of materials. Subsequently, we will continue the development of the database by employing more accurate methods for electronic and phonon transport properties, and also calculating more materials. Based on the present work, one could easily use these data for data mining and thermoelectric material development.

Author Contributions: Conceptualization, methodology, validation, formal analysis, investigation, data curation, Y.F. and H.S.; software, writing—original draft preparation, visualization, Y.F.; resources, writing—review and editing, supervision, project administration, funding acquisition, H.S. All authors have read and agreed to the published version of the manuscript.

Funding: Funding was provided by the National Natural Science Foundation of China (52272006), the Zhejiang Provincial Natural Science Foundation of China (LY22A040001), and the Wenzhou Municipal Natural Science Foundation (G20210016).

Institutional Review Board Statement: Not applicable.

Informed Consent Statement: Not applicable.

Data Availability Statement: Data are contained within the article.

Acknowledgments: We gratefully acknowledge the helpful discussions with Hao Zhang from Fudan University, and the support from Changkun Dong from Wenzhou University.

Conflicts of Interest: The authors declare no conflicts of interest.

References

1. Jain, A.; Ong, S.P.; Hautier, G.; Chen, W.; Richards, W.D.; Dacek, S.; Cholia, S.; Gunter, D.; Skinner, D.; Ceder, G.; et al. Commentary: The Materials Project: A materials genome approach to accelerating materials innovation. *APL Mater* **2013**, *1*, 011002. [CrossRef]
2. Curtarolo, S.; Setyawan, W.; Hart, G.L.W.; Jahnatek, M.; Chepulskii, R.V.; Taylor, R.H.; Wang, S.; Xue, J.; Yang, K.; Levy, O.; et al. AFLOW: An automatic framework for high-throughput materials discovery. *Comput. Mater. Sci.* **2012**, *58*, 218–226. [CrossRef]
3. Saal, J.E.; Kirklin, S.; Aykol, M.; Meredig, B.; Wolverton, C. Materials design and discovery with high-throughput density functional theory: The open quantum materials database (OQMD). *JOM* **2013**, *65*, 1501. [CrossRef]
4. Kirklin, S.; Saal, J.E.; Meredig, B.; Thompson, A.; Doak, J.W.; Aykol, M.; Rühl, S.; Wolverton, C.M. The Open Quantum Materials Database (OQMD): Assessing the accuracy of DFT formation energies. *NPJ Comput. Mater.* **2015**, *1*, 15010. [CrossRef]

5. Pizzi, G.; Cepellotti, A.; Sabatini, R.; Marzari, N.; Kozinsky, B. AiiDA: Automated interactive infrastructure and database for computational science. *Comp. Mat. Sci.* **2016**, *111*, 218–230. [CrossRef]
6. Raccuglia, P.; Elbert, K.C.; Adler, P.D.F.; Falk, C.; Wenny, M.B.; Mollo, A.; Zeller, M.; Friedler, S.A.; Schrier, J.; Norquist, A.J. Machine-learning-assisted materials discovery using failed experiments. *Nature* **2016**, *533*, 73–76. [CrossRef]
7. Selim, S.Z.; Ismail, M.A. K-meanss-class algorithms: A generalized convergence theorem and characterization of local optimality. *IEEE Trans. Pattern Anal. Mach. Intell.* **1984**, *PAMI-6*, 81–87. [CrossRef] [PubMed]
8. Bardeen, J.; Shockley, W. Deformation potentials and mobilities in non-polar crystals. *Phys. Rev.* **1950**, *80*, 72–80. [CrossRef]
9. Lengeling, B.S.; Guzik, A.A. Inverse molecular design using machine learning: Generative models for matter engineering. *Science* **2018**, *361*, 360–365. [CrossRef]
10. Singh, S.; Lang, L.; Dovale-Farelo, V.; Herath, U.; Tavadze, P.; Coudert, F.-X.; Romero, A.H. Mechelastic: A python library for analysis of mechanical and elastic properties of bulk and 2d materials. *Comput. Phys. Commun.* **2021**, *267*, 108068. [CrossRef]
11. Dobson, P.J. Physical properties of crystals—Their representation by tensors and matrices. *Phys. Bull.* **1985**, *36*, 506. [CrossRef]
12. Mouhat, F.; Coudert, F.X. Necessary and sufficient elastic stability conditions in various crystal systems. *Phys. Rev. B* **2014**, *90*, 224104. [CrossRef]
13. Mavko, G.; Mukerji, T.; Dvorkin, J. *The Rock Physics Handbook*; Cambridge University Press: Cambridge, UK, 2020; pp. 220–235.
14. Hill, R. The elastic behaviour of a crystalline aggregate. *Proc. Phys. Society. Sect. A* **1952**, *65*, 349. [CrossRef]
15. Nolas, G.S.; Goldsmid, H.J. Thermal conductivity of semiconductors. In *Thermal Conductivity: Theory, Properties, and Applications*; Tritt, T.M., Ed.; Springer: Boston, MA, USA, 2004; pp. 105–121.
16. Slack, G. Nonmetallic crystals with high thermal conductivity. *J. Phys. Chem. Solids* **1973**, *34*, 321–335. [CrossRef]
17. Kresse, G.; Hafner, J. Ab initio molecular dynamics for liquid metals. *Phys. Rev. B* **1993**, *47*, 558. [CrossRef]
18. Kresse, G.; Furthmüller, J. Efficient iterative schemes for ab initio total-energy calculations using a plane-wave basis set. *Phys. Rev. B* **1996**, *54*, 11169. [CrossRef]
19. Perdew, J.P.; Burke, K.; Ernzerhof, M. Generalized gradient approximation made simple. *Phys. Rev. Lett.* **1996**, *77*, 3865. [CrossRef]
20. Madsen, G.K.; Carrete, J.; Verstraete, M.J. BoltzTraP2, a program for interpolating band structures and calculating semi-classical transport coefficients. *Comput. Phys. Commun.* **2018**, *231*, 140–145. [CrossRef]
21. Pizzi, G.; Vitale, V.; Arita, R.; Bluegel, S.; Freimuth, F.; Géranton, G.; Gibertini, M.; Gresch, D.; Johnson, C.; Koretsune, T.; et al. Wannier90 as a community code: New features and applications. *J. Phys. Condens. Matter* **2020**, *32*, 165902. [CrossRef]
22. McSkimin, H.J.; Andreatch, P., Jr. Elastic moduli of silicon vs hydrostatic pressure at 25.0 °C and −195.8 °C. *J. Appl. Phys.* **1964**, *35*, 2161. [CrossRef]
23. Heyd, J.; Scuseria, G.E.; Ernzerhof, M. Hybrid functionals based on a screened Coulomb potential. *J. Chem. Phys.* **2003**, *118*, 8207–8215; Erratum in *J. Chem. Phys.* **2006**, *124*, 219906.
24. Zhang, X.; Bu, Z.; Shi, X.; Chen, Z.; Lin, S.; Shan, B.; Wood, M.; Snyder, A.H.; Chen, L.; Snyder, G.J.; et al. Electronic quality factor for thermoelectrics. *Sci. Adv.* **2020**, *6*, eabc0726. [CrossRef] [PubMed]

Disclaimer/Publisher's Note: The statements, opinions and data contained in all publications are solely those of the individual author(s) and contributor(s) and not of MDPI and/or the editor(s). MDPI and/or the editor(s) disclaim responsibility for any injury to people or property resulting from any ideas, methods, instructions or products referred to in the content.

Article

First-Principles Study of Doped CdX ($X = Te, Se$) Compounds: Enhancing Thermoelectric Properties

Junfeng Jin [1], Fang Lv [2], Wei Cao [1,2,*] and Ziyu Wang [1,2,*]

[1] The Institute of Technological Sciences, Wuhan University, Wuhan 430060, China; 2021206520035@whu.edu.cn
[2] Key Laboratory of Artificial Micro- and Nano- Structures of Ministry of Education, School of Physics and Technology, Wuhan University, Wuhan 430072, China; fanglv@whu.edu.cn
* Correspondence: wei_cao@whu.edu.cn (W.C.); zywang@whu.edu.cn (Z.W.)

Abstract: Isovalent doping offers a method to enhance the thermoelectric properties of semiconductors, yet its influence on the phonon structure and propagation is often overlooked. Here, we take CdX ($X = Te, Se$) compounds as an example to study the role of isovalent doping in thermoelectrics by first-principles calculations in combination with the Boltzmann transport theory. The electronic and phononic properties of Cd_8Se_8, Cd_8Se_7Te, Cd_8Te_8, and Cd_8Te_7Se are compared. The results suggest that isovalent doping with CdX significantly improves the thermoelectric performance. Due to the similar properties of Se and Te atoms, the electronic properties remain unaffected. Moreover, doping enhances anharmonic phonon scattering, leading to a reduction in lattice thermal conductivity. Our results show that optimized p-type(n-type) ZT values can reach 3.13 (1.33) and 2.51 (1.21) for Cd_8Te_7Se and Cd_8Se_7Te at 900 K, respectively. This research illuminates the potential benefits of strategically employing isovalent doping to enhance the thermoelectric properties of CdX compounds.

Keywords: thermoelectric; doped CdX compounds; first-principles calculations; isovalent doping

Citation: Jin, J.; Lv, F.; Cao, W.; Wang, Z. First-Principles Study of Doped CdX ($X = Te, Se$) Compounds: Enhancing Thermoelectric Properties. *Materials* **2024**, *17*, 1797. https://doi.org/10.3390/ma17081797

Academic Editor: Yaniv Gelbstein

Received: 18 March 2024
Revised: 4 April 2024
Accepted: 4 April 2024
Published: 14 April 2024

Copyright: © 2024 by the authors. Licensee MDPI, Basel, Switzerland. This article is an open access article distributed under the terms and conditions of the Creative Commons Attribution (CC BY) license (https://creativecommons.org/licenses/by/4.0/).

1. Introduction

One of the primary sbyproducts of using various energy forms is heat. The process of converting this excess heat into electrical energy, known as thermoelectricity, is seen as a promising technology for practical energy harvesting applications [1]. The efficiency of thermoelectric conversion is assessed using the dimensionless thermoelectric figure of merit ZT [2]. ZT is defined as

$$ZT = \frac{S^2 \sigma T}{\kappa_e + \kappa_l} \quad (1)$$

where σ, S, T, κ_e and κ_l represent electrical conductivity, Seebeck coefficient, temperature, electronic thermal conductivity, and lattice thermal conductivity, respectively. However, the coupling effects in thermoelectric performance make it challenging to directly enhance the thermoelectric properties [3]. This is due to the intricate interplay between electrical conductivity, thermal conductivity, and Seebeck coefficient, which are often coupled together. Improving one of these properties can inadvertently affect the others, making it difficult to achieve substantial enhancements in overall thermoelectric performance without carefully considering and addressing these interdependent factors. Consequently, targeted strategies that can effectively decouple these properties or optimize their collective interaction are crucial for achieving significant advancements in thermoelectric materials. Various strategies, including doping [4], band engineering [5], phonon engineering [6], nanostructuring [7], and alloying [8], have been proposed to enhance thermoelectric performance.

Doping in thermoelectric materials enables precise tuning of electronic properties, providing flexibility, versatility, and compatibility with other enhancement techniques. There are many types of doping in thermoelectric materials, including cationic doping [9], co-doping [10], ion doping [11] and single-atom doping [12]. Here, we primarily focus on single-atom doping. Based on the doping atoms, doping in thermoelectric materials can be classified into aliovalent doping, which involves introducing impurities of different valences, and isovalent doping, which involves introducing impurities of the same valence. Aliovalent doping is commonly utilized to regulate carrier concentration for optimizing ZT. Research on the effects of Nb doping on the thermoelectric properties of n-type half-Heusler compounds revealed an enhanced power factor and a 20% increase attributed to aliovalent doping-induced decoupling between thermoelectric parameters [13]. Han et al. [14] emphasized the critical impact of aliovalent dopants on controlling the phonon structure and inhibiting the phonon propagation in a heavy-band NbFeSb system. Baranets et al. [15] demonstrated that aliovalent substitutions can alter the dimensionality of the polyanionic sublattice in the resulting quaternary phases, leading to reduced electrical resistivity and a notably enhanced Seebeck coefficient.

Compared to aliovalent doping, isovalent doping ideally decouples and regulates thermoelectric performance by reducing thermal conductivity through phonon scattering while maintaining unchanged electronic properties. Musah et al. [16] summarized a review of isovalent substitution as a method to independently enhance thermoelectric performance and device applications. The substitution of isovalent ions in the anion Te-site of Bi–Sb–Te led to a significant enhancement of the ZT over a wide temperature range, with the ZT being increased by 10% for all measured temperatures and averaging beyond 1.0 between 300 and 520 K, demonstrating the synergetic control of band structure and deformation potential via isovalent substitution [17]. He et al. [18] also demonstrated that isovalent Te substitution effectively reduces κ_l and increases σ in hole carrier concentration.

To thoroughly explore the impact of equiatomic doping on regulating thermoelectric performance, we chose CdX ($X = Se, Te$) as the focus of our research and utilized a first-principles approach. Recent research [19,20] indicated that CdX is commonly used as a dopant in thermoelectric applications. The simple cubic phase structure of CdX provides advantages for first-principles studies due to its well-defined symmetry and straightforward electronic and phononic property calculations. Additionally, Te and Se share similarities in their doping characteristics, owing to their comparable chemical properties and the analogous effects they induce when integrated into host materials. In this paper, we systematically investigated the electronic, phononic, mechanical, bonding, and thermoelectric properties of CdX using first-principles combined with Boltzmann transport theory.

2. Computational Methods

Theoretical computations were conducted using density functional theory (DFT) within the Quantum ESPRESSO v6.2 (QE) code [21,22]. The exchange–correlation functional used is the Generalized Gradient Approximation (GGA) as given by Perdew–Burke–Ernzerhof (PBE) [23], and the corresponding pseudopotential files are sourced from the standard solid-state pseudopotentials (SSSP PBE Efficency v1.3.0) library [24]. A kinetic energy cut-off of 80 Ry was utilized, and all relaxations were carried out until the forces and energy on each atom were reduced to less than 10^{-4} Ry/Bohr and 10^{-10} Ry. The Heyd–Scuseria–Ernzerhof (HSE06) hybrid functional [25,26] was used to obtain a more accurate band structure for the primitive cell of CdX. We constructed the doping structure using a $2 \times 2 \times 2$ supercell of the primitive cell for both CdTe and CdSe. The Brillouin zone was sampled over a uniform Γ-centered k-mesh of $4 \times 4 \times 4$. The projected crystal orbital Hamilton population (COHP) was calculated using the LOBSTER [27,28] package. The mechanical properties were carried out using Voigt–Reuss–Hill approximation [29], as implemented in the ElATools v1.7.0 [30] package. The crystal structure was plotted using VESTA v3.5.7 software [31].

Boltzmann's transport theory was employed to analyze the transport properties of systems using the BoltzTraP code [32]. Under Boltzmann's transport theory, these electronic transport coefficients can be expressed as

$$S_{\alpha\beta}(T,\mu) = \frac{1}{eT} \frac{\int v_\alpha(i,k)v_\beta(i,k)(\varepsilon-\mu)\left[-\frac{\partial f_\mu(T,\varepsilon)}{\partial \varepsilon}\right]d\varepsilon}{\int v_\alpha(i,k)v_\beta(i,k)\left[-\frac{\partial f_\mu(T,\varepsilon)}{\partial \varepsilon}\right]d\varepsilon} \quad (2)$$

$$\frac{\sigma_{\alpha\beta}(T,\mu)}{\tau_e(i,k)} = \frac{1}{V}\int e^2 v_\alpha(i,k)v_\beta(i,k)\left[-\frac{\partial f_\mu(T,\varepsilon)}{\partial \varepsilon}\right]d\varepsilon \quad (3)$$

$$\frac{\kappa^e_{\alpha\beta}(T,\mu)}{\tau(i,k)} = \frac{1}{TV}\int v_\alpha(i,k)v_\beta(i,k)(\varepsilon-\mu)^2\left[-\frac{\partial f_\mu(T,\varepsilon)}{\partial \varepsilon}\right]d\varepsilon \quad (4)$$

where α, β are Cartesian components, μ is the chemical potential of electrons (the Fermi level), V is volume of the unit cell, e is electronic charge, ε is the band eigenvalue, $v_\alpha(i,k)$ is the electron group velocity, and $f_\mu(T,\varepsilon)$ is is the Fermi–Dirac distribution.

The carrier relaxation time (τ_e) under the electron–phonon averaged (EPA) approximation was evaluated using the following equation [33]:

$$\tau_e^{-1}(\varepsilon,\mu,T) = \frac{2\pi\Omega}{g_{sh}}\sum_\nu \Big\{ g_\nu^2(\varepsilon, \varepsilon+\overline{\omega}\nu)[n(\overline{\omega}\nu,T)+f(\varepsilon+\overline{\omega}\nu,\mu,T)] \times \rho(\varepsilon+\overline{\omega}\nu)$$

$$+ g_\nu^2(\varepsilon,\varepsilon-\overline{\omega}\nu)[n(\overline{\omega}\nu,T)+1-f(\varepsilon-\overline{\omega}\nu,\mu,T)]\rho(\varepsilon-\overline{\omega}*\nu) \Big\} \quad (5)$$

Here, ε is the energy of the carriers, μ is the chemical potential, Ω is the volume of the primitive unit cell, $\hbar$ is the reduced Planck's constant, g_s is the spin degeneracy, ν is the phonon mode index, g_ν^2 is the averaged electron–phonon matrix, $\overline{\omega}\nu$ is the averaged phonon mode energy, $n(\overline{\omega}\nu,T)$ is the Bose–Einstein distribution function, $f(\varepsilon+\overline{\omega}_\nu,\mu,T)$ is the Fermi–Dirac distribution function, and ρ is the density of states per unit energy and unit volume.

The lattice thermal conductivity, κ_l, is computed using the Boltzmann transport equation integrated within the ShengBTE code [34], incorporating second- and third-order interatomic force constants (IFCs). The lattice thermal conductivity component $\kappa_l^{\alpha\beta}$ (α, β represents three Cartesian axes) is given by

$$\kappa_l^{\alpha\beta} = \frac{1}{k_B T^2 \Omega N}\sum_\lambda f_0(f_0+1)(\hbar\omega_\lambda)^2 v_\lambda^\alpha v_\lambda^\beta \tau_\lambda^0 \quad (6)$$

where Ω, N, f_0, ω_λ, v_λ, and τ_λ^0 are volume, number of phonon vectors, Bose–Einstein distribution function, frequency, group velocity, and lifetime of phonon mode λ, respectively. The second-order and third-order IFCS were calculated by a $2 \times 2 \times 2$ supercell, including 128 atoms. The third-order IFCS took the 5th nearest neighbor into consideration. The grid mesh for the phonon was set to $20 \times 20 \times 20$ to obtain convergent lattice thermal conductivity.

3. Results and Discussion

3.1. Electronic Properties

CdX adopts a zincblende, sphalerite structure and crystallizes in the cubic $F\overline{4}3m$ space group, as depicted in Figure 1a,d. Each X^{2-} ion is bonded to four equivalent Cd^{2+} atoms to form corner-sharing XCd_4 tetrahedra. To facilitate our study, we constructed a $2 \times 2 \times 2$ supercell of CdX, denoted as Cd_8X_8 as shown in Figure 1b,e, and replaced one X atom. The resulting doped structures are illustrated in Figure 1c,f. The relaxed latice constants are also given in Figure 1. The lattice constant and bond length of Cd_8Te_8 are longer than those of Cd_8Se_8, indicating a stronger bond strength in $Cd-Se$. Upon doping, the lattice constant of Cd_8Se_7Te increases, while that of Cd_8Te_7Se decreases.

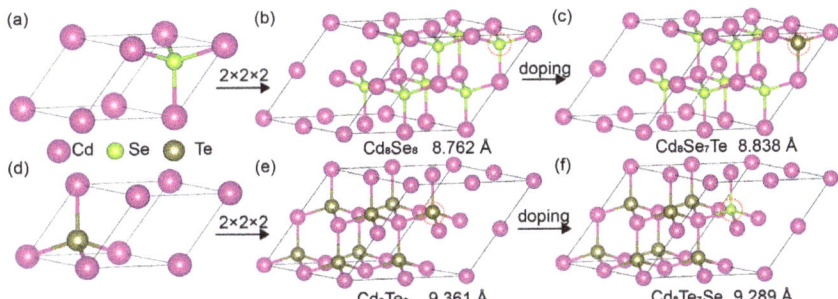

Figure 1. The unit cell of (**a**) $CdSe$ and (**d**) $CdTe$. The $2 \times 2 \times 2$ supercell of (**b**) Cd_8Se_8 and (**e**) Cd_8Te_8. The doped structures of (**c**) Cd_8Se_7Te and (**f**) Cd_8Te_7Se.

The band structures depicted in Figure 2a–f all display a similar band shape, with the only distinguishing factor being the band gap. It is evident that the band structures obtained from the HSE06 method exhibit a similar shape to those obtained from the PBE method, with notable differences observed in the band gaps. Specifically, the band gaps for $CdSe$ and $CdTe$ computed using the PBE method are reported as 0.47 eV and 0.58 eV, respectively, whereas those computed using the HSE06 method are reported as 1.42 eV and 1.34 eV, respectively. These findings closely align with the results reported in Ref. [35]. Following doping, there is a reduction in the band gaps. In the case of Cd_8Se_7Te, the band gap reduced to 0.372 eV, and for Cd_8Te_7Se, it reduced to 0.441 eV. Notably, both the valence band maximum (VBM) and conduction band minimum (CBM) are situated at the Γ point. Furthermore, the valence band demonstrates multiple valleys. Similar trends in band structure changes have been observed in other isovalent doped systems [36]. The relationship between band gaps and composition in these systems can be characterized by the quadratic Vegard's law as [36]

$$E_g^{A_{1-x}B_x}(x) = (1-x)E_g^A + xE_g^B - bx(1-x) \qquad (7)$$

where E_g^A and E_g^B are the band gaps of the host materials, A and B, respectively, x is the composition, and b is a bowing parameter. In our case, A and B represent $CdSe$ and $CdTe$, respectively, with x equal to $1/8$. By fitting bowing parameter b, we found it to be 1.037 eV for Cd_8Se_7Te. b through Cd_8Se_7Te is 1.037 eV. Subsequently, we applied this model and fitted b to Cd_8Te_7Se and obtained a band gap of 0.453 eV, which closely aligns with the calculated value. This suggests that the band gap of doped CdX can be predicted using Vegard's law.

To gain a comprehensive understanding of the band structure, we present the projected band structures of Cd_8Se_7Te and Cd_8Te_7Se in Figure 3. The dot size in the projected band structure represents the contribution of corresponding orbitals. The conduction band is primarily composed of the $5s$ orbitals of Cd. For CdX or doped CdX, their conduction band is the same, while the valence band is dominated by the p and d orbitals of all atoms. Notably, both Se and Te atoms demonstrate similar contributions, with the s orbitals of Se being a little stronger than those of Te, and the p orbitals of Se being a little weaker than those of Te. Although isovalent atoms contribute to the band structure, their effect is relatively subtle.

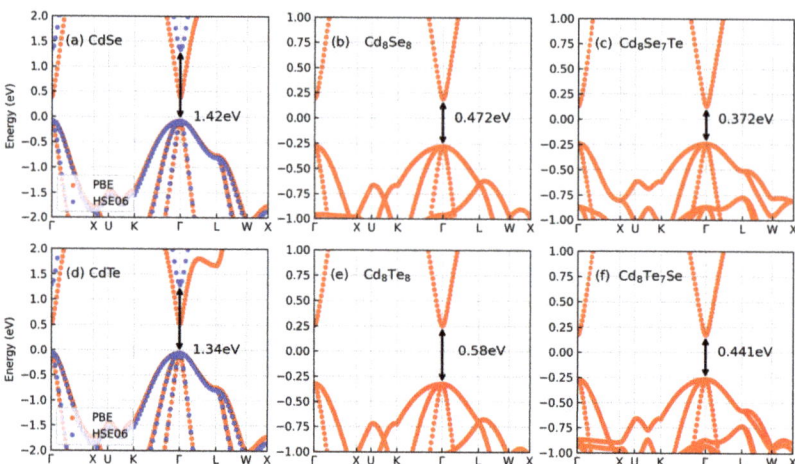

Figure 2. The band structures of (**a**) *CdSe*, (**b**) Cd_8Se_8, (**c**) Cd_8Se_7Te, (**d**) *CdSe*, (**e**) Cd_8Se_8, and (**f**) Cd_8Se_7Te.

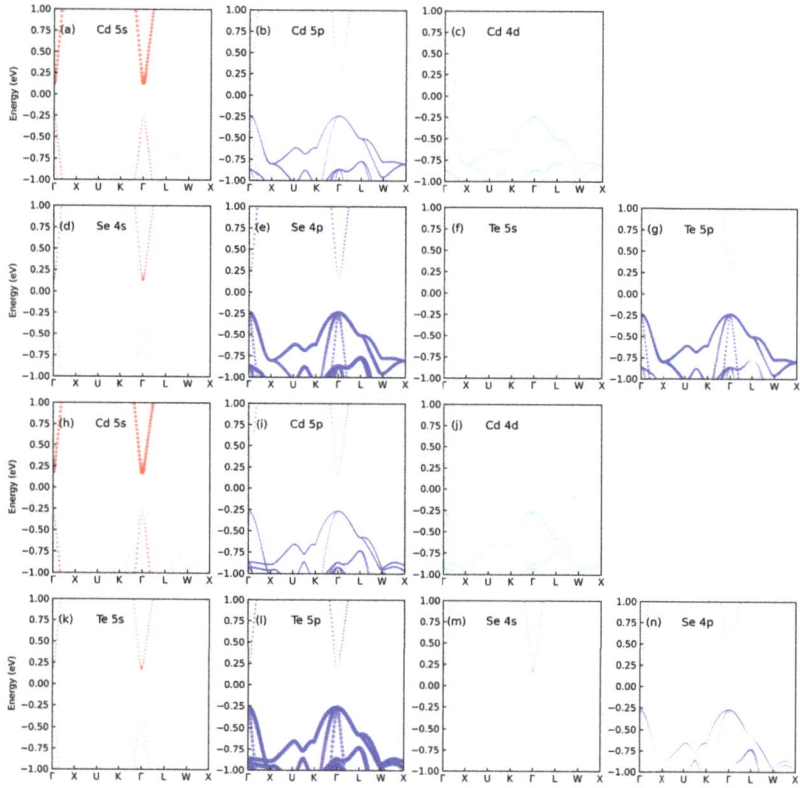

Figure 3. The projected band structure of (**a**) Cd 5s, (**b**) Cd 5p, (**c**) Cd 4d, (**d**) Se 4s, (**e**) Se 4p, (**f**) Te 5s, and (**g**) Te 5p of Cd_8Se_7Te. The projected band structure of (**h**) Cd 5s, (**i**) Cd 5p, (**j**) Cd 4d, (**k**) Te 5s, (**l**) Te 5p, (**m**) Se 4s, and (**n**) Se 4p of Cd_8Te_7Se.

Figure 4 illustrates the calculated values of S, σ, and $S^2\sigma$ at 300 K for different carrier concentrations. Generally, S can be expressed as [37]

$$S = \frac{8\pi^2 k_B^2}{3eh^2} m^* T \left(\frac{\pi}{3n}\right)^{2/3} \quad (8)$$

in which k_B, e, m^*, and h are the Boltzmann constant, electron charge, effective mass, and Planck constant, respectively. Analysis of Figure 4a,d reveals that the absolute values of S all decrease as the carrier concentration increases. Due to their similar band curvatures (m^*), CdX exhibits comparable S values. Notably, the S for hole doping (p-type) is significantly higher than that for electron doping (n-type). For instance, p-type S can reach 400 μV/K, while n-type S is only 100 μV/K at 10^{19} cm^{-3}. This difference can be attributed to the valence band having a much sharper curvature and a larger m^* compared to the conduction band. The behavior of σ as a function of carrier concentration is depicted in Figure 4b,e. In contrast to S, all σ values increase as the carrier concentration rises. N-type σ is higher than p-type σ, especially at low carrier concentrations. When the carrier concentration reaches to 10^{20}–10^{21} cm^{-3}, both n-type and p-type σ reach the 10^5 S/m level. Due to the Wiedemann–Franz relation [38], κ_e exhibits a linear correlation with σ; hence, there is no need to separately display κ_e. In Ref. [39], the electrical parameters of various thin film CdSe samples were investigated. The carrier concentration of thin film CdSe was found to be approximately (10^{20} cm^{-3}) with a Seebeck coefficient of around (-50 μV/K). Our obtained value of (-25 μV/K) aligns closely with this result. It is worth noting that our calculated electrical conductivity (10^5 S/m) significantly exceeds the experimental value (10^2 S/m). This discrepancy can be attributed to the DFT simulation assuming a perfect crystal structure, while experimental samples typically exhibit boundaries, grain effects, and scattering mechanisms that substantially reduce the electrical conductivity.

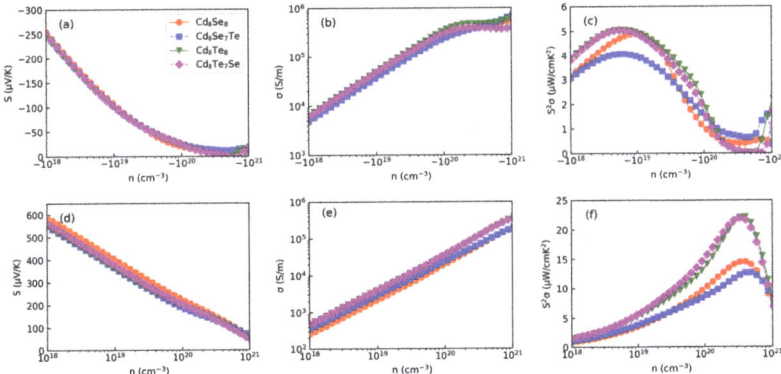

Figure 4. The electronic transport properties of (**a,d**) S, and (**b,e**) σ, and (**c,f**) $S^2\sigma$.

When considering S and σ, it is evident that the n-type $S^2\sigma$ is significantly lower than p-type $S^2\sigma$, as depicted in Figure 4e,f. The optimized carrier concentrations for n-type and p-type are determined to be 10^{19} and 5×10^{20} cm^{-3}, respectively. All CdX materials exhibit similar n-type $S^2\sigma$ values around 4–5 μW/cmK^2. The maximum n-type $S^2\sigma$ for $CdTe$ (20 μW/cmK^2) is twice as much as that of $CdSe$ (10 μW/cmK^2). The p-type $S^2\sigma$ value of CdX is comparable to well-known thermoelectric materials such as Cu_2Se [40] and $SnSe$ [41], indicating their superior p-type thermoelectric properties. This analysis implies that doping can effectively maintain the electronic transport properties of CdX.

3.2. Phononic Properties

Next, we shift our focus to the phonon dispersion in Figure 5. In the low-frequency range, all structures exhibit similar phonon dispersions, as depicted in Figure 5a,c,e,g. All structures show non-negative values in their phonon dispersions, confirming their stability. Upon doping Te atoms into Cd_8Se_8, phonon modes around 130 cm^{-1} moving to a higher frequency. However, for Cd_8Te_8, doping Se atoms leads to phonon modes around 175 cm^{-1} moving to lower frequency. This phenomenon can be seen more clearly in the phonon DOS in Figure 5b,d,f,h. In the phonon DOS, there are two peaks around 130 cm^{-1} in Cd_8Se_8. After doping Te atoms, the peak around 130 cm^{-1} disappears and new peaks at 150 cm^{-1} arise. From the projected phonon DOS, it is evident that the new peaks at 150 cm^{-1} are contributed by Te atoms. The case for Cd_8Te_8 is similar. When doping Se atoms, the peak around 175 cm^{-1} disappears and new peaks contributed by Se atoms at 150 cm^{-1} arise. An increase in the strength of interatomic bonds leads to an increase in the vibration frequency near the atom's position [42,43].

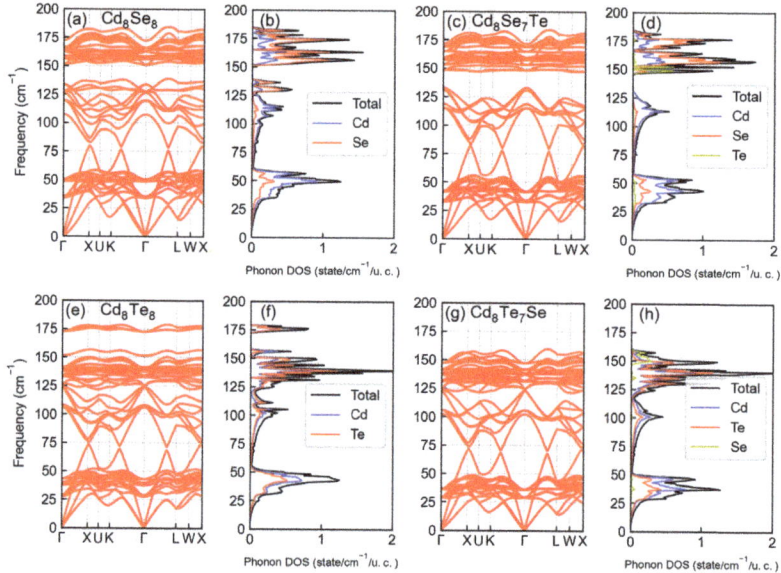

Figure 5. The phonon dispersions for (**a**) Cd_8Se_8, (**c**) Cd_8Se_7Te, (**e**) Cd_8Te_8, and (**g**) Cd_8Te_7Se. The total and projected phononic density of states for (**b**) Cd_8Se_8, (**d**) Cd_8Se_7Te, (**f**) Cd_8Te_8, and (**h**) Cd_8Te_7Se.

We further investigate the thermal transport properties of CdX and their corresponding doping systems. Figure 6a shows κ_l at different temperatures. κ_l decreases with temperature due to stronger phonon–phonon scattering. Cd_8Se_8 exhibit lower κ_l values than Cd_8Te_8, as stronger bonds tend to transfer more heat, leading to higher thermal conductivity in crystal structures with stronger bonds. After doping, κ_l is reduced. Specifically, the κ_l of Cd_8Se_7Te is much lower than that of Cd_8Te_7Se, with the former being 0.5 Wm^{-1}K^{-1} and the latter being 0.8 Wm^{-1}K^{-1} at 300 K. Furthermore, we illustrate the cumulative κ_l as a function of frequency at 300 K in Figure 6b. For undoped CdX, the rate of increase in κ_l begins to decrease at 75 cm^{-1}, while the node at which the rate of increase slows down after doping drops to 50 cm^{-1}. This indicates that doping not only reduces κ_l, but also lowers the frequency at which the maximum rate of κ_l increase occurs, further contributing to the reduction in κ_l.

To further elucidate the reasons for the behavior of κ_l, we calculate the phonon group velocity (v_g), phonon lifetime (τ_{ph}), and Grüneisen parameter (γ) as a function of frequency at 300 K as shown in Figure 7. In thermal transport, γ represents the sensitivity of a material's phonon frequency to changes in volume or pressure, providing insight into the strength of anharmonic scattering. A large value of $|\gamma|$ indicates the potential for strong phonon–phonon anharmonic scattering [44]. From Figure 7a–d, it can be observed that the doped v_g remain largely unchanged, especially in the low-frequency region, consistent with the earlier phonon spectral variations. The speed at which energy is propagated through a material's lattice vibrations correlates with the distribution of vibrational modes across different frequencies. After doping, a significant decrease in the low-frequency τ_{ph} is observed in Figure 7e–h. Figure 6b indicates that thermal conductivity is primarily influenced by low-frequency phonons. Subsequently, we analyze the γ in Figure 7i–l, which describes the strength of anharmonic scattering. It is found that the γ also decreases, indicating an enhancement in the strength of anharmonic scattering. This phenomenon evidently arises from the presence of dopant elements.

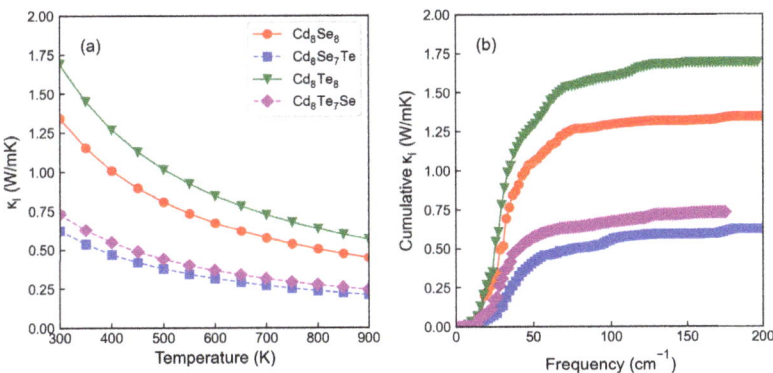

Figure 6. (**a**) The lattice thermal conductivity as a function of temperature. (**b**) Cumulative lattice thermal conductivity as a function of frequency at 300 K.

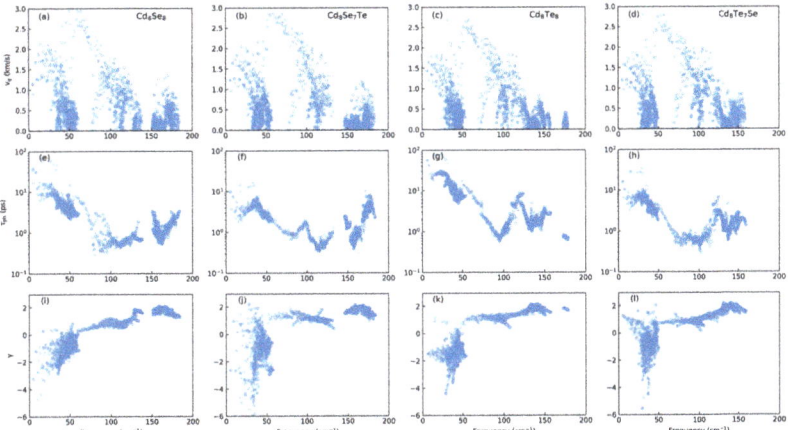

Figure 7. The (**a**–**d**) phonon group velocity, (**e**–**h**) phonon lifetime, and (**i**–**l**) Grüneisen parameter as a function of frequency at 300 K.

3.3. Mechanical Properties and Bonding Analysis

Figure 8a illustrates the calculated Young's modulus and shear modulus. It is evident that both Young's modulus and shear modulus gradually decrease with an increase in the ratio of *Te* content. These mechanical properties are indicative of the averaged bonding strength within the material. As the *Te* content increases, the averaged bonding strength of *CdX* diminishes. Furthermore, we conduct a comprehensive analysis of the bonding in *CdX* using the COHP method, as depicted in Figure 8b,c. In this context, negative COHP values indicate bonding states, while positive COHP values indicate antibonding states.

It is noteworthy that all structures exhibit a similar COHP phenomenon. Upon doping, states for Cd–Se in Cd_8Te_7Se and Cd–Te in Cd_8Se_7Te exhibit slightly enhanced strength compared to pristine states. Doping introduces new antibonding states. Below the Fermi level, bonding and antibonding states appear alternately. Notably, bonding states are stronger than antibonding states, thus ensuring the stability of the structure. Additionally, antibonding states typically possess higher energy levels than bonding states, contributing to a reduction in thermal conductivity [45]. This phenomenon arises from the greater delocalization of electrons in antibonding states, which results in reduced heat-carrying capacity compared to bonding states. Consequently, this accounts for the relatively low lattice thermal conductivity observed in CdX.

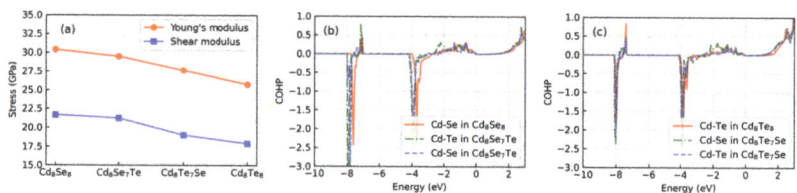

Figure 8. (**a**) Elasctic properties for CdX. COHP for (**b**) Cd_8Se_7Te and (**c**) Cd_8Te_7Se.

3.4. Thermoelectric Properties

Figure 9 illustrates maximum $S^2\sigma$ and ZT values at different temperatures. The $S^2\sigma_{max}$ values for doped and pristine CdX are comparable across temperature ranges as shown in Figure 9a,b. As temperature increases, the n-type $S^2\sigma_{max}$ values exhibit further enhancement, with $S^2\sigma_{max}$ reaching 9.8 μW/cm^{-2} at 900 K for Cd_8Se_7Te. In contrast, there is no clear increasing trend observed for p-type $S^2\sigma_{max}$. P-type $S^2\sigma_{max}$ values are much higher than n-type ones. Notably, for Cd_8Te_8 and Cd_8Te_7Se, p-type $S^2\sigma_{max}$ is approximately 25 μW/cm^{-2}, while for Cd_8Se_8 and Cd_8Se_7Te, it is around 15 μW/cm^{-2}. Doping does not degrade the $S^2\sigma_{max}$ for CdX, as indicated by our findings.

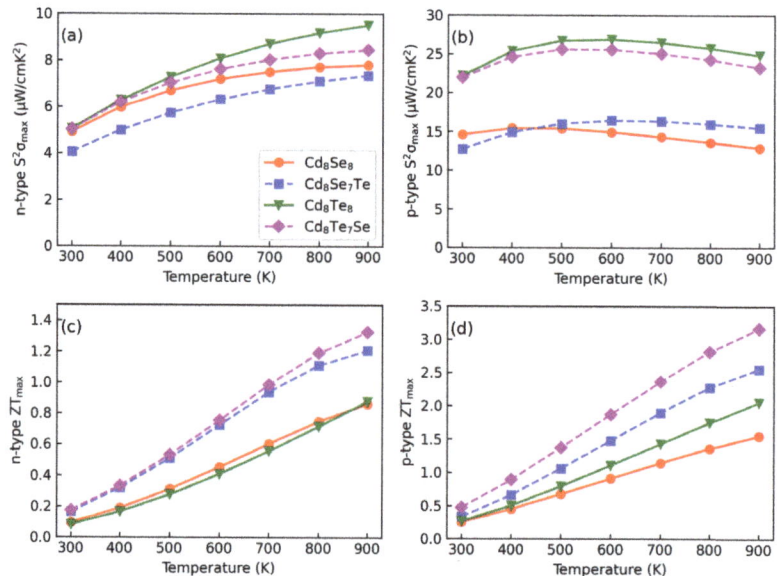

Figure 9. The maximum (**a**) n-type, (**b**) p-type $S^2\sigma$ and (**c**) n-type, (**d**) p-type ZT values as a function of temperatures.

When combined with electronic and phononic transport properties, the ZT_{max} values listed in Figure 9c,d demonstrate high thermoelectric performance for CdX at high temperatures. Despite the lack of increase in n-tpye $S^2\sigma_{max}$ with temperature, there is lower lattice thermal conductivity with temperatures resulting in higher ZT_{max} values for doped systems compared to pristine ones. Furthermore, these values can be further enhanced with increasing temperature, with p-type ZT_{max} values reaching up to 3.13 at 900 K. Our results suggest that doped CdX ($X = Te, Se$) presents potential for realizing both n-type and p-type thermoelectric materials for high-temperature applications.

4. Conclusions

In this study, we investigated the electronic, carrier, phonon transport, and thermoelectric properties of isovalent doped CdX ($X = Te, Se$) compounds using first-principles calculations with the Boltzmann transport equation. Due to the similar properties of Te and Se, the band structures remain nearly unchanged except for the band gaps in doped CdX. The bandgaps are 0.472 eV, 0.372 eV, 0.58 eV, and 0.441 eV for Cd_8Se_8, Cd_8Se_7Te, Cd_8Te_8, and Cd_8Te_7Se, respectively. Electronic transport properties of CdX are comparable for doped and pristine compounds. However, doping significantly reduces lattice thermal conductivity due to the introduction of impurity scattering. The maximum p-type (n-type) ZT values at 900 K are 1.5 (0.84), 2.51 (1.21), 2.1 (0.88), and 3.13 (1.33) for Cd_8Se_8, Cd_8Se_7Te, Cd_8Te_8, and Cd_8Te_7Se, respectively. Our study focuses on investigating the impact of isovalent doping in enhancing the thermoelectric properties of materials. Isovalent doping, such as with selenides and tellurides, can maintain electronic transport properties while effectively scattering phonons and decreasing lattice thermal conductivity. Future investigations could explore the potential of decoupling thermoelectric properties through homoelement doping.

Author Contributions: J.J.: Calculations, Formal analysis, Writing—original draft. F.L.: Review and editing. W.C.: Calculations, Review and editing. Z.W.: Funding acquisition, Supervision, Project administration. All authors have read and agreed to the published version of the manuscript.

Funding: The authors of this work acknowledge the National Natural Science Foundation of China (Grant No. 12122408, 12074292), and the Fundamental Research Funds for the Central Universities (Grant No. 2042023kf0109).

Institutional Review Board Statement: Not applicable.

Informed Consent Statement: Not applicable.

Data Availability Statement: The data provided in this study could be released upon reasonable request.

Conflicts of Interest: The authors declare no conflicts of interest.

References

1. Zoui, M.A.; Bentouba, S.; Stocholm, J.G.; Bourouis, M. A Review on Thermoelectric Generators: Progress and Applications. *Energies* **2020**, *13*, 3606. [CrossRef]
2. Sootsman, J.R.; Chung, D.Y.; Kanatzidis, M.G. New and Old Concepts in Thermoelectric Materials. *Angew. Chem. Int. Ed.* **2009**, *48*, 8616–8639. [CrossRef] [PubMed]
3. Shi, X.L.; Zou, J.; Chen, Z.G. Advanced Thermoelectric Design: From Materials and Structures to Devices. *Chem. Rev.* **2020**, *120*, 7399–7515. [CrossRef] [PubMed]
4. Zhao, W.; Ding, J.; Zou, Y.; Di, C.a.; Zhu, D. Chemical doping of organic semiconductors for thermoelectric applications. *Chem. Soc. Rev.* **2020**, *49*, 7210–7228. [CrossRef] [PubMed]
5. Zheng, Z.H.; Shi, X.L.; Ao, D.W.; Liu, W.D.; Chen, Y.X.; Li, F.; Chen, S.; Tian, X.Q.; Li, X.R.; Duan, J.Y.; et al. Rational band engineering and structural manipulations inducing high thermoelectric performance in n-type $CoSb_3$ thin films. *Nano Energy* **2021**, *81*, 105683. [CrossRef]
6. Hooshmand Zaferani, S.; Ghomashchi, R.; Vashaee, D. Strategies for engineering phonon transport in Heusler thermoelectric compounds. *Renew. Sustain. Energy Rev.* **2019**, *112*, 158–169. [CrossRef]
7. Novak, T.G.; Kim, K.; Jeon, S. 2D and 3D nanostructuring strategies for thermoelectric materials. *Nanoscale* **2019**, *11*, 19684–19699. [CrossRef] [PubMed]

8. Slade, T.J.; Pal, K.; Grovogui, J.A.; Bailey, T.P.; Male, J.; Khoury, J.F.; Zhou, X.; Chung, D.Y.; Snyder, G.J.; Uher, C.; et al. Contrasting SnTe-NaSbTe$_2$ and SnTe-NaBiTe$_2$ Thermoelectric Alloys: High Performance Facilitated by Increased Cation Vacancies and Lattice Softening. *J. Am. Chem. Soc.* **2020**, *142*, 12524–12535. [CrossRef] [PubMed]
9. Liu, Z.; Cheng, H.; Le, Q.; Chen, R.; Li, J.; Ouyang, J. Giant Thermoelectric Properties of Ionogels with Cationic Doping. *Adv. Energy Mater.* **2022**, *12*, 2200858. [CrossRef]
10. Perumal, S.; Samanta, M.; Ghosh, T.; Shenoy, U.S.; Bohra, A.K.; Bhattacharya, S.; Singh, A.; Waghmare, U.V.; Biswas, K. Realization of High Thermoelectric Figure of Merit in GeTe by Complementary Co-doping of Bi and In. *Joule* **2019**, *3*, 2565–2580. [CrossRef]
11. Chen, C.; Jacobs, I.E.; Kang, K.; Lin, Y.; Jellett, C.; Kang, B.; Lee, S.B.; Huang, Y.; BaloochQarai, M.; Ghosh, R.; et al. Observation of Weak Counterion Size Dependence of Thermoelectric Transport in Ion Exchange Doped Conducting Polymers Across a Wide Range of Conductivities. *Adv. Energy Mater.* **2023**, *13*, 2202797. [CrossRef]
12. Zhao, W.; Jin, K.; Fu, L.; Shi, Z.; Xu, B. Mass Production of Pt Single-Atom-Decorated Bismuth Sulfide for n-Type Environmentally Friendly Thermoelectrics. *Nano Lett.* **2022**, *22*, 4750–4757. [CrossRef]
13. Van Du, N.; Rahman, J.U.; Huy, P.T.; Shin, W.H.; Seo, W.S.; Kim, M.H.; Lee, S. X-site aliovalent substitution decoupled charge and phonon transports in XYZ half-Heusler thermoelectrics. *Acta Mater.* **2019**, *166*, 650–657. [CrossRef]
14. Han, S.; Dai, S.; Ma, J.; Ren, Q.; Hu, C.; Gao, Z.; Duc Le, M.; Sheptyakov, D.; Miao, P.; Torii, S.; et al. Strong phonon softening and avoided crossing in aliovalence-doped heavy-band thermoelectrics. *Nat. Phys.* **2023**, *19*, 1649–1657. [CrossRef]
15. Baranets, S.; Balvanz, A.; Darone, G.M.; Bobev, S. On the Effects of Aliovalent Substitutions in Thermoelectric Zintl Pnictides. Varied Polyanionic Dimensionality and Complex Structural Transformations—The Case of Sr$_3$ZnP$_3$ vs. Sr$_3$Al$_x$Zn$_{1-x}$P$_3$. *Chem. Mater.* **2022**, *34*, 4172–4185. [CrossRef]
16. Musah, J.D.; Ilyas, A.M.; Venkatesh, S.; Mensah, S.; Kwofie, S.; Roy, V.A.L.; Wu, C.M.L. Isovalent substitution in metal chalcogenide materials for improving thermoelectric power generation—A critical review. *Nano Res. Energy* **2022**, *1*, e9120034. [CrossRef]
17. Lee, K.H.; Kim, H.S.; Kim, M.; Roh, J.W.; Lim, J.H.; Kim, W.J.; Kim, S.-i.; Lee, W. Isovalent sulfur substitution to induce a simultaneous increase in the effective mass and weighted mobility of a p-type Bi-Sb-Te alloy: An approach to enhance the thermoelectric performance over a wide temperature range. *Acta Mater.* **2021**, *205*, 116578. [CrossRef]
18. He, X.; Zhang, H.; Nose, T.; Katase, T.; Tadano, T.; Ide, K.; Ueda, S.; Hiramatsu, H.; Hosono, H.; Kamiya, T. Degenerated Hole Doping and Ultra-Low Lattice Thermal Conductivity in Polycrystalline SnSe by Nonequilibrium Isovalent Te Substitution. *Adv. Sci.* **2022**, *9*, 2105958. [CrossRef]
19. Gao, B.; Tang, J.; Meng, F.; Li, W. Band manipulation for high thermoelectric performance in SnTe through heavy CdSe-alloying. *J. Mater.* **2019**, *5*, 111–117. [CrossRef]
20. Tao, Q.; Deng, R.; Li, J.; Yan, Y.; Su, X.; Poudeu, P.F.P.; Tang, X. Enhanced Thermoelectric Performance of Bi$_{0.46}$Sb$_{1.54}$Te$_3$ Nanostructured with CdTe. *ACS Appl. Mater. Interfaces* **2020**, *12*, 26330–26341. [CrossRef]
21. Giannozzi, P.; Baroni, S.; Bonini, N.; Calandra, M.; Car, R.; Cavazzoni, C.; Ceresoli, D.; Chiarotti, G.L.; Cococcioni, M.; Dabo, I.; et al. Quantum ESPRESSO: A modular and open-source software project for quantum simulations of materials. *J. Phys. Condens. Matter* **2009**, *21*, 395502. [CrossRef] [PubMed]
22. Giannozzi, P.; Andreussi, O.; Brumme, T.; Bunau, O.; Nardelli, M.B.; Calandra, M.; Car, R.; Cavazzoni, C.; Ceresoli, D.; Cococcioni, M.; et al. Advanced capabilities for materials modelling with Quantum ESPRESSO. *J. Phys. Condens. Matter* **2017**, *29*, 465901. [CrossRef] [PubMed]
23. Perdew, J.P.; Burke, K.; Ernzerhof, M. Generalized Gradient Approximation Made Simple. *Phys. Rev. Lett.* **1996**, *77*, 3865–3868. [CrossRef] [PubMed]
24. Prandini, G.; Marrazzo, A.; Castelli, I.E.; Mounet, N.; Marzari, N. Precision and efficiency in solid-state pseudopotential calculations. *NPJ Comput. Mater.* **2018**, *4*, 72. [CrossRef]
25. Heyd, J.; Scuseria, G.E.; Ernzerhof, M. Hybrid functionals based on a screened Coulomb potential. *J. Chem. Phys.* **2003**, *118*, 8207–8215. [CrossRef]
26. Krukau, A.V.; Vydrov, O.A.; Izmaylov, A.F.; Scuseria, G.E. Influence of the exchange screening parameter on the performance of screened hybrid functionals. *J. Chem. Phys.* **2006**, *125*, 224106. [CrossRef] [PubMed]
27. Maintz, S.; Deringer, V.L.; Tchougréeff, A.L.; Dronskowski, R. Analytic projection from plane-wave and PAW wavefunctions and application to chemical-bonding analysis in solids. *J. Comput. Chem.* **2013**, *34*, 2557–2567. [CrossRef] [PubMed]
28. Maintz, S.; Deringer, V.L.; Tchougréeff, A.L.; Dronskowski, R. LOBSTER: A tool to extract chemical bonding from plane-wave based DFT. *J. Comput. Chem.* **2016**, *37*, 1030–1035. [CrossRef] [PubMed]
29. Hill, R. The Elastic Behaviour of a Crystalline Aggregate. *Proc. Phys. Soc. Sect.* **1952**, *65*, 349. [CrossRef]
30. Yalameha, S.; Nourbakhsh, Z.; Vashaee, D. ElATools: A tool for analyzing anisotropic elastic properties of the 2D and 3D materials. *Comput. Phys. Commun.* **2022**, *271*, 108195. [CrossRef]
31. Momma, K.; Izumi, F. VESTA: A three-dimensional visualization system for electronic and structural analysis. *J. Appl. Crystallogr.* **2008**, *41*, 653–658. [CrossRef]
32. Madsen, G.K.; Singh, D.J. BoltzTraP. A code for calculating band-structure dependent quantities. *Comput. Phys. Commun.* **2006**, *175*, 67–71. [CrossRef]
33. Samsonidze, G.; Kozinsky, B. Accelerated Screening of Thermoelectric Materials by First-Principles Computations of Electron–Phonon Scattering. *Adv. Energy Mater.* **2018**, *8*, 1800246. [CrossRef]

34. Li, W.; Carrete, J.; Katcho, N.A.; Mingo, N. ShengBTE: A solver of the Boltzmann transport equation for phonons. *Comput. Phys. Commun.* **2014**, *185*, 1747–1758. [CrossRef]
35. Gopal, P.; Fornari, M.; Curtarolo, S.; Agapito, L.A.; Liyanage, L.S.I.; Nardelli, M.B. Improved predictions of the physical properties of Zn- and Cd-based wide band-gap semiconductors: A validation of the ACBN0 functional. *Phys. Rev. B* **2015**, *91*, 245202. [CrossRef]
36. Polak, M.P.; Scharoch, P.; Kudrawiec, R. The effect of isovalent doping on the electronic band structure of group IV semiconductors. *J. Phys. Appl. Phys.* **2020**, *54*, 085102. [CrossRef]
37. Snyder, G.J.; Toberer, E.S. Complex thermoelectric materials. *Nat. Mater.* **2008**, *7*, 105–114. [CrossRef] [PubMed]
38. Yadav, A.; Deshmukh, P.; Roberts, K.; Jisrawi, N.; Valluri, S. An analytic study of the Wiedemann–Franz law and the thermoelectric figure of merit. *J. Phys. Commun.* **2019**, *3*, 105001. [CrossRef]
39. Ahamed, S.T.; Kulsi, C.; Kirti.; Banerjee, D.; Srivastava, D.N.; Mondal, A. Synthesis of multifunctional CdSe and Pd quantum dot decorated CdSe thin films for photocatalytic, electrocatalytic and thermoelectric applications. *Surfaces Interfaces* **2021**, *25*, 101149. [CrossRef]
40. Cao, W.; Wang, Z.; Miao, L.; Shi, J.; Xiong, R. Thermoelectric Properties of Strained β-Cu_2Se. *ACS Appl. Mater. Interfaces* **2021**, *13*, 34367–34373. [CrossRef]
41. Cao, W.; Wang, Z.; Miao, L.; Shi, J.; Xiong, R. Extremely Anisotropic Thermoelectric Properties of SnSe Under Pressure. *Energy Environ. Mater.* **2023**, *6*, e12361. [CrossRef]
42. Huang, L.F.; Zeng, Z. Roles of Mass, Structure, and Bond Strength in the Phonon Properties and Lattice Anharmonicity of Single-Layer Mo and W Dichalcogenides. *J. Phys. Chem. C* **2015**, *119*, 18779–18789. [CrossRef]
43. Yang, X.X.; Li, J.W.; Zhou, Z.F.; Wang, Y.; Yang, L.W.; Zheng, W.T.; Sun, C.Q. Raman spectroscopic determination of the length, strength, compressibility, Debye temperature, elasticity, and force constant of the C-C bond in graphene. *Nanoscale* **2012**, *4*, 502–510. [CrossRef] [PubMed]
44. Delaire, O.; Ma, J.; Marty, K.; May, A.F.; McGuire, M.A.; Du, M.H.; Singh, D.J.; Podlesnyak, A.; Ehlers, G.; Lumsden, M.D.; et al. Giant anharmonic phonon scattering in PbTe. *Nat. Mater.* **2011**, *10*, 614–619. [CrossRef]
45. Kundu, A.; Ma, J.; Carrete, J.; Madsen, G.K.H.; Li, W. Anomalously large lattice thermal conductivity in metallic tungsten carbide and its origin in the electronic structure. *Mater. Today Phys.* **2020**, *13*, 100214. [CrossRef]

Disclaimer/Publisher's Note: The statements, opinions and data contained in all publications are solely those of the individual author(s) and contributor(s) and not of MDPI and/or the editor(s). MDPI and/or the editor(s) disclaim responsibility for any injury to people or property resulting from any ideas, methods, instructions or products referred to in the content.

Enhanced Thermoelectric Properties of Nb-Doped Ti(FeCoNi)Sb Pseudo-Ternary Half-Heusler Alloys Prepared Using the Microwave Method

Ruipeng Zhang [1], Jianbiao Kong [2], Yangbo Hou [2], Linghao Zhao [1,3], Junliang Zhu [1], Changcun Li [1] and Degang Zhao [1,*]

1. School of Materials Science and Engineering, University of Jinan, Jinan 250022, China
2. Heze Institute of Product Inspection and Testing, Heze 274000, China
3. Shenzhen Institute of Advanced Electronic Materials, Shenzhen Institute of Advanced Technology, Chinese Academy of Sciences, Shenzhen 518055, China
* Correspondence: mse_zhaodg@ujn.edu.cn; Tel.: +86-531-82769998

Abstract: Pseudo-ternary half-Heusler thermoelectric materials, which are formed by filling the B sites of traditional ternary half-Heusler thermoelectric materials of ABX with equal atomic proportions of various elements, have attracted more and more attention due to their lower intrinsic lattice thermal conductivity. High-purity and relatively dense $Ti_{1-x}Nb_x$(FeCoNi)Sb (x = 0, 0.01, 0.03, 0.05, 0.07 and 0.1) alloys were prepared via microwave synthesis combined with rapid hot-pressing sintering, and their thermoelectric properties are investigated in this work. The Seebeck coefficient was markedly increased via Nb substitution at Ti sites, which resulted in the optimized power factor of 1.45 $\mu Wcm^{-1}K^{-2}$ for n-type $Ti_{0.93}Nb_{0.07}$(FeCoNi)Sb at 750 K. In addition, the lattice thermal conductivity was largely decreased due to the increase in phonon scattering caused by point defects, mass fluctuation and strain fluctuation introduced by Nb-doping. At 750 K, the lattice thermal conductivity of $Ti_{0.97}Nb_{0.03}$(FeCoNi)Sb is 2.37 $Wm^{-1}K^{-1}$, which is 55% and 23% lower than that of TiCoSb and Ti(FeCoNi)Sb, respectively. Compared with TiCoSb, the ZT of the $Ti_{1-x}Nb_x$(FeCoNi)Sb samples were significantly increased. The average ZT values of the Nb-doped pseudo-ternary half-Heusler samples were dozens of times that of the TiCoSb prepared using the same process.

Keywords: thermoelectric; half-Heusler; pseudo-ternary

1. Introduction

The non-renewable nature of traditional mineral energy and the environmental pollution caused by its consumption pose serious challenges to the energy supply of global economic development [1,2]. Thermoelectric materials, as a kind of green energy material that can directly convert heat and electricity into each other, have great potential in waste heat utilization and thermoelectric refrigeration, which can make great contributions to solve the energy crisis [3–5]. In the history of materials science, a variety of performance parameters are used in order to quantify the merits of a material more intuitively. The quantitative parameter used to measure the performance of thermoelectric materials is the dimensionless figure of merit ZT as follows:

$$ZT = S^2 \sigma T / \kappa \quad (1)$$

where $S^2\sigma$ is the power factor (PF), obtained from the Seebeck coefficient (S) and electrical conductivity (σ), T is the absolute temperature and κ is the thermal conductivity, which is the sum of the carrier thermal conductivity (κ_E) and lattice thermal conductivity (κ_L) [6,7]. It can be seen from Formula (1) that the enhancement of performance usually starts from the numerator and denominator aspects. The former is to increase the PF by means of non-equivalent atom doping [8–10], energy band engineering, etc. [11–13], while the latter often uses nano engineering [14–16], defect engineering and other means to reduce the lattice thermal conductivity [17,18]. Due to the coupling relationship between the parameters, an excessive pursuit of the optimization of a single parameter often cannot achieve the enhancement of the ZT of thermoelectric materials [19]. For example, an excessive pursuit of a large carrier concentration to increase the conductivity can inevitably lead to a deterioration of the Seebeck coefficient. Therefore, how to achieve a greater optimization of performance remains a crucial part of thermoelectric research.

The half-Heusler (HH) compound crystallizes into the cubic MgAgAs-type structure, with ABX as the general chemical formula, which should be viewed as the interpenetration of four face-centered cubic lattices. Elements at the X position are usually elements of the main group, while elements at the A and B positions are elements of the transition group [20]. The elements at the A, B and X positions of the HH can be replaced by different elements without destroying their crystal structure. Half-Heusler alloys are superior thermoelectric materials at the middle and high temperature region due to their excellent electrical conductivity and Seebeck coefficient brought about by their highly symmetric crystal structures [20,21]. In addition, their better thermal stability and mechanical properties compared with other types of thermoelectric materials are also indispensable factors [22]. However, the high symmetry of the crystal structure makes HH compounds not only have excellent electrical properties, but also a high thermal conductivity, which is unfavorable to ZT according to the formula. The excessive thermal conductivity of traditional ternary HH thermoelectric materials has become the limit of their commercial application, and researchers have tried various optimization strategies to reduce their thermal conductivity [23–25]. The lattice thermal conductivity, which is the main contribution of κ, is largely determined by the number of atoms in the primitive unit cell (N) [26]. The strategy of reducing the thermal conductivity by increasing N to introduce more disorder and lattice distortion based on the high-entropy core effect has been proven in many thermoelectric material systems such as $Ge_{0.61}Ag_{0.11}Sb_{0.13}Pb_{0.12}Bi_{0.01}Te$ [27], $AgSnSbSe_3$ [28], $BiSbTe_{1.5}Se_{1.5}$ [29], $Sn_{0.25}Pb_{0.25}Mn_{0.25}Ge_{0.25}Te$ [30], etc. The high substitutability of atoms at each position of HH compounds fits well with this strategy. Consequently, pseudo-ternary half-Heusler thermoelectric materials ($N > 3$) with a low intrinsic lattice thermal conductivity caused by disorder scattering and a smaller phonon group velocity have attracted widespread attention in recent years [31]. And many encouraging pseudo-ternary HH results have been achieved [32]. Wang et al. [33] prepared the $Ti_2FeNiSb_2$ double half-Heusler compound by increasing the number of atoms in the primitive unit cell. On the basis of replacing Sb with Sn, the small lattice thermal conductivity of 1.95 $Wm^{-1}K^{-1}$ and the peak ZT value of 0.52 for $Ti_{1.6}Hf_{0.4}FeNiSb_{1.7}Sn_{0.3}$ are obtained at 923 K. Luo et al. [34] and Wang et al. [35] realized the p-n transformation and the thermoelectric performance optimization of $Ti(Fe_{1/3}Co_{1/3}Ni_{1/3})Sb$ by adjusting the ratio of Fe and Ni.

It is commendable that, as verified by previous reports, microwave-synthesized samples have the advantages of refining grains and improving the uniformity of composition [17,36–39]. With the development in recent years, the technique of microwave synthesis has been relatively reliable, and the dense bulk ceramic or metallic samples can be synthesized in a few minutes with the help of carbon, silicon carbide and other dielectric materials with a high dielectric constant, which greatly reduces the preparation cycle and cost [40–42]. Birkel et al. [43] successfully prepared TiNiSn half-Heusler compounds by using 3.6 g of granular carbon as an absorbing material via microwave heating for 1 min. Its performance was tested and compared with that of TiNiSn prepared using the arc melting method. The thermoelectric properties of the two samples are basically the same except for the presence of a trace impurity phase. Lei et al. [44] also successfully synthesized a TiNiSn half-Heusler compound by heating it in a microwave device for 5 min. In addition, excellent thermoelectric material systems such as Mg_2Si [45], PbTe [46] and Bi_2Te_3 [47] have been documented to prove the successful application of a microwave in their preparation. In this work, high-purity and high-density $Ti_{1-x}Nb_x(FeCoNi)Sb$ (x = 0, 0.01, 0.03, 0.05,

0.07 and 0.1) pseudo-ternary half-Heusler alloys were prepared via microwave synthesis combined with the ball-milling and rapid hot-pressing sintering processes, and the effects of Nb substitution at the Ti site on its phase composition, microstructure and thermoelectric properties were studied. As a comparison, a TiCoSb sample was prepared using the same technical route. This work is expected to provide a feasible idea for optimizing the properties of pseudo-ternary HH thermoelectric materials with a more economical synthesis process.

2. Materials and Methods

The original powders of TiCoSb and $Ti_{1-x}Nb_x(FeCoNi)Sb$ (x = 0, 0.01, 0.03, 0.05, 0.07 and 0.1) samples were titanium powder (99.98%, 300 mesh, MACKLIN, Shanghai, China), niobium powder (99.999%, 300 mesh, MACKLIN, Shanghai, China), iron powder (99.9%, 200 mesh, MACKLIN, Shanghai, China), cobalt powder (99.8%, 200 mesh, MACKLIN, Shanghai, China), nickel powder (99.99%, 200 mesh, MACKLIN, Shanghai, China) and antimony powder (99.9%, 800 mesh, MACKLIN, Shanghai, China). The powders, which were weighed in accordance with stoichiometric ratio, were evenly mixed and then loaded into a cold-pressed mold, which was pressed under an axial pressure of 10 MPa for 5 min to obtain cylindrical billet at room temperature. Then, the obtained billet was put into the prepared clean quartz tube for vacuum sealing ($\leq$0.01 Pa) to ensure that the billet would not oxidize during the synthesis process. The sealed quartz tube was placed in an alumina crucible that filled with expanded graphite powder (99.9%, 10–30 μm, MACKLIN, Shanghai, China) as an absorbing material. The whole tube was placed in a self-made microwave synthesis apparatus for microwave synthesis for 5 min. The sketch map of the microwave synthesis device is shown in Figure 1, and its power is 900 W. Then, the block billet obtained via microwave synthesis was mechanically broken and ground for 5 h in a planetary ball mill at a rotating speed of 300 rpm. The compact disc samples were obtained by loading the ball-milled powder in batches into a hot pressing mold with a diameter of 12 mm, and rapid hot-pressing sintering for 20 min at a temperature of 1073 K with a pressure of 80 MPa. In the process of rapid hot-pressing sintering, the heating rate was 100 K/min.

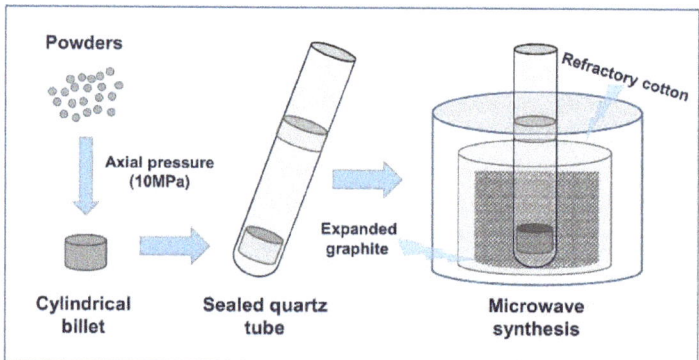

Figure 1. The process of microwave synthesis.

$Ti_{1-x}Nb_x(FeCoNi)Sb$ (x = 0, 0.01, 0.03, 0.05, 0.07 and 0.1) alloys were characterized via X-ray diffraction (XRD) using a Rigaku Ultima IV diffractometer in the 2θ range of 10–90°. The surface morphology and microstructure of samples were analyzed using a scanning electron microscope (SEM, JEOL, JXA-8100, Tokyo, Japan) and energy dispersive spectrometer. The Archimedes method was used to measure the density of $Ti_{1-x}Nb_x(FeCoNi)Sb$ alloys. The electrical conductivity and Seebeck coefficient of alloys at room temperature to 750 K can be measured simultaneously using the ZEM-3 system (ULVAC-RIKO, Yokohama, Japan) in a low-pressure (~10^2 Pa) helium atmosphere. The thermal conductivity (κ) of samples can be calculated according to the formula $\kappa = C_p d \lambda$, where C_p is the specific heat capacity of material, d is the density and λ is the thermal diffusion coefficient. The thermal diffusivity coefficient of $Ti_{1-x}Nb_x(FeCoNi)Sb$ alloys can be obtained using the laser flash thermal analyzer (LFA-457, Netzsch, Bavaria, Germany). The estimated error of measurement of the above physical parameters (d, σ, S and κ) was at most 5%.

3. Results and Discussion
3.1. Phase Analysis

As shown in Table 1, the relative densities of the $Ti_{1-x}Nb_x(FeCoNi)Sb$ (x = 0, 0.01, 0.03, 0.05, 0.07 and 0.1) alloys are above 95%, which were measured over more than three measurements. A relatively high density is usually positively correlated with the electrical properties of thermoelectric materials [48].

Table 1. Room-temperature density (g/cm^3), relative density (%), carrier concentration n (10^{20} cm^{-3}) and carrier mobility μ (cm^{2}v^{-1}s^{-1}) of TiCoSb and $Ti_{1-x}Nb_x(FeCoNi)Sb$ (x = 0, 0.01, 0.03, 0.05, 0.07 and 0.1) alloys.

Composition	Density (g/cm^3)	Relative Density (%)	Carrier Concentration (10^{20} cm^{-3})	Mobility (cm^{2}v^{-1}s^{-1})
TiCoSb	7.133	95.70	4.673	2.885
Ti(FeCoNi)Sb	7.104	95.30	1.471	5.914
Ti$_{0.99}$Nb$_{0.01}$(FeCoNi)Sb	7.093	95.16	1.562	5.451
Ti$_{0.97}$Nb$_{0.03}$(FeCoNi)Sb	7.184	96.38	1.866	4.873
Ti$_{0.95}$Nb$_{0.05}$(FeCoNi)Sb	7.254	97.32	2.185	4.467
Ti$_{0.93}$Nb$_{0.07}$(FeCoNi)Sb	7.315	98.13	2.454	4.239
Ti$_{0.90}$Nb$_{0.10}$(FeCoNi)Sb	7.331	98.39	2.787	3.966

Figure 2 shows the X-ray diffraction results of the $Ti_{1-x}Nb_x(FeCoNi)Sb$ (x = 0, 0.01, 0.03, 0.05, 0.07 and 0.1) pseudo-ternary half-Heusler alloys prepared via microwave synthesis and rapid hot-pressing sintering. The main diffraction peaks of the samples were well indexed to the cubic half-Heusler TiCoSb (PDF#65-5103), which indicated that the pseudo-ternary HH thermoelectric materials with a high density can be successfully prepared via microwave synthesis combined with rapid hot-pressing sintering. As can be seen from Figure 2a, the XRD spectrum of the powder showed no other impurity phase except for the half-Heusler phase. However, the XRD results of the sintered samples in Figure 2b show the presence of trace Fe and FeSb secondary phases. The phase fractions of the $Ti_{1-x}Nb_x(FeCoNi)Sb$ (x = 0, 0.01, 0.03, 0.05, 0.07 and 0.1) alloys are shown in Table 2. The existence of these metallic secondary phases inevitably affected the electrical conductivity and thermal conductivity of the $Ti_{1-x}Nb_x(FeCoNi)Sb$ alloys [23,49–53].

Table 2. The phase fractions of $Ti_{1-x}Nb_x(FeCoNi)Sb$ (x = 0, 0.01, 0.03, 0.05, 0.07 and 0.1) alloys.

Composition	HH (%)	Fe (%)	FeSb (%)
Ti(FeCoNi)Sb	95.1	4.9	~
Ti$_{0.99}$Nb$_{0.01}$(FeCoNi)Sb	94.6	5.4	~
Ti$_{0.97}$Nb$_{0.03}$(FeCoNi)Sb	92.8	7.2	~
Ti$_{0.95}$Nb$_{0.05}$(FeCoNi)Sb	90.3	9.7	~
Ti$_{0.93}$Nb$_{0.07}$(FeCoNi)Sb	84.5	11.3	4.2
Ti$_{0.90}$Nb$_{0.10}$(FeCoNi)Sb	79.1	13.6	7.3

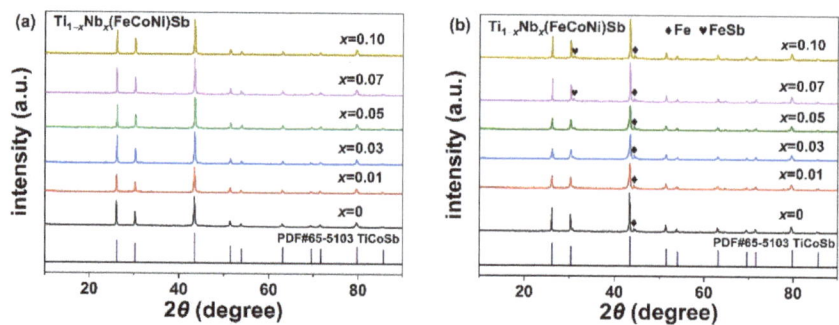

Figure 2. The XRD spectra of powder samples after microwave synthesis (**a**) and compact disc samples after rapid hot-pressing sintering (**b**) of Ti$_{1-x}$Nb$_x$(FeCoNi)Sb (x = 0, 0.01, 0.03, 0.05, 0.07 and 0.1) alloys.

3.2. Microstructural Characterization

The fractural cross-section secondary electron images of the Ti$_{0.9}$Nb$_{0.1}$(FeCoNi)Sb sample in Figure 3a–c show the existence of small grains among large grains. In addition, the back-scattering electron images of the Ti$_{0.9}$Nb$_{0.1}$(FeCoNi)Sb sample also show obvious phase segregation, just as shown in Figure 3d–f. To determine the phase composition, the energy dispersive X-ray (EDX) compositional point analysis and compositional mapping analysis were carried out on the Ti$_{0.9}$Nb$_{0.1}$(FeCoNi)Sb sample. As shown in Figure 4a, the atomic ratio of Fe, Co and Ni obtained via the EDX compositional surface analysis was close to 1:1:1. Moreover, the mapping results of the Ti$_{0.9}$Nb$_{0.1}$(FeCoNi)Sb sample show the constituent element segregation, which verifies the presence of the secondary phases in the XRD results. Based on the results of the compositional point analysis, it can be speculated that the gray regions (point 3, blue) were the Ti$_{0.9}$Nb$_{0.1}$(FeCoNi)Sb HH matrix phase, the dark regions (point 2, yellow) were the Ti-rich phase and Fe-rich phase and the bright regions (point 1, green) were the Sb-rich phase.

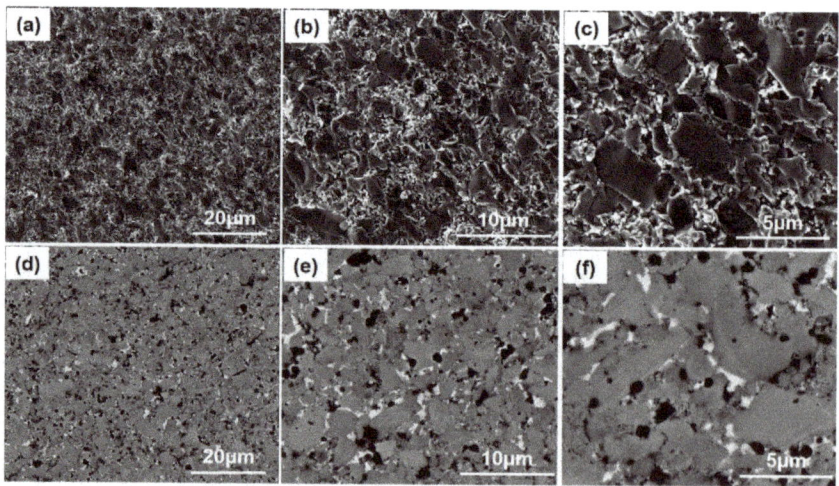

Figure 3. The fractural cross-section secondary electron images (**a**–**c**) and back-scattering electron images (**d**–**f**) of the Ti$_{0.9}$Nb$_{0.1}$(FeCoNi)Sb sample at different magnifications.

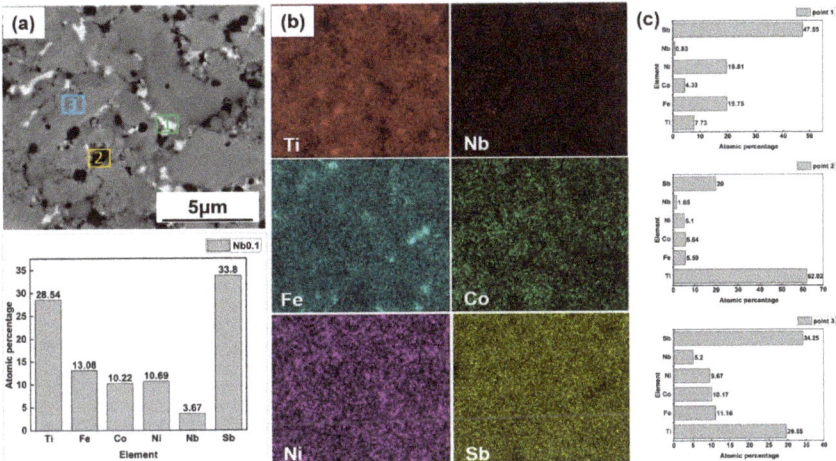

Figure 4. The EDX compositional surface analysis (**a**), mapping (**b**) and point analysis (**c**) of the Ti$_{0.9}$Nb$_{0.1}$(FeCoNi)Sb sample.

3.3. Thermoelectric Transport Properties

The electrical properties of the Ti$_{1-x}$Nb$_x$(FeCoNi)Sb (x = 0, 0.01, 0.03, 0.05, 0.07 and 0.1) alloys as a function of temperature are shown in Figure 5. Except for the TiCoSb and Ti(FeCoNi)Sb samples, the σ of all the samples followed a temperature dependence of $T^{0.5}$, implying that the alloy disorder scattering and ionized impurity scattering caused by the Ti-Nb heteroatomic substitution dominated the charge transport [54,55]. As shown in Figure 5a, except for the decrease in the conductivity of Ti$_{0.99}$Nb$_{0.01}$(FeCoNi)Sb, the electrical conductivity of the other samples increased due to the addition of valence electrons involved in electrical transport [20,56]. The decrease in Ti$_{0.99}$Nb$_{0.01}$(FeCoNi)Sb may be due to the fact that although the incorporation of Nb led to an increase in the carrier concentration (n), it was not enough to compensate for the decrease in the carrier mobility that arose from the introduction of point defects. However, when $x > 0.01$, with the increase in carrier concentration, the adverse impact of ionized impurity scattering on the carrier mobility was weakened due to the screening effect [57]. In addition, the presence of metallic secondary phases constructed the internal high-speed channel of electrons inside the samples, which improved the electrical conductivity.

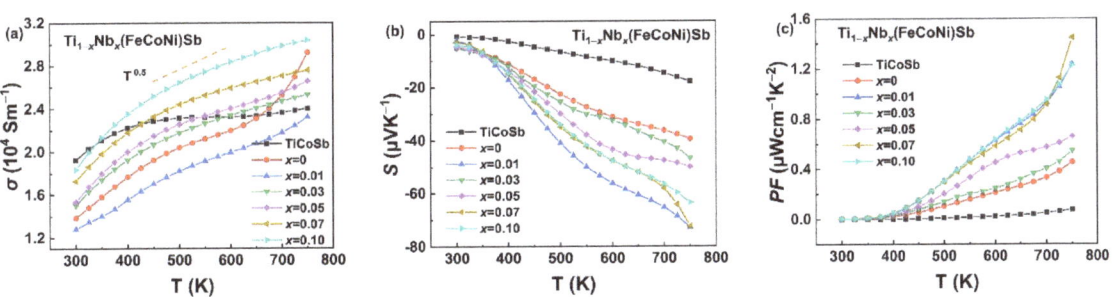

Figure 5. The curves of the electrical properties of TiCoSb and Ti$_{1-x}$Nb$_x$(FeCoNi)Sb (x = 0, 0.01, 0.03, 0.05, 0.07 and 0.1) alloys. Electrical conductivity (**a**), Seebeck coefficient (**b**) and power factor (**c**).

Figure 5b shows that the Seebeck coefficient was negative in the range of 300 K to 750 K, indicating an n-type conducting behavior of the Ti$_{1-x}$Nb$_x$(FeCoNi)Sb samples. For degenerate semiconductors, the Seebeck coefficient S is usually associated with the carrier concentration that can be represented by the Mott equation [19],

$$S = \frac{8\pi^2 k_B^2}{3eh^2} m^* T \left(\frac{\pi}{3n}\right)^{\frac{2}{3}} \quad (2)$$

where k_B, e, h, T, m^* and n are the Boltzmann constant, elementary charge, Planck constant, absolute temperature, density of states (DOSs), effective mass and carrier concentration, respectively. As shown in Figure 6a, the absolute values of S and the average values of $|S|$ of all the Nb-doped samples were larger than that of TiCoSb and Ti(FeCoNi)Sb, indicating that the incorporation of Nb can be beneficial to the improvement of the Seebeck coefficient, which can be due to the increase in the effective mass [34]. However, it can be seen from formula (2) and Table 1 that as the Nb content increased, the coupling relationship between the carrier concentration and the Seebeck coefficient hindered the increase in the Seebeck coefficient [58]. A large m^* of $Ti_{1−x}Nb_x(FeCoNi)Sb$ is conducive to the high Seebeck coefficient, but may, in turn, lead to a deterioration in the carrier mobility (μ) if the band effective mass is also high [59]. Furthermore, the intensified alloy scattering of carriers resulted from the Nb doping was also unfavorable for μ. The power factors (PFs) calculated using the electrical conductivity and Seebeck coefficient of the $Ti_{1−x}Nb_x(FeCoNi)Sb$ alloys are shown in Figure 5c. The PFs of all the pseudo-ternary alloys including the undoped Ti(FeCoNi)Sb were significantly higher than that of the TiCoSb alloy. The maximum PF value was obtained for $Ti_{0.93}Nb_{0.07}(FeCoNi)Sb$ at 750 K, which was 1.45 $\mu Wcm^{−1}K^{−2}$, which is much higher than that of Ti(FeCoNi)Sb. The average power factor of the Nb-doped $Ti_{1−x}Nb_x(FeCoNi)Sb$ samples in Figure 6b also show significant improvement compared with TiCoSb and Ti(FeCoNi)Sb. It is worth noting that compared with other traditional tens or even hundreds of $\mu Wcm^{−1}K^{−2}$ of power factor such as TiCoSb [60], TiNiSn [61] and NbFeSb [50], the result is not enough to support the development of the most outstanding HH thermoelectric materials. Consequently, it is still necessary to further explore the optimization of the electrical properties of the Ti(FeCoNi)Sb pseudo-ternary half-Heusler alloy.

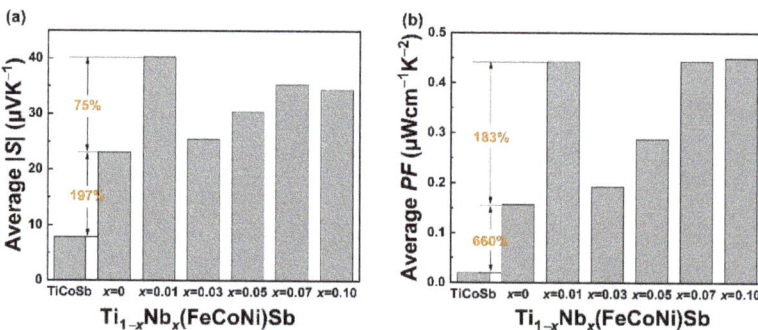

Figure 6. The average Seebeck coefficient (300–750 K) (**a**) and average power factor (300–750 K) (**b**) of TiCoSb and $Ti_{1−x}Nb_x(FeCoNi)Sb$ (x = 0, 0.01, 0.03, 0.05, 0.07 and 0.1) alloys.

The curves of the total thermal conductivity (κ), electronic thermal conductivity (κ_E) and lattice thermal conductivity (κ_L) of the $Ti_{1−x}Nb_x(FeCoNi)Sb$ (x = 0, 0.01, 0.03, 0.05, 0.07 and 0.1) alloys with temperature are shown in Figure 7. The thermal conductivity was significantly reduced with the equal proportional substitution of Fe, Co and Ni at the Co sites over the measured temperature. The electronic thermal conductivity was calculated using the Wiedemann–Franz law as follows:

$$\kappa_E = L\sigma T \quad (3)$$

where L is the Lorenz number. It can be seen from Figure 7b that the variation trend of the electronic thermal conductivity turned out to be consistent with that of the electrical conductivity, and there was no obvious gap between the samples. In contrast to the less noticeable drop in the electronic thermal conductivity, the decreasing trend of κ mainly resulted from the decrease in κ_L. The lattice thermal conductivity κ_L was obtained by subtracting the electronic contribution κ_E from the total thermal conductivity κ. The lattice thermal conductivity of the $Ti_{1-x}Nb_x(FeCoNi)Sb$ samples decreased with the increasing temperature, but this trend was partially slowed down at a high temperature due to the bipolar effect [62]. It can be observed from Figure 7c that the lattice thermal conductivity followed the $T^{-0.5}$ dependence, implying that the alloy disorder scattering of phonons should be dominant [54]. In addition, the contribution of the ionized impurity scattering of phonons cannot be ignored [55,63]. For most HH materials, the simple crystal structure tends to result in a relatively high lattice thermal conductivity ($\approx$10 Wm^{-1}K^{-1} at 300 K) [64], which is more pronounced if a metallic phase is also present in the matrix [65]. Benefiting from the smaller group velocity phonons and disorder scattering caused by a more complex crystal chemistry [31], Ti(FeCoNi)Sb (5.05 Wm^{-1}K^{-1} at 300 K) had the obviously lower κ_L compared with that of TiCoSb (8.07 Wm^{-1}K^{-1} at 300 K). The enhanced phonon scattering mechanism was attributed to several factors, such as point defects introduced by the substitution of Nb at the Ti sites, the mass and strain fluctuation and the lattice distortion generated by the increase in the configuration entropy with the equal proportional substitution of Fe, Co and Ni at the Co sites. Thus, the lattice thermal conductivity of the Nb-doped $Ti_{1-x}Nb_x(FeCoNi)Sb$ alloys was lower than that of Ti(FeCoNi)Sb. However, when Nb > 0.05, this trend was inhibited under the influence of more metallic secondary phases. In the measured temperature range, the lattice thermal conductivity of $Ti_{0.97}Nb_{0.03}MSb$ was decreased by 55% and 23% compared with that of TiCoSb and Ti(FeCoNi)Sb at the same temperature, respectively.

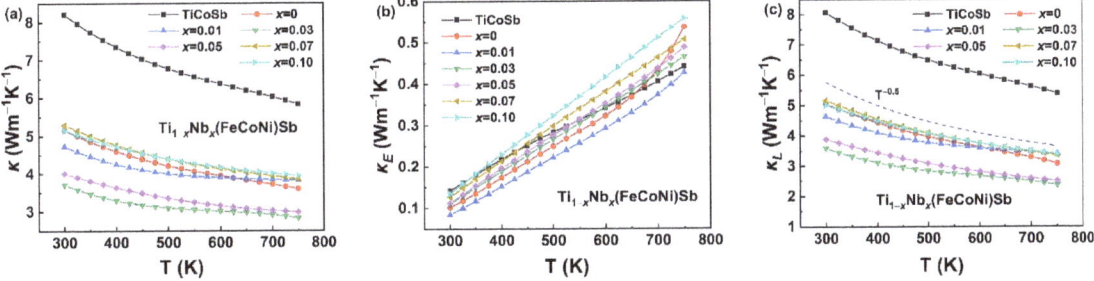

Figure 7. The curves of the thermal properties of TiCoSb and $Ti_{1-x}Nb_x(FeCoNi)Sb$ (x = 0, 0.01, 0.03, 0.05, 0.07 and 0.1) alloys. Total thermal conductivity (**a**), electronic thermal conductivity (**b**) and lattice thermal conductivity (**c**).

Figure 8a,b shows the temperature dependence of the ZT and average ZT values (300–750 K) of the $Ti_{1-x}Nb_x(FeCoNi)Sb$ (x = 0, 0.01, 0.03, 0.05, 0.07 and 0.1) alloys. With the increase in temperature, the ZT of each sample increased, but the gap also widened rapidly. The Nb doping and multi-element substitution at the Co sites resulted in a large gap in the S and κ_L between samples. The maximum ZT value was about 0.03 for the $Ti_{0.93}Nb_{0.07}(FeCoNi)Sb$ sample. However, this is still lower than the ZTs of the p-type $Ti(Fe_{1/3+0.15}Co_{1/3}Ni_{1/3-0.15})Sb$ and n-type $Ti(Fe_{1/3-0.1}Co_{1/3}Ni_{1/3+0.1})Sb$ at 750 K (above 0.2) reported by Luo et al. [34], which is speculated to be due to excessive differences in the electrical properties. Compared with the TiCoSb sample prepared using the same process, the ZT of the $Ti_{1-x}Nb_x(FeCoNi)Sb$ samples were obviously improved due to the increase in the Seebeck coefficient and the reduction in the lattice thermal conductivity. As shown in Figure 8b, the average ZTs in the temperature range of 300–750 K of the

Ti$_{1-x}$Nb$_x$(FeCoNi)Sb pseudo-ternary HH alloys were dozens of times that of TiCoSb, and this trend is further increased after the incorporation of Nb.

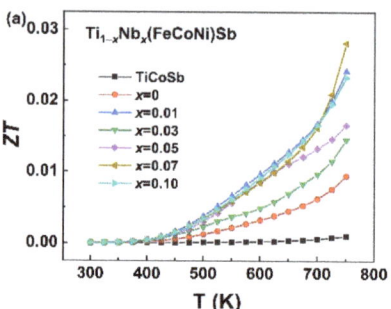
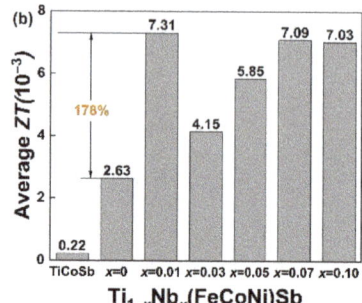

Figure 8. Temperature dependence of ZT (**a**) and average ZT (300–750 K) (**b**) of TiCoSb and Ti$_{1-x}$Nb$_x$(FeCoNi)Sb (x = 0, 0.01, 0.03, 0.05, 0.07 and 0.1) alloys.

4. Conclusions

High-purity and high-density Ti$_{1-x}$Nb$_x$(FeCoNi)Sb (x = 0, 0.01, 0.03, 0.05, 0.07 and 0.1) pseudo-ternary half-Heusler alloys were successfully prepared via microwave synthesis combined with rapid hot-pressing sintering. The multi-element substitution at the Co sites can effectively reduce the lattice thermal conductivity by introducing more phonon scattering mechanisms. In addition, the lattice thermal conductivity was greatly reduced due to the increase in phonon scattering caused by point defects, mass fluctuation and strain fluctuation introduced by Nb doping. The average ZT values of the Nb-doped pseudo-ternary half-Heusler samples were dozens of times that of TiCoSb prepared using the same process. Of course, the factors affecting the electrical properties that are too different from other HH material systems also need to be further studied.

Author Contributions: Conceptualization, R.Z. and D.Z.; methodology, R.Z., J.K. and Y.H.; formal analysis, R.Z. and C.L.; investigation, R.Z. and L.Z.; writing—original draft preparation, R.Z. and J.Z.; writing—review and editing, R.Z.; funding acquisition, D.Z. All authors have read and agreed to the published version of the manuscript.

Funding: This work was funded by the Taishan Scholar Program of Shandong Province (TSQN2023), the Shandong Province Higher Educational Youth Innovative Science and Technology Program (grant no. 2019KJA018) and the Leader of Scientific Research Studio Program of Jinan (grant no. 2021GXRC082).

Institutional Review Board Statement: Not applicable.

Informed Consent Statement: Not applicable.

Data Availability Statement: Not applicable.

Conflicts of Interest: The authors declare no conflict of interest.

References

1. Riffat, S.B.; Ma, X. Thermoelectrics: A review of present and potential applications. *Appl. Therm. Eng.* **2003**, *23*, 913–935. [CrossRef]
2. Shi, X.; He, J. Thermopower and harvesting heat. *Science* **2021**, *371*, 343–344. [CrossRef] [PubMed]
3. Uher, C.; Yang, J.; Hu, S.; Morelli, D.; Meisner, G. Transport properties of pure and doped MNiSn (M = Zr, Hf). *Phys. Rev. B* **1999**, *59*, 8615. [CrossRef]
4. He, J.; Tritt, T.M. Advances in thermoelectric materials research: Looking back and moving forward. *Science* **2017**, *357*, eaak9997. [CrossRef]
5. DiSalvo, F.J. Thermoelectric cooling and power generation. *Science* **1999**, *285*, 703–706. [CrossRef]
6. Gayner, C.; Kar, K.K. Recent advances in thermoelectric materials. *Prog. Mater. Sci.* **2016**, *83*, 330–382. [CrossRef]

7. Shi, X.L.; Zou, J.; Chen, Z.G. Advanced thermoelectric design: From materials and structures to devices. *Chem. Rev.* **2020**, *120*, 7399–7515. [CrossRef]
8. Bhattacharya, S.; Pope, A.; Littleton, R., IV; Tritt, T.M.; Ponnambalam, V.; Xia, Y.; Poon, S. Effect of Sb doping on the thermoelectric properties of Ti-based half-Heusler compounds, TiNiSn$_{1-x}$Sb$_x$. *Appl. Phys. Lett.* **2000**, *77*, 2476–2478. [CrossRef]
9. Yang, X.; Jiang, Z.; Kang, H.; Chen, Z.; Guo, E.; Liu, D.; Yang, F.; Li, R.; Jiang, X.; Wang, T. Enhanced thermoelectric performance of Zr$_{1-x}$Ta$_x$NiSn Half-Heusler alloys by diagonal-rule doping. *ACS Appl. Mater. Interfaces* **2019**, *12*, 3773–3783. [CrossRef]
10. Chen, J.L.; Yang, H.; Liu, C.; Liang, J.; Miao, L.; Zhang, Z.; Liu, P.; Yoshida, K.; Chen, C.; Zhang, Q. Strategy of Extra Zr Doping on the Enhancement of Thermoelectric Performance for TiZr$_x$NiSn Synthesized by a Modified Solid-State Reaction. *ACS Appl. Mater. Interfaces* **2021**, *13*, 48801–48809. [CrossRef]
11. Tan, G.; Liu, W.; Chi, H.; Su, X.; Wang, S.; Yan, Y.; Tang, X.; Wong-Ng, W.; Uher, C. Realization of high thermoelectric performance in *p*-type unfilled ternary skutterudites FeSb$_{2+x}$Te$_{1-x}$ via band structure modification and significant point defect scattering. *Acta Mater.* **2013**, *61*, 7693–7704. [CrossRef]
12. Ning, S.; Huang, S.; Zhang, Z.; Qi, N.; Jiang, M.; Chen, Z.; Tang, X. Band convergence boosted high thermoelectric performance of Zintl compound Mg$_3$Sb$_2$ achieved by biaxial strains. *J. Mater.* **2022**, *8*, 1086–1094. [CrossRef]
13. Guo, S.; Anand, S.; Brod, M.K.; Zhang, Y.S.; Snyder, G.J. Conduction band engineering of half-Heusler thermoelectrics using orbital chemistry. *J. Mater. Chem. A* **2022**, *10*, 3051–3057. [CrossRef]
14. Li, F.; Bo, L.; Zhang, R.; Liu, S.; Zhu, J.; Zuo, M.; Zhao, D. Enhanced thermoelectric properties of Te doped polycrystalline Sn$_{0.94}$Pb$_{0.01}$Se. *Nanomaterials* **2022**, *12*, 1575. [CrossRef]
15. Zhang, K.; Zhang, Q.; Wang, L.; Jiang, W.; Chen, L. Enhanced thermoelectric performance of Se-doped PbTe bulk materials via nanostructuring and multi-scale hierarchical architecture. *J. Alloys Compd.* **2017**, *725*, 563–572. [CrossRef]
16. Kang, H.B.; Poudel, B.; Li, W.J.; Lee, H.; Saparamadu, U.; Nozariasbmarz, A.; Kang, M.; Gupta, A.; Heremans, J.; Priya, S. Decoupled phononic-electronic transport in multi-phase *n*-type Half-Heusler nanocomposites enabling efficient high temperature power generation. *Mater. Today* **2020**, *36*, 63–72. [CrossRef]
17. Bo, L.; Zhang, R.; Zhao, H.; Hou, Y.; Wang, X.; Zhu, J.; Zhao, L.; Zuo, M.; Zhao, D. Achieving High Thermoelectric Properties of Cu$_2$Se via Lattice Softening and Phonon Scattering Mechanism. *ACS Appl. Energy Mater.* **2022**, *5*, 6453–6461. [CrossRef]
18. Sun, Y.X.; Qiu, W.B.; Zhao, L.W.; He, H.; Yang, L.; Chen, L.Q.; Deng, H.; Shi, X.M.; Tang, J. Defects engineering driven high power factor of ZrNiSn-based Half-Heusler thermoelectric materials. *Chem. Phys. Lett.* **2020**, *755*, 137770–137776. [CrossRef]
19. Snyder, G.J.; Toberer, E.S. Complex Thermoelectric Materials. *Nat. Mater.* **2008**, *7*, 105–114. [CrossRef]
20. Huang, L.; He, R.; Chen, S.; Zhang, H.; Dahal, K.; Zhou, H.; Wang, H.; Zhang, Q.; Ren, Z. A new *n*-type half-Heusler thermoelectric material NbCoSb. *Mater. Res. Bull.* **2015**, *70*, 773–778. [CrossRef]
21. Bhattacharya, S.; Skove, M.J.; Russell, M.; Tritt, T.M.; Xia, Y.; Ponnambalam, V.; Poon, S.J.; Thadhani, N. Effect of boundary scattering on the thermal conductivity of TiNiSn-based half-Heusler alloys. *Phys. Rev. B* **2008**, *77*, 184203. [CrossRef]
22. Berry, T.; Fu, C.; Auffermann, G.; Fecher, G.H.; Schnelle, W.; Serrano-Sanchez, F.; Yue, Y.; Liang, H.; Felser, C. Enhancing Thermoelectric Performance of TiNiSn half-Heusler Compounds via Modulation Doping. *Chem. Mater.* **2017**, *29*, 7042–7048. [CrossRef]
23. Hu, C.; Xia, K.; Fu, C.; Zhao, X.; Zhu, T. Carrier Grain Boundary Scattering in Thermoelectric Materials. *Energy Environ. Sci.* **2022**, *15*, 1406–1422. [CrossRef]
24. Ghosh, T.; Dutta, M.; Sarkar, D.; Biswas, K. Insights into low thermal conductivity in inorganic materials for thermoelectrics. *J. Am. Chem. Soc.* **2022**, *144*, 10099–10118. [CrossRef]
25. Zhang, X.L.; Li, S.; Zou, B.; Xu, P.F.; Song, Y.L.; Xu, B.; Wang, Y.F.; Tang, G.D.; Yang, S. Significant enhancement in thermoelectric properties of Half-Heusler compound TiNiSn by grain boundary engineering. *J. Alloys Compd.* **2022**, *901*, 163686–163693. [CrossRef]
26. Toberer, E.S.; Zevalkink, A.; Snyder, G.J. Phonon engineering through crystal chemistry. *J. Mater. Chem.* **2011**, *21*, 15843–15852. [CrossRef]
27. Jiang, B.; Wang, W.; Liu, S.; Wang, Y.; Wang, C.; Chen, Y.; He, J. High figure-of-merit and power generation in high-entropy GeTe-based thermoelectrics. *Science* **2022**, *377*, 208–213. [CrossRef]
28. Luo, Y.B.; Hao, S.Q.; Cai, S.T.; Slade, T.J.; Luo, Z.Z.; Dravid, V.P.; Wolverton, C.; Yan, Q.Y.; Kanatzidis, M.G. High Thermoelectric Performance in the New Cubic Semiconductor AgSnSbSe$_3$ by High-Entropy Engineering. *J. Am. Chem. Soc.* **2020**, *142*, 15187–15198. [CrossRef]
29. Fan, Z.; Wang, H.; Wu, Y.; Liu, X.J.; Lu, Z.P. Thermoelectric high-entropy alloys with low lattice thermal conductivity. *RSC Adv.* **2020**, *6*, 52164–52170. [CrossRef]
30. Wang, X.Y.; Yao, H.H.; Zhang, Z.W.; Li, X.F.; Chen, C.; Yin, L.; Hu, K.N.; Yan, Y.R.; Li, Z.; Yu, B.; et al. Enhanced Thermoelectric Performance in High Entropy Alloys Sn$_{0.25}$Pb$_{0.25}$Mn$_{0.25}$Ge$_{0.25}$Te. *ACS Appl. Mater. Interfaces* **2021**, *13*, 18638–18647. [CrossRef]
31. Anand, S.; Wood, M.; Xia, Y.; Wolverton, C.; Snyder, G.J. Double half-heuslers. *Joule* **2019**, *3*, 1226–1238. [CrossRef]
32. Liu, Z.; Guo, S.; Wu, Y.; Mao, J.; Zhu, Q.; Zhu, H.; Pei, Y.; Sui, J.; Zhang, Y.; Ren, Z. Design of high-performance disordered half-Heusler thermoelectric materials using 18-electron rule. *Adv. Funct. Mater.* **2019**, *29*, 1905044. [CrossRef]
33. Wang, Q.; Li, X.; Chen, C.; Xue, W.; Xie, X.; Cao, F.; Sui, J.; Wang, Y.; Liu, X.; Zhang, Q. Enhanced Thermoelectric Properties in *p*-Type Double Half-Heusler Ti$_{2-y}$Hf$_y$FeNiSb$_{2-x}$Sn$_x$ Compounds. *Phys. Status Solidi A* **2020**, *217*, 2000096. [CrossRef]

34. Luo, P.; Mao, Y.; Li, Z.; Zhang, J.; Luo, J. Entropy engineering: A simple route to both *p*-and *n*-type thermoelectrics from the same parent material. *Mater. Today Phys.* **2022**, *26*, 100745. [CrossRef]
35. Wang, Q.; Xie, X.; Li, S.; Zhang, Z.; Li, X.; Yao, H.; Chen, C.; Cao, F.; Sui, J.; Liu, X. Enhanced thermoelectric performance in Ti(Fe, Co, Ni)Sb pseudo-ternary Half-Heusler alloys. *J. Mater.* **2021**, *7*, 756–765. [CrossRef]
36. Baghurst, D.; Chippindale, A.; Mingos, D.M.P. Microwave syntheses for superconducting ceramics. *Nature* **1988**, *332*, 311. [CrossRef]
37. Agostino, A.; Volpe, P.; Castiglioni, M.; Truccato, M. Microwave synthesis of MgB_2 superconductor. *Mater. Res. Innov.* **2004**, *8*, 75–77. [CrossRef]
38. Wong, W.L.E.; Karthik, S.; Gupta, M. Development of high performance Mg–Al_2O_3 composites containing Al_2O_3 in submicron length scale using microwave assisted rapid sintering. *Mater. Sci. Technol.* **2005**, *21*, 1063–1070. [CrossRef]
39. Wang, L.; Zhang, R.P.; BO, L.; Li, F.J.; Hou, Y.B.; Zuo, M.; Zhao, D.G. Effects of different pressing process on the microstructure and thermoelectric properties of $TiNiSn_{1-x}Te_x$ Half-Heusler alloy prepared by microwave method. *JOM* **2022**, *74*, 4250–4257. [CrossRef]
40. Lekse, J.W.; Stagger, T.J.; Aitken, J.A. Microwave metallurgy: Synthesis of intermetallic compounds via microwave irradiation. *Chem. Mater.* **2007**, *19*, 3601–3603. [CrossRef]
41. Biswas, K.; Muir, S.; Subramanian, M.A. Rapid microwave synthesis of indium filled skutterudites: An energy efficient route to high performance thermoelectric materials. *Mater. Res. Bull.* **2011**, *46*, 2288–2290. [CrossRef]
42. Wang, Y.P.; Wang, W.Y.; Zhao, H.Y.; Bo, L.; Wang, L.; Li, F.J.; Zuo, M.; Zhao, D.G. Rapid microwave synthesis of Cu_2Se thermoelectric material with high conductivity. *Funct. Mater. Lett.* **2021**, *14*, 2151008. [CrossRef]
43. Birkel, C.S.; Zeier, W.G.; Douglas, J.E.; Lettiere, B.R.; Mills, C.E.; Seward, G.; Stucky, G.D. Rapid microwave preparation of thermoelectric TiNiSn and TiCoSb Half-Heusler compounds. *Chem. Mater.* **2012**, *24*, 2558–2565. [CrossRef]
44. Lei, Y.; Li, Y.; Xu, L.; Yang, J.; Wan, R.; Long, H. Microwave synthesis and sintering of TiNiSn thermoelectric bulk. *J. Alloys Compd.* **2016**, *660*, 166–170. [CrossRef]
45. Savary, E.; Gascoin, F.; Marinel, S. Fast synthesis of nanocrystalline Mg_2Si by microwave heating: A new route to nano-structured thermoelectric materials. *Dalton Trans.* **2010**, *39*, 11074–11080. [CrossRef]
46. Hmood, A.; Kadhim, A.; Hassan, H.A. Influence of Yb-doping on the thermoelectric properties of $Pb_{1-x}Yb_xTe$ alloy synthesized using solid-state microwave. *J. Alloys Compd.* **2012**, *520*, 1–6. [CrossRef]
47. Rong, Z.Z.; Fan, X.A.; Yang, F.; Cai, X.Z.; Han, X.W.; Li, G.Q. Microwave activated hot pressing: A new opportunity to improve the thermoelectric properties of *n*-type $Bi_2Te_{3-x}Se_x$ bulks. *Mater. Res. Bull.* **2016**, *83*, 122–127. [CrossRef]
48. Biswas, K.; He, J.; Blum, I.D.; Wu, C.-I.; Hogan, T.P.; Seidman, D.N.; Dravid, V.P.; Kanatzidis, M.G. High-performance bulk thermoelectrics with all-scale hierarchical architectures. *Nature* **2012**, *489*, 414–418. [CrossRef]
49. Yan, J.; Liu, F.; Ma, G.; Gong, B.; Zhu, J.; Wang, X.; Ao, W.; Zhang, C.; Li, Y.; Li, J. Suppression of the lattice thermal conductivity in NbFeSb-based half-Heusler thermoelectric materials through high entropy effects. *Scr. Mater.* **2018**, *157*, 129–134. [CrossRef]
50. Ren, W.; Zhu, H.; Zhu, Q.; Saparamadu, U.; He, R.; Liu, Z.; Mao, J.; Wang, C.; Nielsch, K.; Wang, Z.; et al. Ultrahigh power factor in thermoelectric system $Nb_{0.95}M_{0.05}FeSb$ (M = Hf, Zr, and Ti). *Adv. Sci.* **2018**, *5*, 1800278. [CrossRef]
51. Douglas, J.E.; Birkel, C.S.; Miao, M.S.; Torbet, C.J.; Stucky, G.D.; Pollock, T.M.; Seshadri, R. Enhanced thermoelectric properties of bulk TiNiSn via formation of a $TiNi_2Sn$ second phase. *Appl. Phys. Lett.* **2012**, *101*, 183902. [CrossRef]
52. Schrade, M.; Berland, K.; Eliassen, S.N.; Guzik, M.N.; Echevarria-Bonet, C.; Sørby, M.H.; Jenuš, P.; Hauback, B.C.; Tofan, R.; Gunnæs, A.E. The role of grain boundary scattering in reducing the thermal conductivity of polycrystalline XNiSn (X = Hf, Zr, Ti) half-Heusler alloys. *Sci. Rep.* **2017**, *7*, 13760. [CrossRef] [PubMed]
53. Aversano, F.; Palumbo, M.; Ferrario, A.; Boldrini, S.; Fanciulli, C.; Baricco, M.; Castellero, A. Role of secondary phases and thermal cycling on thermoelectric properties of TiNiSn Half-Heusler alloy prepared by different processing routes. *Intermetallics* **2022**, *127*, 106988. [CrossRef]
54. Xia, K.; Liu, Y.; Anand, S.; Snyder, G.J.; Xin, J.; Yu, J.; Zhao, X.; Zhu, T. Enhanced Thermoelectric Performance in 18-Electron $Nb_{0.8}CoSb$ Half-Heusler Compound with Intrinsic Nb Vacancies. *Adv. Funct. Mater.* **2018**, *28*, 1705845. [CrossRef]
55. Xie, H.; Wang, H.; Fu, C.; Liu, Y.; Snyder, G.J.; Zhao, X.; Zhu, T. The intrinsic disorder related alloy scattering in ZrNiSn half-Heusler thermoelectric materials. *Sci. Rep.* **2014**, *4*, 6888. [CrossRef] [PubMed]
56. Karati, A.; Hariharan, V.; Ghosh, S.; Prasad, A.; Nagini, M.; Guruvidyathri, K.; Mallik, R.C.; Shabadi, R.; Bichler, L.; Murty, B. Thermoelectric properties of Half-Heusler high-entropy $Ti_2NiCoSn_{1-x}Sb_{1+x}$ (x = 0.5, 1) alloys with VEC > 18. *Scr. Mater.* **2020**, *186*, 375–380. [CrossRef]
57. Fistul, V.I. *Heavily Doped Semiconductors*; Springer Science & Business Media: Berlin, Germany, 2012; Volume 1.
58. Tsujii, N.; Mori, T. High thermoelectric power factor in a carrier-doped magnetic semiconductor $CuFeS_2$. *Appl. Phys. Express* **2013**, *6*, 043001. [CrossRef]
59. Zhu, T.; Liu, Y.; Fu, C.; Heremans, J.P.; Snyder, J.G.; Zhao, X. Compromise and synergy in high-efficiency thermoelectric materials. *Adv. Mater.* **2017**, *29*, 1605884. [CrossRef]
60. Wang, R.F.; Li, S.; Xue, W.H.; Chen, C.; Wang, Y.M.; Liu, X.J.; Zhang, Q. Enhanced thermoelectric performance of *n*-type TiCoSb half-Heusler by Ta doping and Hf alloying. *Rare Met.* **2021**, *40*, 40–47. [CrossRef]
61. Karati, A.; Mukherjee, S.; Mallik, R.C.; Shabadi, R.; Murty, B.S.; Varadaraju, U.V. Simultaneous increase in thermopower and electrical conductivity through Ta-doping and nanostructuring in half-Heusler TiNiSn alloys. *Materialia* **2019**, *7*, 100410. [CrossRef]

62. Chen, K.; Zhang, R.; Bos, J.W.G.; Reece, M.J. Synthesis and thermoelectric properties of high-entropy half-Heusler MFe$_{1-x}$Co$_x$Sb (M=equimolar Ti, Zr, Hf, V, Nb, Ta). *J. Alloys Compd.* **2022**, *892*, 162045. [CrossRef]
63. Chauhan, N.S.; Bathula, S.; Vishwakarma, A.; Bhardwaj, R.; Johari, K.K.; Gahtori, B.; Dhar, A. Enhanced thermoelectric performance in *p*-type ZrCoSb based half-Heusler alloys employing nanostructuring and compositional modulation. *J. Mater.* **2019**, *5*, 94–102. [CrossRef]
64. Graf, T.; Felser, C.; Parkin, S.S. Simple rules for the understanding of Heusler compounds. *Prog. Solid State Chem.* **2011**, *39*, 1–50. [CrossRef]
65. Johari, K.K.; Sharma, D.K.; Verma, A.K.; Bhardwaj, R.; Chauhan, N.S.; Kumar, S.; Singh, M.N.; Bathula, S.; Gahtori, B. In Situ Evolution of Secondary Metallic Phases in Off-Stoichiometric ZrNiSn for Enhanced Thermoelectric Performance. *ACS Appl. Mater. Interfaces* **2022**, *14*, 19579–19593. [CrossRef] [PubMed]

Disclaimer/Publisher's Note: The statements, opinions and data contained in all publications are solely those of the individual author(s) and contributor(s) and not of MDPI and/or the editor(s). MDPI and/or the editor(s) disclaim responsibility for any injury to people or property resulting from any ideas, methods, instructions or products referred to in the content.

Review

Copper-Based Diamond-like Thermoelectric Compounds: Looking Back and Stepping Forward

Wenying Wang, Lin Bo, Junliang Zhu and Degang Zhao *

School of Materials Science and Engineering, University of Jinan, Jinan 250022, China
* Correspondence: mse_zhaodg@ujn.edu.cn

Abstract: The research on thermoelectric (TE) materials has a long history. Holding the advantages of high elemental abundance, lead-free and easily tunable transport properties, copper-based diamond-like (CBDL) thermoelectric compounds have attracted extensive attention from the thermoelectric community. The CBDL compounds contain a large number of representative candidates for thermoelectric applications, such as $CuInGa_2$, Cu_2GeSe_3, Cu_3SbSe_4, $Cu_{12}SbSe_{13}$, etc. In this study, the structure characteristics and TE performances of typical CBDLs were briefly summarized. Several common synthesis technologies and effective strategies to improve the thermoelectric performances of CBDL compounds were introduced. In addition, the latest developments in thermoelectric devices based on CBDL compounds were discussed. Further developments and prospects for exploring high-performance copper-based diamond-like thermoelectric materials and devices were also presented at the end.

Keywords: thermoelectric; copper-based diamond-like compounds; zT; lattice conductivity; device

Citation: Wang, W.; Bo, L.; Zhu, J.; Zhao, D. Copper-Based Diamond-like Thermoelectric Compounds: Looking Back and Stepping Forward. *Materials* **2023**, *16*, 3512. https://doi.org/10.3390/ma16093512

Academic Editors: Andres Sotelo and Bertrand Lenoir

Received: 23 March 2023
Revised: 13 April 2023
Accepted: 1 May 2023
Published: 3 May 2023

Copyright: © 2023 by the authors. Licensee MDPI, Basel, Switzerland. This article is an open access article distributed under the terms and conditions of the Creative Commons Attribution (CC BY) license (https://creativecommons.org/licenses/by/4.0/).

1. Introduction

The attractive capability of thermoelectric (TE) materials in actualizing the conversion between temperature gradient and electrical power makes them strong candidates for waste-heat recovery as well as solid-state refrigeration [1–3]. The practical and widespread application of TE technology strongly relies on the development of high-performance TE materials, where the TE performance of materials is evaluated by a dimensionless figure of merit, $zT = \alpha^2 \sigma T / \kappa$. The TE parameters α and σ are the Seebeck coefficient and electrical conductivity which, respectively, constitute the power factor, $PF = \alpha^2 \sigma$, used to evaluate electrical conductivity characteristics. Parameter T is the Kelvin thermodynamic temperature, while κ refers to the total thermal conductivity, which is composed of two major contributions from the charge carriers (κ_E) and the lattice (κ_L), respectively. From a computational perspective, the most ideal high-performance TE material should have a large α, high σ as well as a low κ value. What cannot be avoided is the strong coupling between thermoelectric parameters regarding carrier concentration, such as when a high σ means low α and a high κ_E, limiting the improvement of zT [4–6]. In order to achieve high zT in traditional or emerging TE materials, various methods and approaches have been adopted to reduce the correlation between thermal and electrical properties [7–9], including defect engineering, size effects, alloying effect and high-entropy engineering, etc. In addition to achieving high performance, the exploration of alternative materials consists of earth-abundant and eco-friendly components to meet the sake of clean and environmental protection is also considered as one of the most popular approaches in TE field [10–13]. In recent years, diverse bulk TE materials have been widely researched, including liquid-like $Cu_2(S, Se, Te)$, silver-based chalcogenides, $Sn(Te, S, Se)$, half-heuslers, etc. [14–16].

As an environmentally friendly and promising TE material without precious elements, the performance advantages of copper-based diamond-like TE compounds lie in their high Seebeck coefficient and low thermal conductivity [17–21]. Typical compounds include: Cu_3SbSe_4, with a high zT of 0.89 at 650 K [19]; Cu_2SnSe_3, with α of ~250 μV·K^{-1} in the temperature range of 300–700 K [22]; and $CuInTe_2$, with a κ_L value as low as 0.3 W·m^{-1}·K^{-1} [23], etc. Copper-based diamond-like TE compounds are a type of material that conforms to the concept of "phonon-glass electron-crystal" (PGEC) [17] materials, and their crystal structures are usually composed of two sublattices [23–25], in which one sublattice constitutes a conductive network, while the other acts as a thermal barrier and is sometimes also known as a charge reservoir. In 2011, Skoug et al. [24] summarized the significance of lone-pair electrons in the Cu-Sb-Se diamond-like system and demonstrated that the low intrinsic κ_L in compounds came from the interaction of lone-pair electrons with neighboring atoms. Moreover, Skoug et al. [25] also confirmed that the dominant Cu-Se network controlled the electric transport while the Sn orbitals only compensated the system for electrons. Several diamond-like crystal structures evolved from the cubic zincblende structure are shown in Figure 1a. Simultaneously, a series of advanced CBDL compounds have been discovered since 2009, most of which have presented outstanding TE properties. The timeline of maximum zTs and the temperature dependence of zTs for selected CBDL compounds are shown in Figure 1b,c. Taking the typical diamond-like compounds of Cu(In, Ga)Te$_2$, Cu_3SbSe_4, and Cu_2SnSe_3 as examples, long-term efforts have shown that they all apparently have superior TE transport properties with high zTs that exceed one. For instance, Liu et al. [23] devised a pseudocubic crystal structure in $CuInTe_2$ compounds; thus the highest zT of 1.24 was obtained in Ag-doped $CuInTe_2$ compounds. A peak zT of 1.14 was attained in a $Cu_2Sn_{0.90}In_{0.10}Se_3$ compound at 850 K by replacing Sn sites with In. It is also worth noting that a high average zT (zT_{ave}) value is desirable for overall TE conversion efficiency. For instance, a high zT_{max} of 1.67 at 873 K, and a zT_{ave} of 0.73, were realized in $Cu_{0.7}Ag_{0.3}Ga_{0.4}In_{0.6}Te_2$ [26]. In the latest research of Zhou's group [27,28], record-high zT_{ave} values of 0.73 and 0.77 were achieved in Cu_3SbSe_4–based and Cu_3SbS_4–based materials, respectively, which were also comparable to other state-of-the-art TE compounds. Hence one can see that CBDL compounds are expected to become environmentally friendly candidates for TE applications and to achieve excellent performances.

In this review, the structural origins, and the decoupled transport properties of CBDL thermoelectric compounds, were summarized. The latest advances in different types of CBDL compounds were discussed. Then, several common synthetic methods of CBDL compounds were briefly introduce, typical strategies for optimizing the TE properties of CBDL compounds were described in detail, as well as recent updates on CBDL-based TE devices. Finally, the future development of CBDL thermoelectric compounds was evaluated.

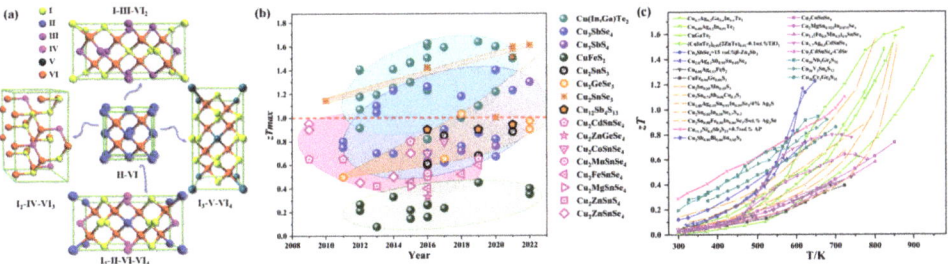

Figure 1. (a) Various crystal structures of CBDL thermoelectric compounds; timeline of zTs (b) and the temperature dependence of zTs (c) for selected copper-based diamond-like thermoelectric compounds, data culled from Cu(In, Ga)Te$_2$ [23,26,29–41], Cu_3SbSe_4 [27,42–49], Cu_3SbS_4 [28,50,51], $CuFeS_2$ [52–55], Cu_2SnS_3 [56–58], Cu_2GeSe_3 [59–61], Cu_2SnSe_3 [62–67], $Cu_{12}SbSe_{13}$ [68,69], $Cu_2CdSnSe_4$ [70–74], $Cu_2ZnGeSe_4$ [75], $Cu_2CoSnSe_4$ [76], $Cu_2MnSnSe_4$ [76], $Cu_2FeSnSe_4$ [76–78], $Cu_2MgSnSe_4$ [79], Cu_2ZnSnS_4 [33], $Cu_2ZnSnSe_4$ [33].

2. Copper-Based Diamond-like Thermoelectric Compounds

CBDL compounds contain a large number of family members, which include ternary I–III–VI$_2$ chalcopyrites, I$_3$–V–VI$_4$ stannites, I$_2$–IV–VI$_3$ stannites, quaternary I$_2$–II–IV–VI$_4$ compounds, and even large-cell Cu$_{10}$B$_2$C$_4$D$_{13}$ tetrahedrites and Cu$_{26}$P$_2$Q$_6$S$_{32}$ colusites. The TE properties of selected typical CBDL compounds including zT_{ave}, zT_{max}, $\alpha^2\sigma$, κ_L, and carrier concentration (n) at room temperature are displayed in Table 1.

Among CBDL compounds, CuGaTe$_2$ and CuInTe$_2$ are typical Cu–III–VI$_2$ (III = In, Ga; VI = Se, S, Te) chalcopyrites structural compounds which have exhibited excellent thermoelectric properties at higher temperatures. In 2012, Plirdpring et al. [29] achieved a record zT of 1.4 in CuGaTe$_2$ compound at 950 K, which indicated that it was a potential material in the field of TE applications. Comparatively, it was found that CuInTe$_2$ possessed a high zT of 1.18 at 850 K [20]. A large number of studies were conducted to optimize the TE transport behaviors of chalcopyrite-based materials in the following years. Through defect engineering, Pei's team obtained a maximum zT of 1.0 at 750 K in the Ag-doped CuGaTe$_2$ compound [40] and identified that vacancy scattering was an active approach to improve TE transport behaviors [80]. Zhang et al. [26] synthesized a quinary alloy compound Cu$_{0.7}$Ag$_{0.3}$Ga$_{0.4}$In$_{0.6}$Te$_2$ with a complex nanosized strain domain structure, which presented excellent TE properties with a peak zT of 1.64 at 873 K and an average zT (zT_{ave}) of 0.73. Through compositing TiO$_2$ nanofibers, Yang et al. [36] achieved a maximum zT of 1.47 at 823 K in a CuInTe$_2$–based TE compound. Moreover, Chen et al. [23] obtained a maximum zT of 1.24 in the Cu$_{0.75}$Ag$_{0.2}$InTe$_2$ compound. The above shows that Cu(In, Ga)Te$_2$ diamond-like TE materials have a higher zTs, comparable to other advanced thermoelectric materials such as PbTe [81–84] and SnTe [85–87]. In addition, a natural chalcopyrite mineral, CuFeS$_2$ [52–55], was also recognized as an advanced CBDL thermoelectric material. It is noteworthy that the CuFeS$_2$ compound is a rare typical n-type TE compound among CBDL thermoelectric materials [88,89].

Cu$_3$–V–VI$_4$ (V = Sb, P, As; VI = Se, S, Te) compounds with a tetragonal diamond-like crystal structure can be approximately regarded as the superposition of four equivalent zincblendes, wherein Cu$_3$SbSe$_4$ is considered as a promising TE candidate owing to its narrow band gap of ~0.3 eV [19,27,47,90]. For improving the TE performance of Cu$_3$SbSe$_4$–based materials, Li et al. [45] coordinately regulated electrical and thermal transport behaviors through the incorporation of Sn-doping and AgSb$_{0.98}$Ge$_{0.02}$Se$_2$ inclusion, and the highest zT of 1.23 was eventually achieved at 675 K. Bo et al. [90] successfully applied the concept of configuration entropy to optimize the TE performance of Cu$_3$SbSe$_4$, and the zT increased by about four times, compared to the initial phase, with the increase of entropy. In their latest report, Zhou's group [27] attained a superior average power factor (PF_{ave}) of 19 μW·cm^{-1}·K^{-2} in 300–723 K by using a small amount of foreign Al atoms as "stabilizers" to supply the high hole concentration, with almost no effect on carrier mobility. Consequently, combined with the reduced κ, a record-high zT of 1.4 and a zT_{ave} of 0.72 were obtained within the Cu$_3$SbSe$_4$–based compounds. A new unconventional doping process that can coordinate the TE properties of materials was also presented. Apart from Cu$_3$SbSe$_4$, Cu$_3$SbS$_4$ is also a promising Cu$_3$–V–VI$_4$–type of TE material [28,50,51], and it has been demonstrated that its PF_{ave} can reach up to 16.1 μW·cm^{-1}·K^{-2} and the zT_{ave} up to 0.77 between 400 and 773 K via its optimization [28].

Different from Cu–III–VI$_2$ and Cu$_3$–V–VI$_4$ compounds, ternary Cu$_2$–IV–VI$_3$ (IV = Sn, Ge, Pb; VI = Se, Te, S) compounds crystallize in more distorted structures that are far from tetragonal, as shown in Figure 1a. Cu$_2$SnSe$_3$ is a kind of CBDL compound with diverse structural phases, which has been found and synthesized successfully, including in cubic, tetragonal, orthogonal, and monoclinic phases involving three variants [22]. Hu et al. [62] improved the TE transport behaviors of Cu$_2$SnSe$_3$ by enhancing the crystal symmetry of it via Mg-doping and intensifying the phonon scattering through the introduction of dislocations and nanoprecipitates. Similarly, Ming et al. [65] obtained a peak zT of 1.51 at 858 K in the Cu$_2$Sn$_{0.82}$In$_{0.18}$Se$_{2.7}$S$_{0.3}$ compound through regulating the band structure and introducing multi-scale defects. In addition, a record-high zT of 1.61 was obtained at 848 K by Qin et al. [66] by constructing the intrinsic point defects, including high-dense stacking faults and endo-grown nanoneedles, to obstruct mid- as well as low-frequency phonons in Cu_2SnSe_3 compounds. Except for Cu_2SnSe_3, $Cu_5A_2B_7$ (A = Si, Ge, Sn; B = S, Se, Te), with a centrosymmetric space group $C2/m$, is also a kind of distorted CBDL compound which has been considered to possess a non-centrosymmetric cubic structure, with the phase crystallized as C-centered, as shown in Figure 2a [91–93]. An undesirable characteristic of $Cu_5A_2B_7$ compounds is that they represent metal-like behaviors, such as the carrier concentration and κ of $Cu_5Sn_2Te_7$ at 300 K are 1.39×10^{21} cm^{-3} and 15.1 $W \cdot m^{-1} \cdot K^{-1}$, respectively [92]. Simultaneously, zinc atoms have been proven to be effective dopants for strengthening the semiconductor properties of $Cu_5Sn_2Te_7$ compounds; Sturm et al. [93] introduced a zinc dopant into $Cu_5Sn_2Se_7$ and $Cu_5Sn_2Te_7$ compounds, which also supports this conclusion. Especially noteworthy is that the effect of zinc doping is not optimal, and the TE performance of the compound still needs further improvement.

Quaternary Cu_2–II–IV–VI_4 (II = Co, Mn, Hg, Mg, Zn, Cd, Fe; IV = Sn, Ge; VI = Se, S, Te) compounds with more complex tetragonal structures have also been widely studied. The distinguishing features of quaternary CBDL compounds are they possess a wider bandgap and a relatively lower carrier mobility compared with the ternary CBDL compounds [68,70,76,94–99]. Taking the orthorhombic enargite-type Cu_2MnGeS_4 as an example [95], the bandgap of it is ~1.0 eV in the initial phase while it only converts to 0.9 eV in the $Cu_{2.5}Mn_{0.5}GeS_4$ by adjusting the ratio of Mn and Cu atoms. The large-cell $Cu_{10}B_2C_4D_{13}$ [100–103] (B = Ag, Cu; C = Co, Ni, Zn, Cu, Mn, Fe, Hg, Cd; Q = Sb, Bi, As; Q = Se, S) tetrahedrites have even more complex crystal structures, as shown in Figure 2b,c, respectively. The featured "PGEC" framework is also displayed in the $Cu_{12}Sb_4S_{13}$ tetrahedrite, where the electric transmission is controlled by a CuS_4 network and the thermal transmission is governed by a cavity polyhedral consisting of CuS_3 and SbS_3 groups [100]. In 2013, Lu et al. [102] achieved an enhanced zT of 0.95 at 720 K in $Cu_{12}Sb_4S_{13}$ utilizing Zn-doping. Moreover, Li et al. [103] attained a high zT of 1.15 at 723 K in a porous $Cu_{12}Sb_4S_{13}$–based material; a segmented single-leg device based on the material was successfully fabricated which realized a high conversion efficiency of 6% when the ΔT reached up to 419 K. $Cu_{26}P_2Q_6S_{32}$ [104–108] (P = V, Ta, Nb, W, Mo; Q = Ge, Sn, As, Sb) colusites are other large-cell examples, which possess 66 atoms in a crystal cell while the tetrahedrites possess 58 atoms. Therefore, the common characteristic of both is their inherent low κ derived from high structural inhomogeneity [108,109]. For instance, Guilmeau's group [105] obtained the lowest κ of 0.4 $W \cdot m^{-1} \cdot K^{-1}$ at 300 K in the $Cu_{26}V_2Sn_6S_{32}$ colusite, which was attributed to the structural complexity of colusite and mass fluctuations among the Cu, V and Sn atoms. In 2018, they further elucidated the potential mechanism related to the fountainhead of intrinsically low κ for a colusite along with the influence of antisite defects and S-vacancies on carrier concentration [105,106].

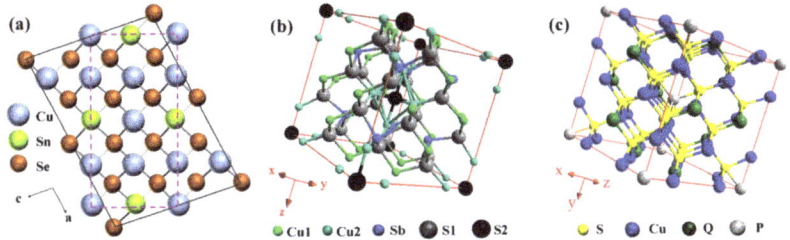

Figure 2. Crystal structure of: (a) $Cu_5Sn_2Se_7$ (reprinted with permission from ref. [91], Copyright 2014 American Chemical Society); (b) $Cu_{12}SbS_{13}$ (reprinted with permission from ref. [110], copyright 2015 American Chemical Society); and (c) $Cu_{26}P_2Q_6S_{32}$ (reprinted with permission from ref. [106], copyright 2018 American Chemical Society).

Table 1. Thermoelectric transport properties of selected CBDL compounds.

Composition	zT_{max}	zT_{ave}	$\alpha^2\sigma$ ($\mu W \cdot cm^{-1} \cdot K^{-2}$)	κ_L ($W \cdot m^{-1} \cdot K^{-1}$)	n@RT ($10^{19} cm^{-3}$)	Synthesis Method *	Ref.
$Cu_{0.75}Ag_{0.2}InTe_2$	1.24@850 K	0.47	7.26	0.3	1.11	M + HP	[23]
Polycrystalline $CuGaTe_2$	1.40@950 K	0.43	8.9	0.45	0.11	M	[29]
$Cu_{0.7}Ag_{0.3}Ga_{0.4}In_{0.6}Te_2$	1.64@873 K	0.73	5.22	0.24	0.007	M + HP	[26]
$(CuInTe_2)_{0.99}(2ZnTe)_{0.01}$–0.1 wt% TiO_2	1.47@823 K	0.50	12.93	0.45	6.01	M	[36]
$Cu_{0.89}Ag_{0.2}In_{0.91}Te_2$	1.60@850 K	0.49	8.81	0.36	0.07	M + SPS	[38]
$Cu_{7.9}In_{8.1}Ga_{0.3}Te_{16}$	1.22@850 K	0.51	11.92	0.55	7.03	M + SPS	[39]
$Cu_{0.7}Ag_{0.3}GaTe_2$	1.00@750 K	0.57	12.26	0.68	4.8	M + HP	[40]
$Cu_{0.8}Ag_{0.2}In_{0.2}Ga_{0.8}Te_2$	1.50@850 K	0.78	14	0.49	0.043	M + SPS	[41]
Cu_3SbSe_4+15 vol% β–Zn_4Sb_3	1.23@648 K	0.43	12.7	0.14	5.5	ST	[45]
$Cu_3Sb_{0.96}Sn_{0.04}Se_4$–5 wt% $AgSb_{0.98}Ge_{0.02}Se_2$	1.23@675 K	0.50	13.8	0.54	8.72	M + SPS	[47]
$Cu_{2.8}Ag_{0.2}Sb_{0.95}Sn_{0.05}Se_4$	1.18@623 K	0.36	9.54	0.27	12.0	MAH + SPS	[48]
$Cu_{2.85}Ag_{0.15}SbSe_4$	0.90@623 K	0.52	10.98	0.66	0.57	M + SPS	[49]
Cu_3SbSe_4–4 wt% $CuAlSe_2$	1.40@723 K	0.72	16	0.35	10	M + HP	[27]
$Cu_{0.92}Zn_{0.08}FeS_2$	0.26@623 K	0.14	5.4	2.24	39.6	M	[52]
$Cu_{0.92}In_{0.08}FeS_2$	0.35@723 K	0.19	4.7	0.79	41.2	M + SPS	[53]
$Cu_{0.88}Ag_{0.12}FeS_2$	0.45@723 K	0.22	7.6	1.15	3.6	M + PAS	[54]
$CuFe_{0.94}Ge_{0.06}S_2$	0.40@723 K	0.17	6	1.04	4.7	M + HP	[55]
$Cu_2Sn_{0.9}In_{0.1}S_3$	0.60@773 K	0.32	6.23	1.01	126	MA + SPS	[56]
$Cu_2Sn_{0.85}Mn_{0.15}S_3$	0.68@723 K	0.28	9.2	0.4	462	M + SPS	[57]
$Cu_2Sn_{0.74}Sb_{0.06}Co_{0.2}S_3$	0.88@773 K	0.43	10.4	0.41	237	M + SPS	[58]
$Cu_{1.85}Ag_{0.15}(Sn_{0.88}Ga_{0.1}Na_{0.02})Se_3$	1.60@823 K	0.50	12.75	0.28	91.8	M	[62]
$Cu_{1.85}Ag_{0.15}Sn_{0.9}In_{0.1}Se_3$	1.42@823 K	0.38	9.70	-	73.4	SHS	[63]
$Cu_{1.85}Ag_{0.15}Sn_{0.91}In_{0.09}Se_3$/4% Ag_2S	1.58@800 K	0.59	12.6	0.12	133.4	SHS + PAS	[64]
$Cu_2Sn_{0.82}In_{0.18}Se_{2.7}S_{0.3}$	1.51@858 K	0.33	9.3	0.35	151	M	[65]
$Cu_2Sn_{0.88}Fe_{0.06}In_{0.06}Se_3$–5 wt% Ag_2Se	1.61@848 K	0.40	7.6	0.2	163	M + MA + HP	[66]
$Cu_{1.9}Ag_{0.1}Ge_{0.997}Ga_{0.003}Se_3$	1.03@768 K	0.58	7.3	0.46	3.5	M + SPS	[60]
$Cu_{1.8}Ag_{0.2}Ge_{0.95}In_{0.05}Se_3$	0.97@723 K	0.44	6.4	0.38	4.6	M + HP	[61]
$Cu_{11.7}Gd_{0.3}Sb_4S_{13}$	0.94@749 K	0.46	16	-	60.3	M + HP	[68]
$Cu_{11.25}Cd_{0.75}Sb_4S_{13}$	0.90@623 K	0.72	12.1	0.33	42	M + HP	[98]
$Cu_{11.5}Ni_{0.5}Sb_4S_{13}$+0.7 vol% AP	1.15@723 K	0.66	12.8	0.17	-	MA + SPS +	[103]
Cu_3SbS_4–9 wt% $CuAlS_2$–1.5 wt% $AgAlS_2$	1.30@773 K	0.77	16.1	0.72	42.2	M + HP	[28]
$Cu_3Sb_{0.95}Sn_{0.05}S_4$	0.72@623 K	0.37	11.3	0.85	41.4	MA + SPS	[50]
$Cu_3Sb_{0.89}Bi_{0.06}Sn_{0.05}S_4$	0.76@623 K	0.38	13.98	0.78	74	MA + SPS	[51]
$Cu_{2.10}Cd_{0.90}SnSe_4$	0.65@700 K	0.27	5.1	0.23	-	M + SPS	[70]
$Cu_2CoSnSe_4$	0.70@850 K	0.31	6.83	0.45	19	M + SPS	[76]
$Cu_2MgSn_{0.925}In_{0.075}Se_4$	0.42@700 K	0.17	5.8	-	14	M + SPS	[79]
$Cu_{2.1}(Fe_{0.5}Mn_{0.5})_{0.9}SnSe_4$	0.60@800 K	0.22	6.1	-	30	M + SPS	[77]
$Cu_{2.1}Fe_{0.9}SnSe_4$	0.52@800 K	0.23	5.9	0.60	23	M + SPS	[78]
$Cu_2CdSnSe_4$	0.50@760 K	0.17	3.1	0.42	1.15	CS + M + SPS	[71]
$Cu_{2.1}Cd_{0.8}SnSe_{3.4}$	0.65@723 K	0.27	6.96	0.42	-	ST + HP	[72]
$Cu_{1.7}Ag_{0.3}CdSnSe_4$	0.80@688 K	0.43	6.5	0.37	-	MAH + SPS	[73]
$Cu_2CdSnSe_4$–CdSe	0.65@725 K	0.34	5.1	0.56	60	M + HP	[74]
$Cu_{26}Nb_2Ge_6S_{32}$	1.00@670 K	0.50	8	0.51	-	M + HP	[104]
$Cu_{26}V_2Sn_6S_{32}$	0.93@675 K	0.55	7.73	0.4	380	MA + SPS	[106]
$Cu_{26}Cr_2Ge_6S_{32}$	1.00@700 K	0.48	19.4	0.48	-	MA + SPS	[108]

* Herein, melting abbreviated to M, hot-pressing abbreviated to HP, spark plasma sintering abbreviated to SPS, microwave-assisted hydrothermal abbreviated to MAH, mechanical alloying abbreviated to MA, plasma-activated sintering abbreviated to PAS, solvothermal abbreviated to ST, self-propagating high-temperature synthesis abbreviated to SHS and colloidal synthesis abbreviated to CS.

3. Material Synthesis Recipes

The synthesis process accompanied by the research and development of the material is a crucial link in obtaining superior TE materials. Therefore, while the performance of TE materials have been improved by leaps and bounds, diverse techniques for synthesizing various TE compounds are also developing vigorously. As shown in Table 1, traditional technologies such as melting, the so-called solid-state reaction, are still widely used in the preparation of high-performance TE materials. Letting nature take its course, the successful application of non-equilibrium formulations, including high-energy ball milling (BM), melt spinning (MS), self-propagating high-temperature synthesis (SHS) and solvothermal (ST) technologies in the TE field provide more options for developing the new generation of TE materials with fine multi-scale microstructures. Simple schematic diagrams of several common synthesis and preparation technologies are shown in Figure 3.

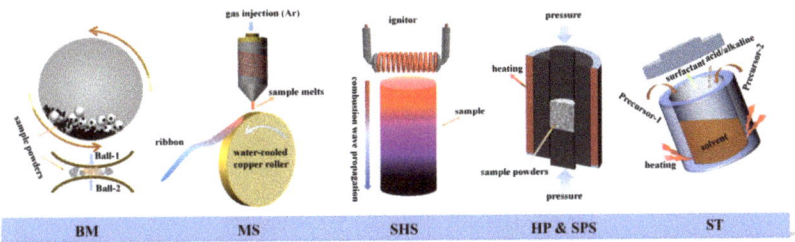

Figure 3. Schematically illustration of BM, MS, SHS, HP&SPS and ST.

High-energy ball milling, also known as mechanical alloying, has been widely adopted to assist in, or directly, synthesize TE compounds with multi-dimensional structures [111–118]. For instance, Nautiyal et al. [117] synthesized a series of polycrystalline Cu_2SnS_3, Cu_2ZnSe_4 and Cu_2ZnSnS_4 TE compounds through MA, which proved that the introduction of nanostructures into the material stabilized the disordered phase structure at low temperatures was conducive to optimizing the TE transport performance of the material. The mechanism of high-energy reaction is achieved by using the inertia between the grinding balls to cause a high-energy impact on the material particles, resulting in cold welding, fracture and re-welding between the particles, leading to further crushing [114]. In addition, most of the BM process involves dry grinding in protective gas to ensure that the collision energy among balls can be effectively applied to the ground powders, and sometimes ethanol and other solvents are used as grinding media. After BM, the fine structure and even nano-powders existing in the material can effectively enhance the phonon scattering and significantly reduce the κ. BM has the advantages of high synthesis efficiency, easy operation, high cost-efficiency and the ability to synthesize thermoelectric materials in large quantities. It is usually used to produce multi-dimensional structure [115], synthetic compound [111,116,117] and mix composites [118] in the TE industry.

Melt-spinning technology is an effective approach to achieve rapid solidification by injecting a molten alloy flow into a rotating and internally cooled roller [119], as shown in Figure 3. When the melt contacts the roller, the melt will undergo rapid solidification or even amorphous transformation accompanying the rapid transfer of heat and will be produced in the form of thin strips or ribbons [119–121]. The microstructure, that depends on local temperature and cooling rate, can be easily controlled by adjusting the machining parameters in the process of MS [119,122]. Previous studies have shown that a large number of refined microstructures and nano-grains can be introduced in TE compounds by MS, such as SnTe, BiSbTe, PbTe and skutterudite, etc., [120,123–125]. In 2019, Zhao's group [121] successfully prepared Cu-Te alloy ribbons with nanocrystalline structures using MS, and achieved the lowest κ of 0.22 $W \cdot m^{-1} \cdot K^{-1}$ in the Cu_2SnSe_3–based composite.

The self-propagating high-temperature synthesis starts with the heating of a small part of the sample at a point, and then the combustion wave spreads along the material to gradually realize the synthesis of the material in an extremely short amount of time [64,126], as shown in Figure 3. In 2014, Su et al. [126] successfully applied SHS to the preparation of various TE compounds for the first time, including Cu_2SnSe_3, $CoSb_3$, $Bi_2(Te, Se)_3$, SnTe, $Mg_2(Sn, Si)$, etc. As the combustion wave spreads across the whole sample, it plays a role in purifying the material and maintaining its stoichiometry [126,127]. The most attractive aspect of SHS is its rapid one-step process, which can be expanded and completed with minimal energy. This feature makes it popular in the synthesis of a variety of CBDL compounds [63,64,127,128]. The main shortcoming of the self-propagating high-temperature synthesis process is that the reaction is so rapid that the sintering size of the sample is difficult to control, requiring secondary processing to ensure the quality of the materials [64,126]. For subsequent measurements and characterizations, dense block TE materials are generally manufactured using sintering technology, including HP and SPS (also known as plasma-activated sintering (PAS)). In most cases, the procedure of sintering is the last step of fabrication, as shown in Table 1, which can strengthen the densification of products and further purify the phases.

In addition, hydrothermal as well as solvothermal reactions are very efficient approaches to preparing refined materials with controllable dimensions and morphologies through the chemical synthesis process [42,45,48,129–133]. In the process of ST, the stoichiometric precursor material required for the synthetic material is first dissolved in the aqueous solution, and then the internal reaction conditions, such as the pressure, pH value and additive concentration, are strictly controlled to make it react in a sealed autoclave [130,131,134–136]. Although the operation is more complex compared to the physical methods mentioned above, controllable thermoelectric compound nanostructures can be synthesized through a wet process, which has the advantages of a low synthesis temperature and fine grain size. It is also worth noting that some the morphologies and sizes of the products can be greatly modified by external conditions, such as the ultrasonic mixture pretreatment time, and the reaction temperature and time [131]. For instance, Wang et al. [136] synthesized the monodispersed Cu_2SnTe_3 nanocrystals (~25 nm) using hot-injection synthesis for the first time, in which the Te precursor was selected by dissolving TeO_2 in 1-dodecanethiol and the reaction solvent was a Cu-Sn complex solution. Moreover, Wei et al. [135] synthesized Cu_3SbSe_4 hollow microspheres dispersed with TiO_2 by a procedure of microwave-assisted hydrothermal synthesis. One advantage of chemical synthesis is that it can control the doping of foreign ions and optimize the grain orientations of nanostructures, which has an important impact on adjusting the carrier concentration and improving phonon scattering [43,129,132,135,136].

In recent years, additive manufacturing [137,138] and machine learning [14,139–144], as emerging intelligent industries, have gradually entered the thermoelectric field, which opens a novel and convenient means to exploring multi-phase space. Referring to diverse indicators closely related to material properties, a series of high-performance CBDL compounds have been discovered. For instance, Zhang's group [139] investigated and predicted the electronic structures and the TE transport behaviors of ABX_2 materials using a high-throughput (HTP) framework, as shown in Figure 4a. Taking the energy position of the band edge as an indicator, Chen's team [140] verified the HTP strategy with the bandgap as an indicator by screening out the potential high-performance n-type TE compounds from Cu-containing chalcogenides, as shown in Figure 4b. In addition, Shi et al. [145] proposed a new performance indicator, shown in Figure 4c,d, for guiding the discovery of TE compounds with low κ. The new indicator referred to the number mismatch (δ) between anions and cations. It should also be noted that since the difference of atomic mass was not considered, the indicator was applicable to compound families with the same elements but different compositions. It was well demonstrated in the Cu-Sn-S systems shown in Figure 4d.

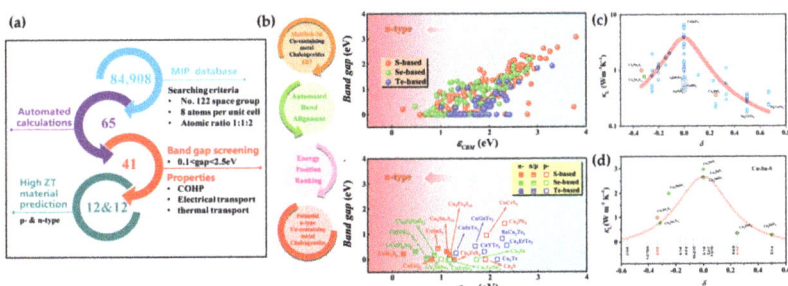

Figure 4. (a) The high-performance thermoelectric material screening workflow for ternary compounds ABX$_2$ with diamond-like structures, reprinted with permission from ref. [139], copyright 2019 American Chemical Society; (b) workflow of the HTP screening process in Cu–containing metal chalcogenides, reprinted with permission from ref. [140], copyright 2022 American Chemical Society. Room temperature κ_L varying with number mismatch in (c) ternary Cu– and Ag–based chalcogenides; and (d) Cu–Sn–S compounds, reprinted with permission from ref. [145], copyright 2020 the Springer Nature.

4. Strategies for Optimizing the TE Performances of CBDL Compounds

There are two main basic principles for achieving high-performance TE materials, one of which is to maximize the *PF* while the other is to minimize the κ_L. One of the typical characteristics of CBDL compounds is that the highly degenerated valence band results in the compound possessing a high Seebeck coefficient. The common defect for most CBDL compounds is that they generally have low carrier concentrations at low temperatures and high κ_L in their initial form. Therefore, trying to promote or maintain *PF* is the critical issue in the development of high-performance CBDL compounds while reducing the κ_L.

4.1. Several Representative Strategies for Enhancing Power Factor

4.1.1. Optimization in Carrier Concentration

As most TE materials have an optimal carrier concentration in the range of 10^{19} to 10^{21} cm^{-3}, one of the most common approaches to maximizing the PFs of TE materials is tuning the carrier concentration [4,5]. For optimizing the carrier concentration in CBDL compounds, a quantity of impurities with different functions has been introduced into pristine compounds. Successful cases among Cu(In, Ga)$_{1-x}$N$_x$Te$_2$ (M = Ag, Zn, Ni, Mn, Cd, Hg, Gd) [23,34,146–151], Cu$_{1-x}$Fe$_{1+x}$S$_2$ [152], Cu$_3$Sb$_{1-x}$N$_x$Se$_4$ (N = As, Zr, Hf, Al, In, Sn, Ge, Bi, La) [42,43,113,115,132,153,154] and Cu$_2$Cd$_{1-x}$In$_x$SnSe$_4$ [155] have demonstrated that doping towards a higher charge-carrier density can effectively improve the electrical performances of the compounds. In addition, introducing vacancies is also another available approach to optimizing the electrical transport properties as well as minimizing the κ_L. On the one hand, as the most common form of *p*-type doping, Cu vacancy has been widely created in CBDL compounds owing to the small formation energy of defects, as seen in Cu$_{12-x}$N$_x$Sb$_4$S$_{13}$ (N = Cd, Mn, Ge, Fe, Co, Sn, Ni, Bi, Zn) [98,156], Cu$_{1-x}$(In, Ga)Te$_2$ [40,157,158] and Cu$_{3-x}$SbSe$_4$ [159]. On the other hand, it is feasible to use anion vacancies for donor doping, as displayed in Cu$_2$ZnGeSe$_{4-x}$S$_x$ [160], CuFeS$_{2-x}$ [161], Cu$_{12}$Sb$_4$S$_{13-x}$Se$_x$ [100] and Cu$_2$FeSnS$_{4-x}$ [162].

4.1.2. Modulation Doping

It should be pointed out that the effect of traditional doping by substituting host atoms by alien ones is fettered by the solubility limit, and worse still, it is easy to cause intense charge-carrier scattering at room temperature, resulting in a loss of electrical transport performance [7]. Compared with traditional doping, modulation doping can effectively avoid the above problems. It is usually designed as a composite composed of two kinds of nanoparticles, and only one of them contains a doping agent [86]. Recently, an unconventional doping (UDOP) strategy was proposed by Zhou et al. [27,28,163] supported this view, where the increase in the vacancies concentration was obtained from an Sb vacancy stabilized by Al rather than alien atoms. Combined with an optimized hole concentration ($3.1 \times 10^{20} \cdot cm^{-3}$) and a maintained carrier mobility, a considerably high average PF of 19 $\mu W \cdot cm^{-1} \cdot K^{-2}$ was obtained in the temperature range of 300–723 K [27]. In contrast to the conventional doping method (Figure 5a), the carrier concentration and carrier mobility decouple by vacancies in the route of UDOP (Figure 5b). In other words, it can be considered that in the purposeful doping process, the doping additive itself does not provide carriers, but acts as a "stabilizer" of the cationic vacancy (Figure 5c–e), which actually offers additional holes for p-type conductive semiconductors. It has been proved that the modulation-doping strategy can be used to not only improve the PF of CBDL compounds, but also to maintain the carrier mobility of various compounds requiring a high carrier concentration.

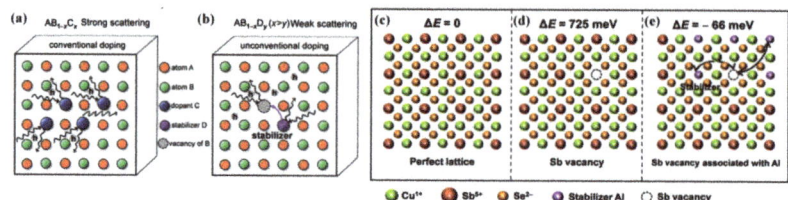

Figure 5. Schematic diagrams of (**a**) conventional; and (**b**)unconventional doping; Cu_3SbSe_4 with (**c**) perfect lattice; (**d**) Sb vacancy; and (**e**) Sb vacancy surrounded by Al as a stabilizer. Reprinted with permission from ref. [27], copyright 2022 Wiley-VCH GmbH.

4.1.3. Pseudocubic Structure

Apart from obtaining an optimal carrier concentration, the regulation of PF is also linked with the electronic band structure [5,7,8]. The high band convergence (N_v) originating from high symmetric crystal structures is beneficial for obtaining large α and high σ. Similarly, the CBDL compounds derived from the high-symmetry cubic phase, ensures that they possess highly degenerate valence bands [18,32,33]. In particular, Zhang et al. [31] proposed a pseudocubic strategy were the PF could be optimized to the greatest extent by pruning the band split-off, which was also considered as an efficient approach to exploring and screening high-performance non-cubic TE compounds. As shown in Figure 6a,b, when the valence band-splitting energy Δ_{CF} approximates to zero, this means that the distortion parameter $\eta = c/2a$ approaches one; in other words, the bands are in a degenerate state at this time, which can trigger the maximum PF. The pseudocubic approach, also known as the unity-η rule ($\eta = c/2a$), has been successfully applied to screen out high-performance tetragonal CBDL compounds [23,35,76,97]. For instance, Li et al. [97] found that the Δ_{CF} of $Cu_2ZnSnSe_4$ could be appropriately tuned by applying the proper strain, which provides an alternative way to improve the thermoelectric properties of the compound. As a systematic strategy, the unity-η rule is used to qualitatively guide the evaluation and manipulation of TE diamond-like lattices. For instance, the distortion parameter η, as a function of the cell parameter a, for tetragonal diamond-like chalcogenides [32] is shown in Figure 6c. It should be noted that the pseudocubic approach is limited to low-symmetry material with an ideal bandgap and a low κ_L [7].

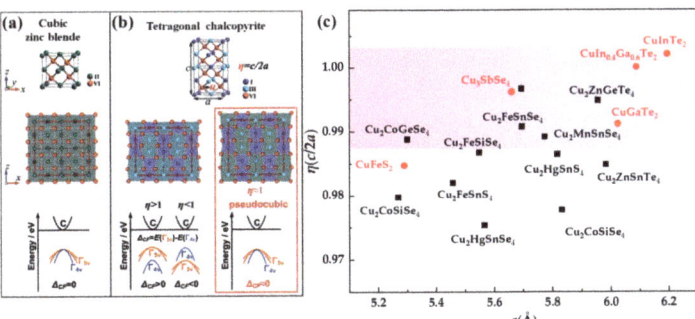

Figure 6. Band convergence in (**a**) cubic zincblende structure; and (**b**) pseudocubic ternary chalcopyrites, reprinted with permission from ref. [31], copyright 2014 WILEY-VCH Verlag GmbH & Co. KGaA, Weinheim. The c and a are the lattice constants. Γ_{4v} is a nondegenerate band, and Γ_{5v} is a doubly degenerate band. Δ_{CF} is the crystal field-induced energy split at the top of the Γ_{4v} and Γ_{5v} bands; (**c**) distortion parameter η as a function of the lattice parameter a, reprinted with permission from ref. [32], copyright 2018 Science China Press and Springer-Verlag GmbH Germany, part of Springer Nature.

4.1.4. Softening p-d Hybridization

It is well known that the p-d hybridization in CBDL compounds is very strongly attributable to the quite small separation energy among the atomic levels of chalcogen p-orbitals and Cu-3d states [91,99,164]. For most CBDL compounds, the electric transport channel (mostly the valence band maxima, VBM) is regarded as being constructed by Cu-X bonds [18,22]. The chemical bonding and the electronic structure in TE materials are closely linked to their internal charge carriers and phonon transport behaviors. Therefore, the regulation of p-d hybrid strength potentially serves as an adjustable critical parameter in adjusting the properties of CBDL compounds. Taking Cu$_2$SnSe$_3$ [25] as an example in Figure 7a, the VBM is mostly occupied by the p-d hybridization from Cu-Se bonds, which acts as the charge-conduction pathway as well as a structural retainer. In contrast, the p-orbitals of Sn atoms contribute little to the occupied states while the conductive band is primarily dominant. Similarly, the Cu-X conduction channel has also been demonstrated in some other CBDL compounds, such as CuGaS$_2$ [164], Cu$_3$SbSe$_4$ [24], Cu$_2$CdZnSe$_4$ [70], etc. The Cu$_3$SbSe$_4$ diamond-like compound with a small bandgap is also an example influenced by the strong relativistic orbital-contraction effect [164,165]. Softening p-d hybridization, as an active strategy, has been adopted to synergistically improve the electronic and thermal transport performance of Cu$_3$SbSe$_4$ via Ag-doping [49,166], as shown in Figure 7b. Zhang et al. [49] discovered that the *PF* of Cu$_3$SbSe$_4$ was significantly enhanced by changes in the bandgap and the density of states caused by the softening of p-d hybridization, which, accompanied by Ag-doping, induced large strain fluctuations in some local structural distortions and resulted in greatly reduced κ_L. In addition, Ge et al. [53] introduced an abnormally high concentration of indium in CuFeS$_2$ compound, as shown in Figure 7c–e; the indium was not fully ionized to In^{3+} cation when on the Cu sublattice and existed mainly in the In$^+$ oxidation state. The latter, with 5s^2 lone-pair electrons, could cause strong local bond distortions, thereby softening the In-S and Cu-S bonds and introducing localized low-frequency vibrations [89]. Therefore, a low κ_L value of 0.79 W·m^{-1}·K^{-1} (Figure 7f) and a high zT value of 0.36 were recorded at 723 K in Cu$_{1-x}$In$_x$FeS$_2$ samples.

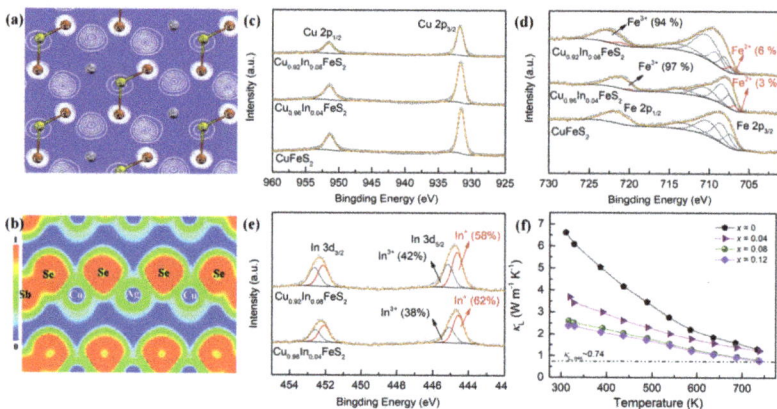

Figure 7. (**a**) Schematic diagrams of the partial charge density of the states close to upper valence-band in Cu_2SnSe_3 on (100) crystal face, reprinted with permission from ref. [25], copyright 2010 American Chemical Society; (**b**) the calculated electron-localization function of Ag-doped Cu_3SbSe_4 on (101) crystal face, reproduced from ref. [49], copyright 2019 the Royal Society of Chemistry. The X-ray photoelectron spectra (XPS) of (**c**) Cu 2p; (**d**) Fe 2p; and (**e**) In 3d for the $Cu_{1-x}In_xFeS_2$ samples; and (**f**) the κ_L of $Cu_{1-x}In_xFeS_2$ samples, reproduced from ref. [53], copyright 2022 Elsevier Ltd.

4.2. Strategies for Reducing Lattice Thermal Conductivity

4.2.1. Point-Defect Scattering

In the thermoelectric field, defect and nanostructuring engineering have been widely adopted to optimize the thermoelectric performance enhancement of TE materials, especially the dislocations and nanostructured interfaces which involve the scattering of low- and mid-frequency phonons, have received more attention [6]. In the process of improving the TE performance for CBDL compounds, the existence of point defects plays a more important and beneficial role in phonon scattering than in affecting the electrical behavior. There are two main types of influence on κ_L originating from point defects in TE materials: the mass fluctuation (Figure 8a) and strain field fluctuation (Figure 8b) among the host and guest atoms. Shen et al. [40] testified that substitutional defects of Ag_{Cu} in $CuGaTe_2$ could reduce the κ_L more efficiently than substitutional defects of Zn_{Ga} or In_{Ga} at the equivalent concentration, which was attributable to the larger mass fluctuation. When the dominant point defects are vacancies, the types of scattering inflected by the strain and mass fluctuations can be maximized [5]. Thus, the compounds with an intrinsic high concentration of cation vacancies, such as In_2Te_3 and Ga_2Te_3, were introduced in $CuGaTe_2$ to depress the κ_L of the matrix phase by constituting solid solutions [80]. Additionally, an elaborate investigation about the room temperature κ for cation-substituted $Cu_2ZnGeSe_{4-x}S_x$ compounds displayed a reduction of 42% for κ_L, where the reduction caused by mass contrast accounted for 34% and the remaining 8% was caused by strain fluctuations [160]. In their latest study, Xie et al. [151] observed the off-centering effect (Figure 8c) of an Ag atom by investigating the thermal transmission behaviors in $Cu_{1-x}Ag_xGaTe_2$ as well as in $CuGa_{1-x}In_xTe_2$. It is obvious that the off-centering behavior of the Ag atom means a new phonon scattering mechanism is brought about by point defects, where the Ag-alloyed solid solutions resulted in an extremely low κ_L, which was attributed to crystallographic distortion and extra-strong acoustic-optical phonon scattering, as shown in Figure 8d. Moreover, it can also be seen that a modified Klemens model was developed by integrating the off-centering effect and alloy-scattering with the crystallographic distortion parameter (η), which can be used as an indicator to predict the κ of diamondoid solid solutions.

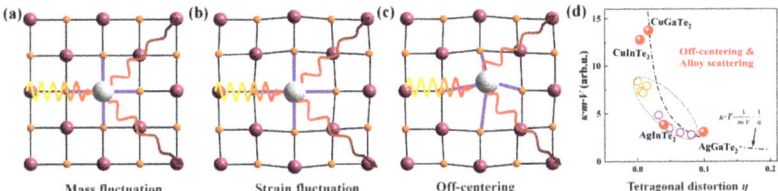

Figure 8. Schematic diagrams of (**a**) mass fluctuation; (**b**) strain fluctuation; (**c**) off-centering effect in phonon transport; and (**d**) relationship between tetragonal distortion and thermal conductivity for different Ag-based and Cu-based diamondoid compounds, the V is crystal volume, m is the formula weight. Reprinted with permission from ref. [151], copyright 2023 American Chemical Society.

4.2.2. Nanostructure Engineering

Controlling the nanostructures of TE materials is also an effective approach to enhancing phonon scattering through realizing an all-scale hierarchical architecture in TE materials. Zhang et al. [26] adopted a quinary alloy compound system with a complex nanosized strain-domain structure in $CuGaTe_2$ (Figure 9a), which made the room-temperature κ decline from 6.1 $W \cdot m^{-1} \cdot K^{-1}$ for the initial compound to 1.5 $W \cdot m^{-1} \cdot K^{-1}$ for the Ag and In co-doped sample. Wang et al. [167] achieved low κ values of 0.491 $W \cdot m^{-1} \cdot K^{-1}$ and 0.481 $W \cdot m^{-1} \cdot K^{-1}$ in $Cu_3Sb_{0.92}Sc_{0.08}Se_4$ and $Cu_3Sb_{0.92}Y_{0.08}Se_4$ at 623 K, respectively, with a constructed multiscale heterostructure. In 2021, Hu et al. [103] designed pore networks for tetrahedrite $Cu_{12}Sb_4S_{13}$–based TE materials using a BiI_3 sublimation technique, as shown in Figure 9b, which led to a hierarchical structure which contained pores, pore interfaces, point defects, and granular precipitates. The effect of various scattering mechanisms on phonon-transport behaviors for $Cu_{12}Sb_4S_{13}$–based samples are shown in Figure 9c,d. First, the existence of specially designed pores and pore interfaces reduced the κ_L of samples with 0.7 vol% annealed pores (AP) by about 36%. Furthermore, $Cu_{1.8}S$ precipitates, point defects involved Ni-alloying and Bi-doping, dislocations, the solid solution of impurity Cu_3SbS_4 phase as well as volume expansion also contributed to the reduction of κ_L because they realized full-scale phonon scattering in the TE sample. Consequently, a ~72% reduction in the κ_L was obtained for samples with 0.7 vol% AP with the addition of a small amount of BiI_3. Moreover, previous works demonstrated that high-density stacking faults (SFs) could be realized in doped Cu_2SnSe_3 [62,66,168], as shown in Figure 9e–g, which also caused strong scattering of phonons as a phonon-scattering center. In addition, solvothermal synthesis [43,134,135,153] and ball milling [113,115] are effective and convenient approaches to constructing nanostructures for TE materials.

4.2.3. Nanocomposite

Compositing with uniformly dispersed nanoinclusions, secondary phases or nanoparticles has been widely considered as a predominant and effective strategy to optimize TE performance in CBDL compounds [27,47,67,118,128,169–172]. Nanoparticles (NPs) introduced in composites can be effectively used as intermediate frequency phonons scatter centers and diminish κ_L [5]. Sun et al. successfully incorporated ZnO [173] and Nb_2O_5 [174] NPs into the grain boundaries of $Cu_{11.5}Ni_{0.5}Sb_4S_{13}$ compounds via mechanical alloying and spark plasma sintering, respectively, and the both composites achieved a reduced κ and high zTs. In our previous work, we also introduced graphene nanosheets or SnTe NPs into Cu_3SbSe_4 through ball milling and realized the optimization of thermoelectric properties. Hu et al. [175] obtained a relatively low κ of 0.9 $W \cdot m^{-1} \cdot K^{-1}$ at all temperatures in Fe_2O_3–dispersed $Cu_{12}Sb_4S_{13}$ tetrahedrite via the combination of nanostructuring and defect engineering (Figure 10a–e). As shown in Figure 10a–d, dislocations along with diverse nanostructures, such as NPs, nanotwins and nanoprecipitates, were introduced in $Cu_{11.5}Ni_{0.5}Sb_4S_{13}$ by compositing magnetic γ-Fe_2O_3 NPs, which realized all-scale hierarchical phonon scattering in the samples, making the zT reached up to ~1.0 (Figure 10f). For reducing κ_L, Li et al. [39] synthesized $CuInTe_2$–based compounds with in-situ formed InTe nanostrips, which wrapped the nanodomains (Figure 10g–j) and resulted in the reduction of κ_L by a factor of ~2 compared to parent compound. It is notably anticipated that the content, dimensions and especially distribution of nano-additives in composites have an important impact the effective regulation of TE performances.

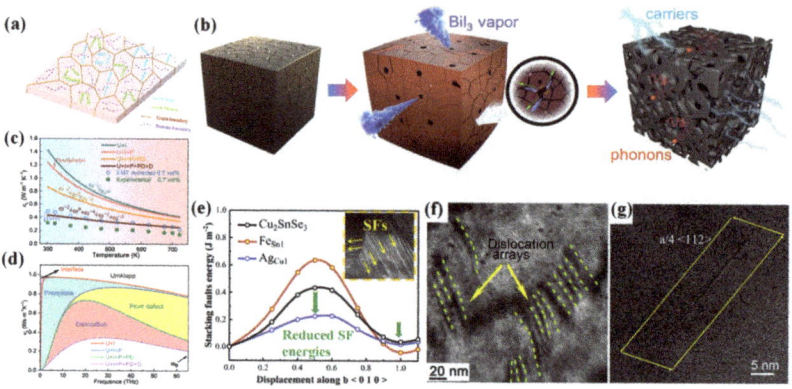

Figure 9. (**a**) Schematic illustration of the transport behaviors for phonons and holes in CuGaTe$_2$, reprinted with permission from ref. [26], copyright 2019 WILEY-VCH Verlag GmbH & Co. KGaA, Weinheim; (**b**) schematic illustration showing the formation of a porous network during BiI$_3$ sublimation; (**c**) κ_L of sample with 0.7 vol% AP, which took Umklapp process (U), porous interfaces (I), precipitates (P), point defects (PD), dislocation cores (DC), and strains (DS, D = DC + DS) into account; (**d**) frequency-dependent accumulative reduction in the lattice thermal conductivity of the EMT-corrected sample with 0.7 vol% AP due to various scattering mechanisms. Reprinted with permission from ref. [103], copyright 2021 Wiley-VCH GmbH; (**e**) calculated generalized stacking fault energies as a function of normalized Burger's vector b <010> in Cu$_2$SnSe$_3$–based system, the insert was the high-dense stacking faults (SFs) in (Fe, Ag, In)-doped Cu$_2$SnSe$_3$. Reprinted with permission from ref. [66], copyright 2022 Elsevier Ltd. High-dense SFs in (**f**) (Ag, Ga, Na)-doped (reprinted with permission from ref. [62], copyright 2021 Wiley-VCH GmbH); and (**g**) Ni-doped (reprinted with permission from ref. [168], copyright 2021 American Chemical Society) Cu$_2$SnSe$_3$.

4.2.4. Lattice Softening Effects

The internal strain fluctuation induced by lattice defects, such as nanoprecipitates and dislocations, can locally shift the phonon frequencies in the TE material, which in principle can bring about lattice-softening accompanied by phonon scattering owing to changes in phonon speed, as shown in Figure 11a. In several cases, improvements in TE performance ascribed to lattice-softening through the introduction of vacancies or alloying have been presented [176–178], such as SnTe with AgSbTe$_2$ alloying, and the lattice-softening effect in Cu$_2$Se, as shown in Figure 11b. In 2019, Hanus et al. [179] authenticated that the changes of thermal transport behavior in the PbTe system were attributable to the lattice-softening through alloying or lattice defects, and pointed out that the modulation of lattice stiffness had a significant impact on the phonon transport in some states. In addition, Muchtar et al. [176] introduced lattice-softening into SnTe by inserting Ti and Zr atoms, which effectively suppressed the phonon group velocities and reduced the κ. Moreover, Snyder et al. [180] found the lattice-softening effect induced by charge-carrier-mediated in several high-performing ($zT > 1$) TE materials (such as SnTe, PbTe, Nb$0_{.8+x}$CoSb, etc.) contributed more than 20% to zT. Simultaneously, the results shown in Figure 11c indicate that a strong dependence of sound velocities v_s on Hall charge-carrier concentration n_H was observed in each compound in which the measured v_s significantly decreased with increasing n_H. Lattice-softening effects also have been successfully used to improve the TE performances of CBDL compounds. Pöhls et al. [181] demonstrated that the Li-induced phonon-softening effect was feasible to enhance the TE performance of chalcopyrite $CuGaTe_2$. Xie et al. [38] obtained an extremely low κ_L of 0.47 W·m^{-1}·K^{-1} at 850 K in Ag-doped $CuInTe_2$ compound that was attributed to strong interactions among low-frequency optical phonons derived from the weakened Ag-Te bonds, as shown in Figure 11d.

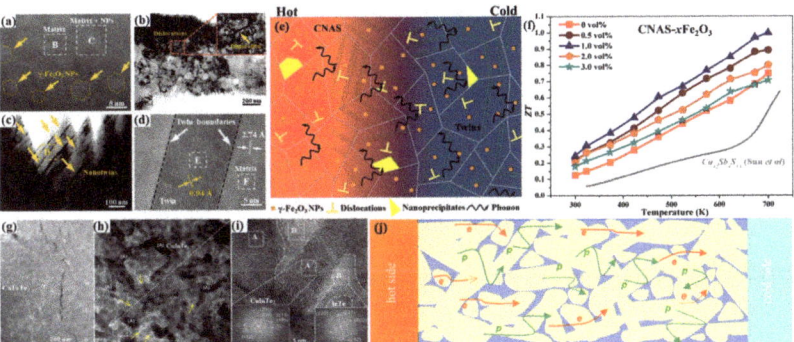

Figure 10. Microstructure of the $Cu_{11.5}Ni_{0.5}Sb_4S_{13}$–1.0% Fe_2O_3 sample including: (**a**) HRTEM image; (**b**) dislocation; (**c**) nanotwins; and (**d**) HRTEM images of the area D; (**f**) schematic diagram of phonon scattering in γ-Fe_2O_3 dispersed $Cu_{11.5}Ni_{0.5}Sb_4S_{13}$ (CNAS), (**e**) zTs for all CNAS-xFe_2O_3 samples. Reprinted with permission from ref. [175], copyright 2020 American Chemical Society; (**g**) high-angle annular dark field; (**h**) high-resolution TEM image; (**i**) magnified TEM image and the fast Fourier transform of the $CuInTe_2$:23 wt% InTe bulk sample; (**j**) schematic illustrating of the transport in both the phonons p and electrons-e. Reprinted with permission from ref. [39], copyright 2020 WILEY-VCH Verlag GmbH & Co. KGaA, Weinheim.

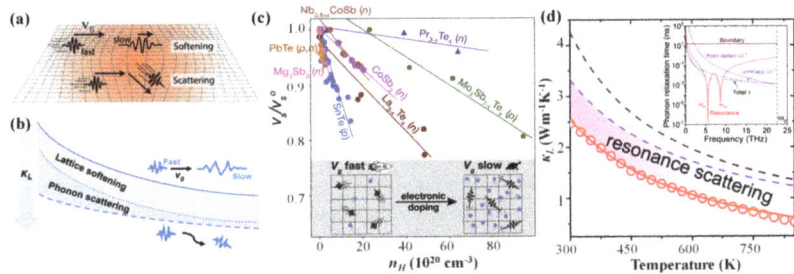

Figure 11. (**a**) Schematic illustration of lattice-softening effects and phonon scattering originated from internal-strain fields. Reprinted with permission from ref. [179], copyright 2019 WILEY-VCH Verlag GmbH & Co. KGaA, Weinheim; (**b**) schematic illustration of lattice-softening in Cu_2Se. Reprinted with permission from ref. [177], copyright 2022, American Chemical Society; (**c**) sound velocities plotted against measured Hall charge-carrier concentration for SnTe, PbTe, $Nb_{0.8+x}CoSb$, $CoSb_3$, $La_{3-x}Te_4$, $Pr_{3-x}Te_4$, and Mo_3Sb_7. Reprinted with permission from ref. [180], copyright 2021 Elsevier Inc.; and (**d**) contribution of distinct scattering mechanism to the κ_L of $Cu_{0.8}Ag_{0.2}InTe_2$. Here the U, B, P and R represent Umklapp scattering, grain-boundaries scattering, point-defect scattering, and phonon-resonance scattering, respectively; the insert shows the calculated phonon relaxation times τ versus phonon frequency ω for $Cu_{0.8}Ag_{0.2}InTe_2$ with different scattering mechanisms. Reprinted with permission from ref. [38], copyright 2020 The Royal Society of Chemistry.

4.3. Synergistic Regulation

4.3.1. Entropy Engineering

In the process of optimizing the electrical and thermal transport properties of TE materials, it is never just to adjust one of them individually. To some extent, the above optimization process can realize the decoupling of electron and phonon transmission. Entropy engineering provides a new pathway to synergistically optimize the electrical, thermal, and mechanical properties for promoting the development of CBDL compounds [15,41,90,182,183]. Through synergistic regulation, Xie et al. [41] achieved a maximum zT of 1.5 at 850 K in the quinary $(Cu_{0.8}Ag_{0.2})(In_{0.2}Ga_{0.8})Te_2$ compound, in which Ga-substituted In and Ag-substituted Cu effectively optimized the electrical and thermal transport properties, respectively. In addition, Cai et al. [183] obtained a high zT of 1.02 in $CuInTe_2$ compound, which was attributed to the reduction of κ by devising a high-entropy structure as well as by improving the carrier mobility by one order of magnitude. In many cases long before that, Liu et al. [15] utilized the entropy attribute as the comprehensive gene-like performance indicator to screen and devise TE materials with high zT. As can be seen in Figure 12a,b, a special example can be noted that when multi-component alloy elements are adopted in compounds, the configurational entropy can especially be changed. For a given multi-component material, the maximum entropy lies on the solubility parameter δ of the whole material, which is linked to the mismatch of the atomic radius, shear modulus and lattice constant in the material, as shown in Figure 12c. Instructing with δ-criterion (Figure 12d), representative multi-component $(Cu/Ag)(In/Ga)Te_2$–based CBDL compounds with zTs approaches to 1.6 were screened out owing to the optimization of entropy.

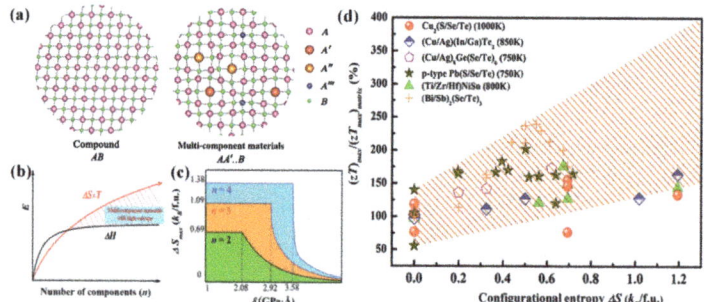

Figure 12. (a) Schematic diagram of the lattice framework in multicomponent materials compared to a simple binary compound; (b) schematic diagram of the entropy engineering with multicomponent TE materials; (c) the maximum configurational entropy (in units of k_B per formula unit) as a function of a material's solubility parameter δ for given multicomponent TE materials, where n is the number of components; and (d) plots of maximum zT versus the configurational entropy in several selected TE systems. Reprinted with permission from ref. [15], copyright 2017 WILEY-VCH Verlag GmbH & Co. KGaA, Weinheim.

4.3.2. Progressive Regulation Strategy

The progressive regulation strategy can be realized via integrating point defects and microstructure engineering. Luo et al. [30,36] successfully acquired high-performance $CuInTe_2$ compounds by integrating the cation/anion substitution and in-situ oxidation, as shown in Figure 13a. Taking the in-situ substitution reaction between $CuInTe_2$ and ZnO additive as a case [36], the priority generation of acceptor defects Zn_{In}^- significantly optimized the PF while the In_2O_3 nanoinclusions incurred by the in-situ reaction led to a low κ of $CuInTe_2$. Through triple doping in Cu_2SnSe_3, Hu et al. [62] obtained an excellent zT of 1.6 at 823 K in cubic $Cu_{1.85}Ag_{0.15}(Sn_{0.88}Ga_{0.1}Na_{0.02})Se_3$ and a decent zT_{ave} of 0.7 from 475 to 823 K in $Cu_{1.85}Ag_{0.15}(Sn_{0.93}Mg_{0.06}Na_{0.01})Se_3$ via synergistic effects. As shown in Figure 13b, during the management process from the initial phase to (Ag, Ga, Na)-doped Cu_2SnSe_3, the gradually improved zT originated from symmetry enhancing, alloying scattering and dislocation/nanoprecipitate construction, respectively. Similarly, synergistically optimized $CuGaTe_2$ [135] (Figure 13c), Cu_3SbSe_4 nanocrystals with $Cu_{2-x}Se$ in-situ inclusions [48], $CuIn_{1-x}Ga_xTe_2$:y$InTe$ with in situ formed nanoscale phase $InTe$ [39],

Cu$_2$SnSe$_3$ with CuInSe$_2$ alloying [184], *etc*, demonstrated that the progressive and collaborative optimization strategies have been widely applied in CBDL materials.

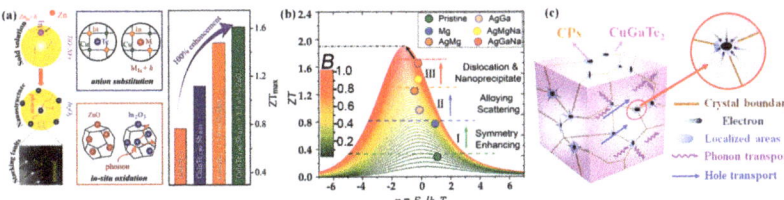

Figure 13. (a) Synergistic strategies of point defects and microstructure engineering in CuInTe$_2$. Reprinted with permission from ref. [30,36], copyright 2015 Elsevier Ltd. All rights reserved and 2016 WILEY-VCH Verlag GmbH & Co. KGaA, Weinheim; (b) quality factor analysis on the relationship of chemical potential η versus zT in Cu$_2$SnSe$_3$–based compounds. Reprinted with permission from ref. [62], copyright 2021 Wiley-VCH GmbH; and (c) schematic diagram illustrating various phonon scattering mechanisms and the electron localized region near carbon particles (CPs) within the CuGaTe$_2$+x wt% CPs sample. Reprinted with permission from ref. [135], copyright 2020 The Royal Society of Chemistry.

5. CBDL-Based TE Devices

For practical TE applications, moving from high-performance materials to high-efficiency devices is of great significance. CBDL compounds conform to the concept of green environmental protection and have great practical application value while the absence of n-type conductive compounds greatly hinders the manufacture and application of CBDL-based TE devices. During the journey of device development, researchers have made a lot of efforts. In 2017, Qiu et al. [185] manufactured a CBDL-based TE module via integrating high-performance n-type Ag$_{0.9}$Cd$_{0.1}$InSe$_2$ and p-type Cu$_{0.99}$In$_{0.6}$Ga$_{0.4}$Te$_2$ leg, respectively, as shown in Figure 14a. The output power of module reached 0.06 W under a temperature difference of 520 K (Figure 14b), demonstrating that diamond-like compounds are also potential candidates for TE applications. On the foundation of obtaining high-performance in (Sn, Bi)-codoped nanocrystalline Cu$_3$SbSe$_4$ materials, Liu et al. [153] fabricated a hot pipe integrated by a series of ring-shaped Cu$_3$Sb$_{0.88}$Sn$_{0.10}$Bi$_{0.02}$Se$_4$–based TE modules (Figure 14c), which can be used for the purpose of retrieving the waste heat from exhaust gas pipes in vehicles. Moreover, Li et al. [103] synthetized a segmented Cu$_{12}$Sb$_4$S$_{13}$-based single-leg module, which had a superior conversion efficiency η of 6% at ΔT = 419 K, as shown in Figure 14d,e. Recently, the Cu$_3$SbS$_4$–based single-leg module synthetized by Zhang et al. [28] approached a conversion efficiency η of 2% with ΔT = 375 K, which reached to 5.5% predicted by the COMSOL simulation analysis (Figure 14f,g). Apart from realizing excellent TE efficiency, good thermal stability is also crucial for the manufacturing of TE devices. In practice, the volatilization induced softening and decomposition is the core issue for thermoelectric selenides and sulfides working at elevated temperatures. In the latest research from Zhou's group [163] demonstrated that the compositing of CuAlS$_2$ significantly optimized the thermal stability of Cu$_3$SbSe$_4$–based compounds by pushing the decomposition temperature to a higher value, while also greatly improving the mechanical properties of the material. Eventually, a maximum η over 3% was achieved at a ΔT = of 367 K and an I = 0.8 A. Based on the above research, it seems that CBDL has considerable TE performance and has gradually attracted researchers' attention in the field of practical TE applications.

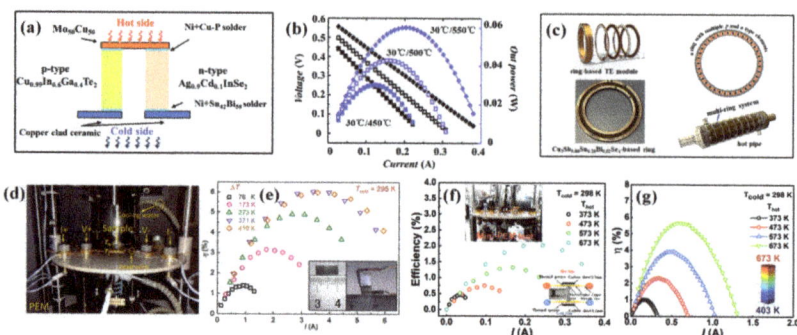

Figure 14. (**a**) Schematic diagram of the fabricated diamond-like module, and (**b**) plots of output voltage and power versus current for TE module based on diamond-like materials. Reprinted with permission from ref. [185], copyright 2018 The Royal Society of Chemistry. (**c**) Schematic diagrams of annular $Cu_3Sb_{0.88}Sn_{0.10}Bi_{0.02}Se_4$–based TE modules, reprinted with permission from ref. [153], copyright 2017 The Royal Society of Chemistry. (**d**) The Mini-PEM used to measure the conversion efficiency of a segmented $Cu_{12}Sb_4S_{13}$–based single-leg, and (**e**) experimental power generation efficiency for the segmented leg. The insets are the fabricated TE single-leg. Reprinted with permission from ref. [103], copyright 2021 Wiley-VCH GmbH. (**f**) Experimental TE conversion efficiency η and (**g**) simulated η by COMSOL Multiphysics software for Cu_3SbS_4–based single-leg module, the inset is a photo of mini-PEM test. Reprinted with permission from ref. [28], copyright 2023 Wiley-VCH GmbH.

6. Conclusions and Perspectives

By reviewing the research on copper-based diamond-like thermoelectric materials, it has been found that diverse compounds appear to have excellent TE performances as well as possessing zT higher than unity and an even approach to two. Advanced approaches to guide the development of new high-performance CBDL materials have been found, such as machine learning, high-throughput and union-η rules. There are also various approaches to improving the TE properties of CBDL compounds. It is worth considering that, during the process of optimizing electrical and thermal transport behaviors of TE materials, the regulation is never carried out separately, but that coordination and unification of the two are sought. Based on the efforts of researchers, the CBDL compounds have been greatly developed. There is no escaping the fact that the softening and decomposition of Cu-based compounds occurs when the compounds are exposed to high temperatures. Therefore, compared with practical materials, the CBDL compounds still have great room for improvement.

Considering practical applications, it is of great significance to shift our focus from high-performance TE materials to highly efficient devices. The integration for TE equipment requires both high-performance n-type and p-type legs. Currently, CBDL compounds are mostly p-type materials, while the further development of n-type CBDL compounds is beneficial for its TE application. In addition, in the research and development process of TE devices, it is also necessary to consider the comprehensive properties such as thermal stability, processability and self-compatibility. Therefore, the feasibility of manufacturing efficient TE devices based on CBDL materials remains a highly challenging issue.

The exploration of material properties is still ongoing, and the practical application of devices also needs to be developed. There has been a deep understanding of the transport mechanism of TE materials with the iteration and update of characterization methods, accompanied by the assistance of more advanced manufacturing technologies, and that the development of high-performance TE materials and devices based on CBDL compounds has a bright future.

Author Contributions: Conceptualization, W.W. and D.Z.; methodology, W.W. and L.B.; writing—original draft preparation, W.W. and J.Z.; writing—review and editing, W.W.; funding acquisition, D.Z. All authors have read and agreed to the published version of the manuscript.

Funding: This work was funded by the National Natural Science Foundation of China (grant no. 51772132); Shandong Province Higher Educational Youth Innovative Science and Technology Program (grant no. 2019KJA018); the leader of scientific research studio program of Jinan (grant no. 2021GXRC082).

Institutional Review Board Statement: Not applicable.

Informed Consent Statement: Not applicable.

Data Availability Statement: Not applicable.

Conflicts of Interest: The authors declare no conflict of interest.

References

1. Bell, L.E. Cooling, heating, generating power, and recovering waste heat with thermoelectric systems. *Science* **2008**, *321*, 1457–1461. [CrossRef] [PubMed]
2. Shi, X.L.; Zou, J.; Chen, Z.G. Advanced Thermoelectric Design: From Materials and Structures to Devices. *Chem. Rev.* **2020**, *120*, 7399–7515. [CrossRef] [PubMed]
3. DiSalvo, F.J. Thermoelectric cooling and power generation. *Science* **1999**, *285*, 703–706. [CrossRef] [PubMed]
4. Snyder, G.J.; Toberer, E.S. Complex thermoelectric materials. *Nat. Mater.* **2008**, *7*, 105–114. [CrossRef] [PubMed]
5. Tan, G.; Zhao, L.D.; Kanatzidis, M.G. Rationally Designing High-Performance Bulk Thermoelectric Materials. *Chem. Rev.* **2016**, *116*, 12123–12149. [CrossRef]
6. Mukherjee, M.; Srivastava, A.; Singh, A.K. Recent Advances in Designing Thermoelectric Materials. *J. Mater. Chem. C* **2022**, *10*, 12524–12555. [CrossRef]
7. He, J.; Tritt, T.M. Advances in Thermoelectric Materials Research: Looking Back and Moving Forward. *Science* **2017**, *357*, eaak9997. [CrossRef]
8. Wu, Z.; Zhang, S.; Liu, Z.; Mu, E.; Hu, Z. Thermoelectric Converter: Strategies from Materials to Device Application. *Nano Energy* **2022**, *91*, 106692. [CrossRef]
9. Pei, Y.Z.; Shi, X.Y.; LaLonde, A.; Wang, H.; Chen, L.D.; Snyder, G.J. Convergence of Electronic Bands for High Performance Bulk Thermoelectrics. *Nature* **2011**, *473*, 66–69. [CrossRef]
10. Liu, W.; Yin, K.; Zhang, Q.J.; Uher, C.; Tang, X.F. Eco-Friendly High-Performance Silicide Thermoelectric Materials. *Nat. Sci. Rev.* **2017**, *4*, 611–626. [CrossRef]
11. Liu, W.S.; Jie, Q.; Kim, H.S.; Ren, Z.F. Current Progress and Future Challenges in Thermoelectric Power Generation: From Materials to Devices. *Acta Mater.* **2015**, *87*, 357–376. [CrossRef]
12. Li, J.F.; Pan, Y.; Wu, C.H.; Sun, F.H.; Wei, T.R. Processing of Advanced Thermoelectric Materials. *Sci. China Technol. Sci.* **2017**, *60*, 1347–1364. [CrossRef]
13. He, Y.; Day, T.; Zhang, T.; Liu, H.L.; Shi, X.; Chen, L.D.; Snyder, G.J. High Thermoelectric Performance in Non-Toxic Earth-Abundant Copper Sulfide. *Adv. Mater.* **2014**, *26*, 3974–3978. [CrossRef]
14. Cao, Y.; Sheng, Y.; Li, X.; Xi, L.L.; Yang, J. Application of Materials Genome Methods in Thermoelectrics. *Fron. Mater.* **2022**, *9*, 861817. [CrossRef]
15. Liu, R.H.; Chen, H.Y.; Zhao, K.P.; Qin, Y.T.; Jiang, B.B.; Zhang, T.S.; Sha, G.; Shi, X.; Uher, C.; Zhang, W.Q.; et al. Entropy as a Gene-Like Performance Indicator Promoting Thermoelectric Materials. *Adv. Mater.* **2017**, *29*, 1702712. [CrossRef]
16. Tee, S.Y.; Ponsford, D.; Lay, C.L.; Wang, X.; Wang, X.; Neo, D.C.J.; Wu, T.; Thitsartarn, W.; Yeo, J.C.C.; Guan, G.; et al. Thermoelectric Silver-Based Chalcogenides. *Adv. Sci.* **2022**, *9*, 2204624. [CrossRef]
17. Slack, G.A. New Materials and Performance Limits for Thermoelectric Cooling. In *CRC Handbook of Thermoelectrics*; Rowe, D.M., Ed.; CRC Press: Boca Raton, FL, USA, 1995; pp. 407–440.
18. Qiu, P.F.; Shi, X.; Chen, L.D. Cu-Based Thermoelectric Materials. *Energy Storage Mater.* **2016**, *3*, 85–97. [CrossRef]
19. Skoug, E.J.; Cain, J.D.; Morelli, D.T. High Thermoelectric Figure of Merit in the Cu_3SbSe_4–Cu_3SbS_4 solid Solution. *Appl. Phys. Lett.* **2011**, *98*, 261911. [CrossRef]
20. Liu, R.H.; Xi, L.L.; Liu, H.L.; Shi, X.; Zhang, W.Q.; Chen, L.D. Ternary Compound $CuInTe_2$: A Promising Thermoelectric Material with Diamond-Like Structure. *Chem. Commun.* **2012**, *48*, 3818–3820. [CrossRef]
21. Suekuni, K.; Takabatake, T. Research Update: Cu-S Based Synthetic Minerals as Efficient Thermoelectric Materials at Medium Temperatures. *APL Mater.* **2016**, *4*, 104503. [CrossRef]
22. Fan, J.; Carrillo-Cabrera, W.; Akselrud, L.; Antonyshyn, I.; Chen, L.D.; Grin, Y.R. New Monoclinic Phase at the Composition Cu_2SnSe_3 and Its Thermoelectric Properties. *Inorg. Chem.* **2013**, *52*, 11067–11074. [CrossRef] [PubMed]
23. Liu, R.H.; Qin, Y.T.; Cheng, N.; Zhang, J.W.; Shi, X.; Grin, Y.R.; Chen, L.D. Thermoelectric Performance of $Cu_{1-x-\delta}Ag_xInTe_2$ Diamond-Like Materials with a Pseudocubic Crystal Structure. *Inorg. Chem. Front.* **2016**, *3*, 1167–1177. [CrossRef]

24. Skoug, E.J.; Morelli, D.T. Role of Lone-Pair Electrons in Producing Minimum Thermal Conductivity in Nitrogen-Group Chalcogenide Compounds. *Phys. Rev. Lett.* **2011**, *107*, 235901. [CrossRef] [PubMed]
25. Shi, X.; Xi, L.L.; Fan, J.; Zhang, W.Q.; Chen, L.D. Cu-Se Bond Network and Thermoelectric Compounds with Complex Diamondlike Structure. *Chem. Mater.* **2010**, *22*, 6029–6031. [CrossRef]
26. Zhang, J.; Huang, L.L.; Zhu, C.; Zhou, C.J.; Jabar, B.; Li, J.; Zhu, X.G.; Wang, L.; Song, C.J.; Xin, H.X.; et al. Design of Domain Structure and Realization of Ultralow Thermal Conductivity for Record-High Thermoelectric Performance in Chalcopyrite. *Adv. Mater.* **2019**, *31*, e1905210. [CrossRef]
27. Huang, Y.L.; Zhang, B.; Li, J.W.; Zhou, Z.Z.; Zheng, S.K.; Li, N.H.; Wang, G.W.; Zhang, D.; Zhang, D.L.; Han, G.; et al. Unconventional Doping Effect Leads to Ultrahigh Average Thermoelectric Power Factor in Cu_3SbSe_4-Based Composites. *Adv. Mater.* **2022**, *34*, 2109952. [CrossRef]
28. Zhang, D.; Wang, X.C.; Wu, H.; Huang, Y.L.; Zheng, S.K.; Zhang, B.; Fu, H.X.; Cheng, Z.E.; Jiang, P.F.; Han, G.; et al. High Thermoelectric Performance in Earth-Abundant Cu_3SbS_4 by Promoting Doping Efficiency via Rational Vacancy Design. *Adv. Funct. Mater.* **2023**, *33*, 2214163. [CrossRef]
29. Plirdpring, T.; Kurosaki, K.; Kosuga, A.; Day, T.; Firdosy, S.; Ravi, V.; Snyder, G.J.; Harnwunggmoung, A.; Sugahara, T.; Ohishi, Y.; et al. Chalcopyrite $CuGaTe_2$: A High-Efficiency Bulk Thermoelectric Material. *Adv. Mater.* **2012**, *24*, 3622–3626. [CrossRef]
30. Luo, Y.B.; Yang, J.Y.; Jiang, Q.H.; Li, W.X.; Zhang, D.; Zhou, Z.W.; Cheng, Y.D.; Ren, Y.Y.; He, X. Progressive Regulation of Electrical and Thermal Transport Properties to High-Performance $CuInTe_2$ thermoelectric Materials. *Adv. Energy Mater.* **2016**, *6*, 160007. [CrossRef]
31. Zhang, J.W.; Liu, R.H.; Cheng, N.; Zhang, Y.B.; Yang, J.H.; Uher, C.; Shi, X.; Chen, L.D.; Zhang, W. High-Performance Pseudocubic Thermoelectric Materials from Non-Cubic Chalcopyrite Compounds. *Adv. Mater.* **2014**, *26*, 3848–3853. [CrossRef]
32. Wei, T.R.; Qin, Y.T.; Deng, T.T.; Song, Q.F.; Jiang, B.B.; Liu, R.H.; Qiu, P.F.; Shi, X.; Chen, L.D. Copper Chalcogenide Thermoelectric Materials. *Sci. China Mater.* **2018**, *62*, 8–24. [CrossRef]
33. Zhang, D.; Bai, H.C.; Li, Z.L.; Wang, J.L.; Fu, G.S.; Wang, S.F. Multinary Diamond-Like Chalcogenides for Promising Thermoelectric Application. *Chin. Phys. B* **2018**, *27*, 047206. [CrossRef]
34. Zhang, Z.P.; Gao, Y.; Wu, Y.; Wang, B.Y.; Sun, W.L.; Yu, L.; Wei, S.T.; Zheng, S.Q. P-Type Doping of Transition Metal Elements to Optimize the Thermoelectric Properties of $CuGaTe_2$. *Chem. Eng. J.* **2022**, *427*, 131807. [CrossRef]
35. Qin, Y.T.; Qiu, P.F.; Liu, R.H.; Li, Y.L.; Hao, F.; Zhang, T.S.; Ren, D.D.; Shi, X.; Chen, L.D. Optimized Thermoelectric Properties in Pseudocubic Diamond-Like $CuGaTe_2$ Compounds. *J. Mater. Chem. A* **2016**, *4*, 1277–1289. [CrossRef]
36. Luo, Y.B.; Yang, J.Y.; Jiang, Q.H.; Li, W.X.; Xiao, Y.; Fu, L.W.; Zhang, D.; Zhou, Z.W.; Cheng, Y.D. Large Enhancement of Thermoelectric Performance of $CuInTe_2$ via a Synergistic Strategy of Point Defects and Microstructure Engineering. *Nano Energy* **2015**, *18*, 37–46. [CrossRef]
37. Yan, Y.C.; Lu, X.; Wang, G.Y.; Zhou, X.Y. Zt = 1.1 in $CuInTe_2$ Solid Solutions Enabled by Rational Defect Engineering. *ACS Appl. Energy Mater.* **2019**, *3*, 2039–2048. [CrossRef]
38. Xie, H.Y.; Hao, S.Q.; Cai, S.T.; Bailey, T.P.; Uher, C.; Wolverton, C.; Dravid, V.P.; Kanatzidis, M.G. Ultralow Thermal Conductivity in Diamondoid Lattices: High Thermoelectric Performance in Chalcopyrite $Cu_{0.8+Y}Ag_{0.2}In_{1-Y}Te_2$. *Energy Environ. Sci.* **2020**, *13*, 3693–3705. [CrossRef]
39. Li, M.; Luo, Y.; Hu, X.J.; Cai, G.M.; Han, Z.K.; Du, Z.L.; Cui, G.L. Synergistic Regulation of Phonon and Electronic Properties to Improve the Thermoelectric Performance of Chalcogenide $CuIn_{1-x}Ga_xTe_2$:yInTe (x = 0–0.3) with In Situ Formed Nanoscale Phase InTe. *Adv. Electron. Mater.* **2020**, *6*, 190114. [CrossRef]
40. Shen, J.W.; Zhang, X.Y.; Chen, Z.W.; Lin, S.Q.; Li, J.; Li, W.; Li, S.S.; Chen, Y.; Pei, Y.Z. Substitutional Defects Enhancing Thermoelectric $CuGaTe_2$. *J. Mater. Chem. A* **2017**, *5*, 5314–5320. [CrossRef]
41. Xie, H.Y.; Hao, S.Q.; Bailey, T.P.; Cai, S.T.; Zhang, Y.Y.; Slade, T.J.; Snyder, G.J.; Dravid, V.P.; Uher, C.; Wolverton, C.; et al. Ultralow Thermal Conductivity in Diamondoid Structures and High Thermoelectric Performance in $(Cu_{1-x}Ag_x)(In_{1-y}Ga_y)Te_2$. *J. Am. Chem. Soc.* **2021**, *143*, 5978–5989. [CrossRef]
42. Wang, B.Y.; Zheng, S.Q.; Chen, Y.X.; Wu, Y.; Li, J.; Ji, Z.; Mu, Y.N.; Wei, Z.B.; Liang, Q.; Liang, J.X. Band Engineering for Realizing Large Effective Mass in Cu_3SbSe_4 by Sn/La Codoping. *J. Phys. Chem. C* **2020**, *124*, 10336–10343. [CrossRef]
43. Wang, B.Y.; Zheng, S.Q.; Wang, Q.; Li, Z.L.; Li, J.; Zhang, Z.P.; Wu, Y.; Zhu, B.S.; Wang, S.Y.; Chen, Y.X.; et al. Synergistic Modulation of Power Factor and Thermal Conductivity in Cu_3SbSe_4 Towards High Thermoelectric Performance. *Nano Energy* **2020**, *71*, 104658. [CrossRef]
44. Kumar, A.; Dhama, P.; Saini, D.S.; Banerji, P. Effect of Zn Substitution at a Cu Site on The Transport Behavior and Thermoelectric Properties in Cu_3SbSe_4. *RSC Adv.* **2016**, *6*, 5528–5534. [CrossRef]
45. Zou, T.H.; Qin, X.Y.; Li, D.; Li, L.L.; Sun, G.L.; Wang, Q.Q.; Zhang, J.; Xin, H.X.; Liu, Y.F.; Song, C.J. Enhanced Thermoelectric Performance of β-Zn_4Sb_3 Based Composites Incorporated with Large Proportion of Nanophase Cu_3SbSe_4. *J. Alloys Compd.* **2014**, *588*, 568–572. [CrossRef]
46. Li, J.M.; Li, D.; Song, C.J.; Wang, L.; Xin, H.X.; Zhang, J.; Qin, X.Y. Realized High Power Factor and Thermoelectric Performance in Cu_3SbSe_4. *Intermetallics* **2019**, *109*, 68–73. [CrossRef]
47. Li, J.M.; Ming, H.W.; Zhang, B.L.; Song, C.J.; Wang, L.; Xin, H.X.; Zhang, J.; Qin, X.Y.; Li, D. Ultra-Low Thermal Conductivity and High Thermoelectric Performance Realized in a Cu_3SbSe_4 Based System. *Mater. Chem. Front.* **2021**, *5*, 324–332. [CrossRef]

48. Xie, D.D.; Zhang, B.; Zhang, A.; Chen, Y.; Yan, Y.; Yang, H.; Wang, G.; Wang, G.; Han, X.; Han, G.; et al. High Thermoelectric Performance of Cu$_3$SbSe$_4$ Nanocrystals with Cu$_{2-x}$Se in Situ Inclusions Synthesized by a Microwave-Assisted Solvothermal Method. *Nanoscale* **2018**, *10*, 14546–14553. [CrossRef]
49. Zhang, D.; Yang, J.Y.; Bai, H.C.; Luo, Y.B.; Wang, B.; Hou, S.H.; Li, Z.L.; Wang, S. Significant Average Zt Enhancement in Cu$_3$SbSe$_4$-Based Thermoelectric Material via Softening P-D Hybridization. *J. Mater. Chem. A* **2019**, *7*, 17648–17654. [CrossRef]
50. Chen, K.; Di Paola, C.; Du, B.; Zhang, R.Z.; Laricchia, S.; Bonini, N.; Weber, C.; Abrahams, I.; Yan, H.; Reece, M. Enhanced Thermoelectric Performance of Sn-Doped Cu$_3$SbS$_4$. *J. Mater. Chem. C* **2018**, *6*, 8546–8552. [CrossRef]
51. Shen, M.J.; Lu, S.Y.; Zhang, Z.F.; Liu, H.Y.; Shen, W.X.; Fang, C.; Wang, Q.Q.; Chen, L.C.; Zhang, Y.W.; Jia, X.P. Bi and Sn Co-Doping Enhanced Thermoelectric Properties of Cu$_3$SbS$_4$ Materials with Excellent Thermal Stability. *ACS Appl. Mater. Interfaces* **2020**, *12*, 8271–8279. [CrossRef]
52. Xie, H.Y.; Su, X.L.; Zheng, G.; Zhu, T.; Yin, K.; Yan, Y.G.; Uher, C.; Kanatzidis, M.G.; Tang, X.F. The Role of Zn in Chalcopyrite CuFeS$_2$: Enhanced Thermoelectric Properties of Cu$_{1-x}$Zn$_x$FeS$_2$ with in Situ Nanoprecipitates. *Adv. Energy Mater.* **2016**, *7*, 601299.
53. Ge, B.Z.; Lee, H.; Zhou, C.J.; Lu, W.Q.; Hu, J.B.; Yang, J.; Cho, S.P.; Qiao, G.J.; Shi, Z.Q.; Chung, I. Exceptionally Low Thermal Conductivity Realized in the Chalcopyrite CuFeS$_2$ via Atomic-Level Lattice Engineering. *Nano Energy* **2022**, *94*, 106941. [CrossRef]
54. Ge, B.Z.; Shi, Z.Q.; Zhou, C.J.; Hu, J.B.; Liu, G.W.; Xia, H.Y.; Xu, J.T.; Qiao, G.J. Enhanced Thermoelectric Performance of N-Type Eco-Friendly Material Cu$_{1-x}$Ag$_x$FeS$_2$ (X=0–0.14) via Bandgap Tuning. *J. Alloys Compd.* **2019**, *809*, 151717. [CrossRef]
55. Tippireddy, S.; Azough, F.; Vikram; Bhui, A.; Chater, P.; Kepaptsoglou, D.; Ramasse, Q.; Freer, R.; Grau-Crespo, R.; Biswas, K.; et al. Local Structural Distortions and Reduced Thermal Conductivity in Ge-Substituted Chalcopyrite. *J. Mater. Chem. A* **2022**, *10*, 23874–23885. [CrossRef]
56. Tan, Q.; Sun, W.; Li, Z.L.; Li, J.F. Enhanced Thermoelectric Properties of Earth-Abundant Cu$_2$SnS$_3$ via in Doping Effect. *J. Alloys Compd.* **2016**, *672*, 558–563. [CrossRef]
57. Zhang, Z.; Zhao, H.W.; Wang, Y.F.; Hu, X.H.; Lyu, Y.; Cheng, C.C.; Pan, L.; Lu, C.H. Role of Crystal Transformation on the Enhanced Thermoelectric Performance in Mn-Doped Cu$_2$SnS$_3$. *J. Alloys Compd.* **2019**, *780*, 618–625. [CrossRef]
58. Zhao, Y.Q.; Gu, Y.; Zhang, P.; Hu, X.H.; Wang, Y.F.; Zong, P.; Pan, L.; Lyu, Y.; Koumoto, K. Enhanced Thermoelectric Performance in Polymorphic Heavily Co-Doped Cu$_2$SnS$_3$ through Carrier Compensation by Sb Substitution. *Sci Technol. Adv. Mater.* **2021**, *22*, 363–372. [CrossRef]
59. Huang, T.Y.; Yan, Y.C.; Peng, K.L.; Tang, X.D.; Guo, L.J.; Wang, R.F.; Lu, X.; Zhou, X.Y.; Wang, G.Y. Enhanced Thermoelectric Performance in Copper-Deficient Cu$_2$GeSe$_3$. *J. Alloys Compd.* **2017**, *723*, 708–713. [CrossRef]
60. Wang, R.F.; Li, A.; Huang, T.; Zhang, B.; Peng, K.L.; Yang, H.Q.; Lu, X.; Zhou, X.Y.; Han, X.D.; Wang, G.Y. Enhanced Thermoelectric Performance in Cu$_2$GeSe$_3$ via (Ag,Ga)-Co-Doping on Cation Sites. *J. Alloys Compd.* **2018**, *769*, 218–225. [CrossRef]
61. Yang, J.; Lu, B.B.; Song, R.F.; Hou, H.G.; Zhao, L.J.; Zhang, X.Z.; Liu, G.W.; Qiao, G.J. Realizing Enhanced Thermoelectric Properties in Cu$_2$GeSe$_3$ via a Synergistic Effect of in and Ag Dual-Doping. *J. Eur. Ceram. Soc.* **2022**, *42*, 169–174. [CrossRef]
62. Hu, L.; Luo, Y.B.; Fang, Y.W.; Qin, F.Y.; Cao, X.; Xie, H.Y.; Liu, J.W.; Dong, J.F.; Sanson, A.; Giarola, M.; et al. High Thermoelectric Performance through Crystal Symmetry Enhancement in Triply Doped Diamondoid Compound Cu$_2$SnSe$_3$. *Adv. Energy Mater.* **2021**, *11*, 2100661. [CrossRef]
63. Li, Y.Y.; Liu, G.H.; Cao, T.F.; Liu, L.M.; Li, J.T.; Chen, K.X.; Li, L.F.; Han, Y.M.; Zhou, M. Enhanced Thermoelectric Properties of Cu$_2$SnSe$_3$ by (Ag, In)-Co-Doping. *Adv. Funct. Mater.* **2016**, *26*, 6025–6032. [CrossRef]
64. Cheng, X.; Yang, D.W.; Su, X.L.; Xie, H.Y.; Liu, W.; Zheng, Y.; Tang, X.F. Synergistically Enhanced Thermoelectric Performance of Cu$_2$SnSe$_3$-Based Composites via Ag Doping Balance. *ACS Appl. Mater. Interfaces* **2021**, *13*, 55178–55187. [CrossRef]
65. Ming, H.W.; Zhu, G.F.; Zhu, C.; Qin, X.Y.; Chen, T.; Zhang, J.; Li, D.; Xin, H.X.; Jabar, B. Boosting Thermoelectric Performance of Cu$_2$SnSe$_3$ via Comprehensive Band Structure Regulation and Intensified Phonon Scattering by Multidimensional Defects. *ACS Nano* **2021**, *15*, 10532–10541. [CrossRef]
66. Ming, H.W.; Zhu, C.; Chen, T.; Yang, S.H.; Chen, Y.; Zhang, J.; Li, D.; Xin, H.X.; Qin, X.Y. Creating High-Dense Stacking Faults and Endo-Grown Nanoneedles to Enhance Phonon Scattering and Improve Thermoelectric Performance of Cu$_2$SnSe$_3$. *Nano Energy* **2022**, *100*, 107510. [CrossRef]
67. Ming, H.W.; Zhu, C.; Qin, X.Y.; Zhang, J.; Li, D.; Zhang, B.L.; Chen, T.; Li, J.M.; Lou, X.N.; Xin, H.X. Improved Figure of Merit of Cu$_2$SnSe$_3$ via Band Structure Modification and Energy-Dependent Carrier Scattering. *ACS Appl. Mater. Interfaces* **2020**, *12*, 19693–19700. [CrossRef]
68. Zhu, C.; Chen, Q.; Ming, H.W.; Qin, X.Y.; Yang, Y.; Zhang, J.; Peng, D.; Chen, T.; Li, D.; Kawazoe, Y. Improved Thermoelectric Performance of Cu$_{12}$Sb$_4$S$_{13}$ through Gd-Substitution Induced Enhancement of Electronic Density of States and Phonon Scattering. *ACS Appl. Mater. Interfaces* **2021**, *13*, 25092–25101. [CrossRef]
69. Zhu, C.; Ming, H.; Huang, L.; Zhang, B.; Lou, X.; Li, D.; Jabar, B.; Xin, H.; Zhang, J.; Qin, X. Achieving High Power Factor and Thermoelectric Performance through Dual Substitution of Zn and Se in Tetrahedrites Cu$_{12}$Sb$_4$S$_{13}$. *Appl. Phys. Lett.* **2019**, *115*, 182102. [CrossRef]
70. Liu, M.L.; Chen, I.W.; Huang, F.Q.; Chen, L.D. Improved Thermoelectric Properties of Cu-Doped Quaternary Chalcogenides of Cu$_2$CdSnSe$_4$. *Adv. Mater.* **2009**, *21*, 3808–3812. [CrossRef]
71. Chen, Q.F.; Yan, Y.C.; Zhan, H.; Yao, W.; Chen, Y.; Dai, J.Y.; Sun, X.N.; Zhou, X.Y. Enhanced Thermoelectric Performance of Chalcogenide Cu$_2$CdSnSe$_4$ by Ex-Situ Homogeneous Nanoinclusions. *J. Mater.* **2016**, *2*, 179–186.

72. Fan, F.J.; Yu, B.; Wang, Y.X.; Zhu, Y.L.; Liu, X.J.; Yu, S.H.; Ren, Z.F. Colloidal Synthesis of $Cu_2CdSnSe_4$ Nanocrystals and Hot-Pressing to Enhance the Thermoelectric Figure-of-Merit. *J. Am. Chem. Soc.* **2011**, *133*, 15910–15913. [CrossRef] [PubMed]
73. Chen, Q.F.; Wang, G.W.; Zhang, A.J.; Yang, D.F.; Yao, W.; Peng, K.L.; Yan, Y.C.; Sun, X.N.; Liu, A.P.; Wang, G.Y.; et al. Colloidal Synthesis of $Cu_{2-x}Ag_xCdSnSe_4$ Nanocrystals: Microstructures Facilitate High Performance Thermoelectricity. *J. Mater. Chem. C* **2015**, *3*, 12273–12280. [CrossRef]
74. Basu, R.; Mandava, S.; Bohra, A.; Bhattacharya, S.; Bhatt, R.; Ahmad, S.; Bhattacharyya, K.; Samanta, S.; Debnath, A.K.; Singh, A.; et al. Improving the Thermoelectric Performance of Tetrahedrally Bonded Quaternary Selenide $Cu_2CdSnSe_4$ Using Cdse Precipitates. *J. Electron. Mater.* **2019**, *48*, 2120–2130. [CrossRef]
75. Chetty, R.; Bali, A.; Femi, O.E.; Chattopadhyay, K.; Mallik, R.C. Thermoelectric Properties of in-Doped $Cu_2ZnGeSe_4$. *J. Electron. Mater.* **2015**, *45*, 1625–1632. [CrossRef]
76. Song, Q.F.; Qiu, P.F.; Hao, F.; Zhao, K.P.; Zhang, T.S.; Ren, D.D.; Shi, X.; Chen, L.D. Quaternary Pseudocubic $Cu_2TmSnSe_4$(Tm = Mn, Fe, Co) Chalcopyrite Thermoelectric Materials. *Adv. Electron. Mater.* **2016**, *2*, 1600312. [CrossRef]
77. Song, Q.F.; Qiu, P.F.; Zhao, K.; Deng, T.T.; Shi, X.; Chen, L.D. Crystal Structure and Thermoelectric Properties of $Cu_2Fe_{1-x}Mn_xSnSe_4$ Diamond-Like Chalcogenides. *ACS Appl. Energy Mater.* **2019**, *3*, 2137–2146. [CrossRef]
78. Song, Q.F.; Qiu, P.F.; Chen, H.; Zhao, K.; Guan, M.; Zhou, Y.; Wei, T.R.; Ren, D.D.; Xi, L.; Yang, J.; et al. Enhanced Carrier Mobility and Thermoelectric Performance in $Cu_2FeSnSe_4$ Diamond-Like Compound via Manipulating the Intrinsic Lattice Defects. *Mater. Today Phys.* **2018**, *7*, 45–53. [CrossRef]
79. Pavan Kumar, V.; Guilmeau, E.; Raveau, B.; Caignaert, V.; Varadaraju, U.V. A New Wide Band Gap Thermoelectric Quaternary Selenide $Cu_2MgSnSe_4$. *J. Appl. Phys.* **2015**, *118*, 155101. [CrossRef]
80. Shen, J.W.; Zhang, X.Y.; Lin, S.Q.; Li, J.; Chen, Z.W.; Li, W.; Pei, Y.Z. Vacancy Scattering for Enhancing the Thermoelectric Performance of $CuGaTe_2$ Solid Solutions. *J. Mater. Chem. A* **2016**, *4*, 15464–15470. [CrossRef]
81. Zhong, Y.; Tang, J.; Liu, H.T.; Chen, Z.W.; Lin, L.; Ren, D.; Liu, B.; Ang, R. Optimized Strategies for Advancing N-Type PbTe Thermoelectrics: A Review. *ACS Appl. Mater. Interfaces* **2020**, *12*, 49323–49334. [CrossRef]
82. Shi, H.N.; Qin, Y.X.; Qin, B.C.; Su, L.Z.; Wang, Y.P.; Chen, Y.J.; Gao, X.; Liang, H.; Ge, Z.H.; Hong, T.; et al. Incompletely Decomposed In_4SnSe_4 Leads to High-Ranged Thermoelectric Performance in N-Type PbTe. *Adv. Energy Mater.* **2022**, *12*, 2202539. [CrossRef]
83. Jia, B.H.; Huang, Y.; Wang, Y.; Zhou, Y.; Zhao, X.D.; Ning, S.T.; Xu, X.; Lin, P.J.; Chen, Z.Q.; Jiang, B.B.; et al. Realizing High Thermoelectric Performance in Non-Nanostructured N-Type PbTe. *Energy Environ. Sci.* **2022**, *15*, 1920–1929. [CrossRef]
84. Wang, S.Q.; Chang, C.; Bai, S.L.; Qin, B.C.; Zhu, Y.C.; Zhan, S.P.; Zheng, J.Q.; Tang, S.W.; Zhao, L.D. Fine Tuning of Defects Enables High Carrier Mobility and Enhanced Thermoelectric Performance of N-Type PbTe. *Chem. Mater.* **2023**, *35*, 755–763. [CrossRef]
85. Xu, H.H.; Wan, H.; Xu, R.; Hu, Z.Q.; Liang, X.L.; Li, Z.; Song, J.M. Enhancing the Thermoelectric Performance of $SnTe-CuSbSe_2$ with an Ultra-Low Lattice Thermal Conductivity. *J. Mater. Chem. A* **2023**, *11*, 4310–4318. [CrossRef]
86. Chen, Z.Y.; Guo, X.M.; Zhang, F.J.; Shi, Q.; Tang, M.J.; Ang, R. Routes for Advancing SnTe Thermoelectrics. *J. Mater. Chem. A* **2020**, *8*, 16790–16813. [CrossRef]
87. Li, W.; Wu, Y.X.; Lin, S.Q.; Chen, Z.W.; Li, J.; Zhang, X.Y.; Zheng, L.L.; Pei, Y.Z. Advances in Environment-Friendly SnTe Thermoelectrics. *ACS Energy Lett.* **2017**, *2*, 2349–2355. [CrossRef]
88. Tippireddy, S.; Azough, F.; Vikram; Tompkins, F.T.; Bhui, A.; Freer, R.; Grau-Crespo, R.; Biswas, K.; Vaqueiro, P.; Powell, A.V. Tin-Substituted Chalcopyrite: An N-Type Sulfide with Enhanced Thermoelectric Performance. *Chem. Mater.* **2022**, *34*, 5860–5873. [CrossRef]
89. Xie, H.Y.; Su, X.L.; Hao, S.Q.; Zhang, C.; Zhang, Z.K.; Liu, W.; Yan, Y.G.; Wolverton, C.; Tang, X.F.; Kanatzidis, M.G. Large Thermal Conductivity Drops in the Diamondoid Lattice of $CuFeS_2$ by Discordant Atom Doping. *J. Am. Chem. Soc.* **2019**, *141*, 18900–18909. [CrossRef]
90. Bo, L.; Li, F.J.; Hou, Y.B.; Wang, L.; Wang, X.L.; Zhang, R.P.; Zuo, M.; Ma, Y.Z.; Zhao, D.G. Enhanced Thermoelectric Properties of Cu_3SbSe_4 via Configurational Entropy Tuning. *J. Mater. Sci.* **2022**, *57*, 4643–4651. [CrossRef]
91. Fan, J.; Carrillo-Cabrera, W.; Antonyshyn, I.; Prots, Y.; Veremchuk, I.; Schnelle, W.; Drathen, C.; Chen, L.D.; Grin, Y. Crystal Structure and Physical Properties of Ternary Phases around the Composition $Cu_5Sn_2Se_7$ with Tetrahedral Coordination of Atoms. *Chem. Mater.* **2014**, *26*, 5244–5251. [CrossRef]
92. Adhikary, A.; Mohapatra, S.; Lee, S.H.; Hor, Y.S.; Adhikari, P.; Ching, W.Y.; Choudhury, A. Metallic Ternary Telluride with Sphalerite Superstructure. *Inorg. Chem.* **2016**, *55*, 2114–2122. [CrossRef] [PubMed]
93. Sturm, C.; Macario, L.R.; Mori, T.; Kleinke, H. Thermoelectric Properties of Zinc-Doped $Cu_5Sn_2Se_7$ and $Cu_5Sn_2Te_7$. *Dalton Trans.* **2021**, *50*, 6561–6567. [CrossRef] [PubMed]
94. Hasan, S.; San, S.; Baral, K.; Li, N.; Rulis, P.; Ching, W.Y. First-Principles Calculations of Thermoelectric Transport Properties of Quaternary and Ternary Bulk Chalcogenide Crystals. *Materials* **2022**, *15*, 2843–2871. [CrossRef] [PubMed]
95. Pavan Kumar, V.; Passuti, S.; Zhang, B.; Fujii, S.; Yoshizawa, K.; Boullay, P.; Le Tonquesse, S.; Prestipino, C.; Raveau, B.; Lemoine, P.; et al. Engineering Transport Properties in Interconnected Enargite-Stannite Type $Cu_{2+x}Mn_{1-x}GeS_4$ Nanocomposites. *Angew. Chem. Int. Ed. Engl.* **2022**, *61*, e202210600. [CrossRef]
96. Shi, X.Y.; Huang, F.Q.; Liu, M.L.; Chen, L.D. Thermoelectric Properties of Tetrahedrally Bonded Wide-Gap Stannite Compounds $Cu_2ZnSn_{1-x}In_xSe_4$. *Appl. Phys. Lett.* **2009**, *94*, 122103. [CrossRef]

97. Zou, D.F.; Nie, G.Z.; Li, Y.; Xu, Y.; Lin, J.; Zheng, H.; Li, J.G. Band Engineering via Biaxial Strain for Enhanced Thermoelectric Performance in Stannite-Type $Cu_2ZnSnSe_4$. *RSC Adv.* **2015**, *5*, 24908–24914. [CrossRef]
98. Prem Kumar, D.S.; Chetty, R.; Rogl, P.; Rogl, G.; Bauer, E.; Malar, P.; Mallik, R.C. Thermoelectric Properties of Cd Doped Tetrahedrite: $Cu_{12-x}Cd_xSb_4S_{13}$. *Intermetallics* **2016**, *78*, 21–29. [CrossRef]
99. Chen, S.Y.; Gong, X.G.; Walsh, A.; Wei, S.H. Electronic Structure and Stability of Quaternary Chalcogenide Semiconductors Derived from Cation Cross-Substitution of II-VI And I-III-VI_2 compounds. *Phys. Rev. B* **2009**, *79*, 165211. [CrossRef]
100. Lu, X.; Morelli, D.T.; Wang, Y.X.; Lai, W.; Xia, Y.; Ozolins, V. Phase Stability, Crystal Structure, and Thermoelectric Properties of $Cu_{12}Sb4S_{13-x}Se_x$ Solid Solutions. *Chem. Mater.* **2016**, *28*, 1781–1786. [CrossRef]
101. Chetty, R.; Bali, A.; Mallik, R.C. Tetrahedrites as Thermoelectric Materials: An Overview. *J. Mater. Chem. C* **2015**, *3*, 12364–12378. [CrossRef]
102. Lu, X.; Morelli, D.T.; Xia, Y.; Zhou, F.; Ozolins, V.; Chi, H.; Zhou, X.Y.; Uher, C. High Performance Thermoelectricity in Earth-Abundant Compounds Based on Natural Mineral Tetrahedrites. *Adv. Energy Mater.* **2013**, *3*, 342–348. [CrossRef]
103. Hu, H.H.; Zhuang, H.L.; Jiang, Y.L.; Shi, J.L.; Li, J.W.; Cai, B.W.; Han, Z.N.; Pei, J.; Su, B.; Ge, Z.H.; et al. Thermoelectric $Cu_{12}Sb_4S_{13}$-Based Synthetic Minerals with a Sublimation-Derived Porous Network. *Adv. Mater.* **2021**, *33*, e2103633. [CrossRef]
104. Bouyrie, Y.; Ohta, M.; Suekuni, K.; Kikuchi, Y.; Jood, P.; Yamamoto, A.; Takabatake, T. Enhancement in the Thermoelectric Performance of Colusites $Cu_{26}A_2E_6S_{32}$(A=Nb, Ta; E=Sn, Ge) Using E-Site Non-Stoichiometry. *J. Mater. Chem. C* **2017**, *5*, 4174–4184. [CrossRef]
105. Bourgès, C.; Gilmas, M.; Lemoine, P.; Mordvinova, N.E.; Lebedev, O.I.; Hug, E.; Nassif, V.; Malaman, B.; Daou, R.; Guilmeau, E. Structural Analysis and Thermoelectric Properties of Mechanically Alloyed Colusites. *J. Mater. Chem. C* **2016**, *4*, 7455–7463. [CrossRef]
106. Bourgès, C.; Bouyrie, Y.; Supka, A.R.; Al Rahal Al Orabi, R.; Lemoine, P.; Lebedev, O.I.; Ohta, M.; Suekuni, K.; Nassif, V.; Hardy, V.; et al. High-Performance Thermoelectric Bulk Colusite by Process Controlled Structural Disordering. *J. Am. Chem. Soc.* **2018**, *140*, 2186–2195. [CrossRef]
107. Kim, F.S.; Suekuni, K.; Nishiate, H.; Ohta, M.; Tanaka, H.I.; Takabatake, T. Tuning the Charge Carrier Density in the Thermoelectric Colusite. *J. Appl. Phys.* **2016**, *119*, 175105. [CrossRef]
108. Pavan Kumar, V.; Supka, A.R.; Lemoine, P.; Lebedev, O.I.; Raveau, B.; Suekuni, K.; Nassif, V.; Al Rahal Al Orabi, R.; Fornari, M.; Guilmeau, E. High Power Factors of Thermoelectric Colusites $Cu_{26}T_2Ge_6S_{32}$ (T= Cr, Mo, W): Toward Functionalization of the Conductive "Cu-S" Network. *Adv. Energy Mater.* **2018**, *9*, 1803249. [CrossRef]
109. Guélou, G.; Lemoine, P.; Raveau, B.; Guilmeau, E. Recent Developments in High-Performance Thermoelectric Sulphides: An Overview of the Promising Synthetic Colusites. *J. Mater. Chem. C* **2021**, *9*, 773–795. [CrossRef]
110. Bouyrie, Y.; Candolfi, C.; Dauscher, A.; Malaman, B.; Lenoir, B. Exsolution Process as a Route toward Extremely Low Thermal Conductivity in $Cu_{12}Sb_{4-x}Te_xS_{13}$ Tetrahedrites. *Chem. Mater.* **2015**, *27*, 8354–8361. [CrossRef]
111. Vaidya, M.; Muralikrishna, G.M.; Murty, B.S. High-Entropy Alloys by Mechanical Alloying: A Review. *J. Mater. Res.* **2019**, *34*, 664–686. [CrossRef]
112. Murty, B.S.; Ranganathan, S. Novel materials synthesis by mechanical alloying/milling. *Int. Mater. Rev.* **1998**, *43*, 101–141. [CrossRef]
113. Wei, T.R.; Wang, H.; Gibbs, Z.M.; Wu, C.F.; Snyder, G.J.; Li, J.F. Thermoelectric Properties of Sn-Doped P-Type Cu_3SbSe_4: A Compound with Large Effective Mass and Small Band Gap. *J. Mater. Chem. A* **2014**, *2*, 13527–13533. [CrossRef]
114. Suryanarayana, C.; Ivanov, E.; Boldyrev, V.V. The Science and technology of mechanical Alloying. *Mater. Sci. Eng. A* **2001**, *304*, 151–158.
115. Zhang, D.; Yang, J.Y.; Jiang, Q.H.; Zhou, Z.W.; Li, X.W.; Xin, J.; Basit, A.; Ren, Y.Y.; He, X.; Chu, W.J.; et al. Combination of Carrier Concentration Regulation and High Band Degeneracy for Enhanced Thermoelectric Performance of Cu_3SbSe_4. *ACS Appl. Mater. Interfaces* **2017**, *9*, 28558–28565. [CrossRef]
116. Chen, K.; Du, B.; Bonini, N.; Weber, C.; Yan, H.X.; Reece, M.J. Theory-Guided Synthesis of an Eco-Friendly and Low-Cost Copper Based Sulfide Thermoelectric Material. *J. Phys. Chem. C* **2016**, *120*, 27135–27140. [CrossRef]
117. Nautiyal, H.; Lohani, K.; Mukherjee, B.; Isotta, E.; Malagutti, M.A.; Ataollahi, N.; Pallecchi, I.; Putti, M.; Misture, S.T.; Rebuffi, L.; et al. Mechanochemical Synthesis of Sustainable Ternary and Quaternary Nanostructured Cu_2SnS_3, Cu_2ZnSnS_4, and $Cu_2ZnSnSe_4$ Chalcogenides for Thermoelectric Applications. *Nanomaterials* **2023**, *13*, 366–387. [CrossRef]
118. Wang, W.Y.; Bo, L.; Wang, Y.P.; Wang, L.; Li, F.J.; Zuo, M.; Zhao, D.G. Enhanced Thermoelectric Properties of Graphene /Cu_3SbSe_4 Composites. *J. Electron. Mater.* **2021**, *50*, 4880–4886. [CrossRef]
119. Wang, S.Y.; Xie, W.J.; Li, H.; Tang, X.F.; Zhang, Q.J. Effects of Cooling Rate on Thermoelectric Properties of N-Type $Bi_2(Se_{0.4}Te_{0.6})_3$ Compounds. *J. Electron. Mater.* **2011**, *40*, 1150–1157. [CrossRef]
120. Xie, W.J.; Tang, X.F.; Yan, Y.G.; Zhang, Q.J.; Tritt, T.M. Unique Nanostructures and Enhanced Thermoelectric Performance of Melt-Spun Bisbte Alloys. *Appl. Phys. Lett.* **2009**, *94*, 102111. [CrossRef]
121. Zhao, D.G.; Wang, L.; Wu, D.; Bo, L. Thermoelectric Properties of $CuSnSe_3$-Based Composites Containing Melt-Spun Cu-Te. *Metals* **2019**, *9*, 971–980. [CrossRef]
122. Zheng, Y.; Xie, H.Y.; Zhang, Q.; Suwardi, A.; Cheng, X.; Zhang, Y.F.; Shu, W.; Wan, X.J.; Yang, Z.L.; Liu, Z.H.; et al. Unraveling the Critical Role of Melt-Spinning Atmosphere in Enhancing the Thermoelectric Performance of P-Type $Bi_{0.52}Sb_{1.48}Te_3$ Alloys. *ACS Appl. Mater. Interfaces* **2020**, *12*, 36186–36195. [CrossRef]

123. Ding, G.C.; Si, J.X.; Wu, H.F.; Yang, S.D.; Zhao, J.; Wang, G.W. Thermoelectric Properties of Melt Spun PbTe with Multi-Scaled Nanostructures. *J. Alloys Compd.* **2016**, *662*, 368–373. [CrossRef]
124. Yang, B.; Li, S.M.; Li, X.; Liu, Z.P.; Zhong, H.; Feng, S.K. Ultralow Thermal Conductivity and Enhanced Thermoelectric Properties of SnTe Based Alloys Prepared by Melt Spinning Technique. *J. Alloys Compd.* **2020**, *837*, 155568. [CrossRef]
125. Geng, H.Y.; Zhang, J.L.; He, T.H.; Zhang, L.X.; Feng, J.C. Microstructure Evolution and Mechanical Properties of Melt Spun Skutterudite-Based Thermoelectric Materials. *Materials* **2020**, *13*, 984–998. [CrossRef]
126. Su, X.L.; Fu, F.; Yan, Y.G.; Zheng, G.; Liang, T.; Zhang, Q.; Cheng, X.; Yang, D.W.; Chi, H.; Tang, X.F.; et al. Self-Propagating High-Temperature Synthesis for Compound Thermoelectrics and New Criterion for Combustion Processing. *Nat. Commun.* **2014**, *5*, 4908–4915. [CrossRef]
127. Cheng, X.; You, Y.H.; Fu, J.F.; Hu, T.Z.; Liu, W.; Su, X.L.; Yan, Y.G.; Tang, X.F. Self-Propagating High-Temperature Synthesis and Thermoelectric Performances of $CuSnSe_3$. *J. Alloys Compd.* **2018**, *750*, 965–971. [CrossRef]
128. Cheng, X.; Zhu, B.; Yang, D.W.; Su, X.L.; Liu, W.; Xie, H.Y.; Zheng, Y.; Tang, X.F. Enhanced Thermoelectric Properties of Cu_2SnSe_3-Based Materials with Ag_2Se Addition. *ACS Appl. Mater. Interfaces* **2022**, *14*, 5439–5446. [CrossRef]
129. Wei, S.T.; Wang, B.Y.; Zhang, Z.P.; Li, W.H.; Yu, L.; Wei, S.K.; Ji, Z.; Song, W.Y.; Zheng, S.Q. Achieving High Thermoelectric Performance through Carrier Concentration Optimization and Energy Filtering in Cu_3SbSe_4-Based Materials. *J. Mater.* **2022**, *8*, 929–936. [CrossRef]
130. Nandihalli, N.; Gregory, D.H.; Mori, T. Energy-Saving Pathways for Thermoelectric Nanomaterial Synthesis: Hydrothermal/Solvothermal, Microwave-Assisted, Solution-Based, and Powder Processing. *Adv. Sci.* **2022**, *9*, e2106052. [CrossRef]
131. Shi, X.L.; Tao, X.Y.; Zou, J.; Chen, Z.G. High-Performance Thermoelectric SnSe: Aqueous Synthesis, Innovations, and Challenges. *Adv. Sci.* **2020**, *7*, 1902923. [CrossRef]
132. Balow, R.B.; Tomlinson, E.P.; Abu-Omar, M.M.; Boudouris, B.W.; Agrawal, R. Solution-Based Synthesis and Characterization of Earth Abundant $Cu_3(As,Sb)Se_4$ Nanocrystal Alloys: Towards Scalable Room-Temperature Thermoelectric Devices. *J. Mater. Chem. A* **2016**, *4*, 2198–2204. [CrossRef]
133. Xiong, Q.H.; Xie, D.D.; Wang, H.; Wei, Y.Q.; Wang, G.W.; Wang, G.Y.; Liao, H.J.; Zhou, X.Y.; Lu, X. Colloidal Synthesis of Diamond-Like Compound Cu_2SnTe_3 and Thermoelectric Properties of $(Cu_{0.96}InTe_2)_{1-x}(Cu_2SnTe_3)_x$ Solid Solutions. *Chem. Eng. J.* **2021**, *422*, 129985. [CrossRef]
134. Wang, B.Y.; Wang, Y.L.; Zheng, S.Q.; Liu, S.C.; Li, J.; Chang, S.Y.; An, T.; Sun, W.L.; Chen, Y.X. Improvement of Thermoelectric Properties of Cu_3SbSe_4 Hierarchical with In-Situ Second Phase Synthesized by Microwave-Assisted Solvothermal Method. *J. Alloys Compd.* **2019**, *806*, 676–682. [CrossRef]
135. Huang, L.L.; Zhang, J.; Zhu, C.; Ge, Z.H.; Li, Y.Y.; Li, D.; Qin, X.Y. Synergistically Optimized Electrical and Thermal Properties by Introducing Electron Localization and Phonon Scattering Centers in $CuGaTe_2$ with Enhanced Mechanical Properties. *J. Mater. Chem. C* **2020**, *8*, 7534–7542. [CrossRef]
136. Wang, W.L.; Feng, W.L.; Ding, T.; Yang, Q. Phosphine-Free Synthesis and Characterization of Cubic-Phase Cu_2SnTe_3 Nanocrystals with Optical and Optoelectronic Properties. *Chem. Mater.* **2015**, *27*, 6181–6184. [CrossRef]
137. Zhang, D.W.; Lim, W.Y.S.; Duran, S.S.F.; Loh, X.J.; Suwardi, A. Additive Manufacturing of Thermoelectrics: Emerging Trends and Outlook. *ACS Energy Lett.* **2022**, *7*, 720–735. [CrossRef]
138. Oztan, C.; Welch, R.; LeBlanc, S. Additive Manufacturing of Bulk Thermoelectric Architectures: A Review. *Energies* **2022**, *15*, 3121–3137. [CrossRef]
139. Li, R.X.; Li, X.; Xi, L.L.; Yang, J.; Singh, D.J.; Zhang, W.Q. High-Throughput Screening for Advanced Thermoelectric Materials: Diamond-Like Abx(2) Compounds. *ACS Appl. Mater. Interfaces* **2019**, *11*, 24859–24866. [CrossRef]
140. Xiong, Y.F.; Jin, Y.Q.; Deng, T.T.; Mei, K.L.; Qiu, P.F.; Xi, L.L.; Zhou, Z.Y.; Yang, J.; Shi, X.; Chen, L.D. High-Throughput Screening for Thermoelectric Semiconductors with Desired Conduction Types by Energy Positions of Band Edges. *J. Am. Chem. Soc.* **2022**, *144*, 8030–8037. [CrossRef]
141. Xi, L.L.; Pan, S.S.; Li, X.; Xu, Y.L.; Ni, J.Y.; Sun, X.; Yang, J.; Luo, J.; Xi, J.; Zhu, W.H.; et al. Discovery of High-Performance Thermoelectric Chalcogenides through Reliable High-Throughput Material Screening. *J. Am. Chem. Soc.* **2018**, *140*, 10785–10793. [CrossRef]
142. Recatala-Gomez, J.; Suwardi, A.; Nandhakumar, I.; Abutaha, A.; Hippalgaonkar, K. Toward Accelerated Thermoelectric Materials and Process Discovery. *ACS Appl. Energy Mater.* **2020**, *3*, 2240–2257. [CrossRef]
143. Wang, T.; Zhang, C.; Snoussi, H.; Zhang, G. Machine Learning Approaches for Thermoelectric Materials Research. *Adv. Funct. Mater.* **2019**, *30*, 1906041. [CrossRef]
144. Sparks, T.D.; Gaultois, M.W.; Oliynyk, A.; Brgoch, J.; Meredig, B. Data Mining Our Way to the Next Generation of Thermoelectrics. *Scr. Mater.* **2016**, *111*, 10–15. [CrossRef]
145. Deng, T.T.; Wei, T.R.; Huang, H.; Song, Q.F.; Zhao, K.P.; Qiu, P.F.; Yang, J.; Chen, L.D.; Shi, X. Number Mismatch between Cations and Anions as an Indicator for Low Lattice Thermal Conductivity in Chalcogenides. *npj Comput. Mater.* **2020**, *6*, 81. [CrossRef]
146. Cheng, N.S.; Liu, R.H.; Bai, S.; Shi, X.; Chen, L.D. Enhanced Thermoelectric Performance in Cd Doped $CuInTe_2$ Compounds. *J. Appl. Phys.* **2014**, *115*, 163705. [CrossRef]
147. Zhang, J.; Qin, X.Y.; Li, D.; Xin, H.X.; Song, C.J.; Li, L.L.; Wang, Z.M.; Guo, G.L.; Wang, L. Enhanced Thermoelectric Properties of Ag-Doped Compounds $CuAg_xGa_{1-x}Te_2$ ($0 \leq x \leq 0.05$). *J. Alloys Compd.* **2014**, *586*, 285–288. [CrossRef]

148. Ahmed, F.; Tsujii, N.; Mori, T. Thermoelectric Properties of CuGa$_{1-x}$Mn$_x$Te$_2$: Power Factor Enhancement by Incorporation of Magnetic Ions. *J. Mater. Chem. A* **2017**, *5*, 7545–7554. [CrossRef]
149. Kucek, V.; Drasar, C.; Kasparova, J.; Plechacek, T.; Navratil, J.; Vlcek, M.; Benes, L. High-Temperature Thermoelectric Properties of Hg-Doped CuInTe$_2$. *J. Appl. Phys.* **2015**, *118*, 125105. [CrossRef]
150. Shen, J.W.; Chen, Z.W.; Lin, S.Q.; Zheng, L.L.; Li, W.; Pei, Y.Z. Single Parabolic Band Behavior of Thermoelectric P-Type CuGaTe$_2$. *J. Mater. Chem. C* **2016**, *4*, 209–214. [CrossRef]
151. Xie, H.Y.; Li, Z.; Liu, Y.; Zhang, Y.K.; Uher, C.; Dravid, V.P.; Wolverton, C.; Kanatzidis, M.G. Silver Atom Off-Centering in Diamondoid Solid Solutions Causes Crystallographic Distortion and Suppresses Lattice Thermal Conductivity. *J. Am. Chem. Soc.* **2023**, *145*, 3211–3220. [CrossRef]
152. Li, Y.L.; Zhang, T.S.; Qin, Y.T.; Day, T.; Jeffrey Snyder, G.; Shi, X.; Chen, L.D. Thermoelectric Transport Properties of Diamond-Like Cu$_{1-x}$Fe$_{1+x}$S$_2$ Tetrahedral Compounds. *J. Appl. Phys.* **2014**, *116*, 203705. [CrossRef]
153. Liu, Y.; García, G.; Ortega, S.; Cadavid, D.; Palacios, P.; Lu, J.Y.; Ibáñez, M.; Xi, L.L.; De Roo, J.; López, A.M.; et al. Solution-Based Synthesis and Processing of Sn- and Bi-Doped Cu$_3$SbSe$_4$ nanocrystals, Nanomaterials and Ring-Shaped Thermoelectric Generators. *J. Mater. Chem. A* **2017**, *5*, 2592–2602. [CrossRef]
154. Chang, C.H.; Chen, C.L.; Chiu, W.T.; Chen, Y.Y. Enhanced Thermoelectric Properties of Cu$_3$SbSe$_4$ by Germanium Doping. *Mater. Lett.* **2017**, *186*, 227–230. [CrossRef]
155. Chetty, R.; Bali, A.; Mallik, R.C. Thermoelectric Properties of Indium Doped Cu$_2$CdSnSe$_4$. *Intermetallics* **2016**, *72*, 17–24. [CrossRef]
156. Ohta, M.; Jood, P.; Murata, M.; Lee, C.H.; Yamamoto, A.; Obara, H. An Integrated Approach to Thermoelectrics: Combining Phonon Dynamics, Nanoengineering, Novel Materials Development, Module Fabrication, and Metrology. *Adv. Energy Mater.* **2018**, *9*, 1801304. [CrossRef]
157. Kucek, V.; Drasar, C.; Navratil, J.; Plechacek, T.; Benes, L. Thermoelectric Properties of Ni-Doped CuInTe$_2$. *J. Phys. Chem. Solids* **2015**, *83*, 18–23. [CrossRef]
158. Zhong, Y.H.; Wang, P.D.; Mei, H.Y.; Jia, Z.Y.; Cheng, N.P. Elastic, Vibration and Thermodynamic Properties of Cu$_{1-x}$Ag$_x$InTe$_2$ (x = 0, 0.25, 0.5, 0.75 and 1) Chalcopyrite Compounds via First Principles. *Semicond. Sci. Technol.* **2018**, *33*, 065014. [CrossRef]
159. Wei, T.R.; Li, F.; Li, J.F. Enhanced Thermoelectric Performance of Nonstoichiometric Compounds Cu$_{3-x}$SbSe$_4$ by Cu Deficiencies. *J. Electron. Mater.* **2014**, *43*, 2229–2238. [CrossRef]
160. Heinrich, C.P.; Day, T.W.; Zeier, W.G.; Snyder, G.J.; Tremel, W. Effect of Isovalent Substitution on the Thermoelectric Properties of the Cu$_2$ZnGeSe$_{4-x}$S$_x$ Series of Solid Solutions. *J. Am. Chem. Soc.* **2014**, *136*, 442–448. [CrossRef]
161. Li, J.H.; Tan, Q.; Li, J.F. Synthesis and Property Evaluation of CuFeS$_{2-x}$ as Earth-Abundant and Environmentally-Friendly Thermoelectric Materials. *J. Alloys Compd.* **2013**, *551*, 143–149. [CrossRef]
162. Goto, Y.; Naito, F.; Sato, R.; Yoshiyasu, K.; Itoh, T.; Kamihara, Y.; Matoba, M. Enhanced Thermoelectric Figure of Merit in Stannite-Kuramite Solid Solutions Cu$_{2+x}$Fe$_{1-x}$SnS$_{4-y}$ (x = 0–1) with Anisotropy Lowering. *Inorg. Chem.* **2013**, *52*, 9861–9866. [CrossRef]
163. Huang, Y.L.; Shen, X.C.; Wang, G.W.; Zhang, B.; Zheng, S.K.; Yang, C.C.; Hu, X.; Gong, S.K.; Han, G.; Wang, G.Y.; et al. High Thermoelectric Performance and Compatibility in Cu$_3$SbSe$_4$–CuAlS$_2$ Composites. *Energy Environ. Sci.* **2023**, *16*, 1763–1772. [CrossRef]
164. Zhang, Y.B.; Xi, L.L.; Wang, Y.W.; Zhang, J.W.; Zhang, P.H.; Zhang, W.Q. Electronic Properties of Energy Harvesting Cu-Chalcogenides: P-D Hybridization and D-Electron Localization. *Comp. Mater. Sci.* **2015**, *108*, 239–249. [CrossRef]
165. Do, D.; Ozolins, V.; Mahanti, S.D.; Lee, M.S.; Zhang, Y.S.; Wolverton, C. Physics of Bandgap Formation in Cu-Sb-Se Based Novel Thermoelectrics: The Role of Sb Valency and Cu D Levels. *J. Phys. Condens. Matter* **2012**, *24*, 415502. [CrossRef]
166. Ai, L.; Ming, H.W.; Chen, T.; Chen, K.; Zhang, J.H.; Zhang, J.; Qin, X.Y.; Li, D. High Thermoelectric Performance of Cu$_3$SbSe$_4$ Obtained by Synergistic Modulation of Power Factor and Thermal Conductivity. *ACS Appl. Energy Mater.* **2022**, *5*, 13070–13078. [CrossRef]
167. Wang, B.Y.; Zheng, S.Q.; Chen, Y.; Wang, Q.; Li, Z.; Wu, Y.; Li, J.; Mu, Y.; Xu, S.; Liang, J. Realizing Ultralow Thermal Conductivity in Cu$_3$SbSe$_4$ via All-Scale Phonon Scattering by Co-Constructing Multiscale Heterostructure and Iiib Element Doping. *Mater. Today Energy* **2021**, *19*, 100620. [CrossRef]
168. Li, C.; Song, H.L.; Cheng, Y.; Qi, R.J.; Huang, R.; Cui, C.Q.; Wang, Y.F.; Zhang, Y.; Miao, L. Highly Suppressed Thermal Conductivity in Diamond-Like Cu$_2$SnS$_3$ by Dense Dislocation. *ACS Appl. Energy Mater.* **2021**, *4*, 8728–8733. [CrossRef]
169. Deng, S.; Jiang, X.; Chen, L.; Zhang, Z.; Qi, N.; Wu, Y.; Chen, Z.; Tang, X. The Reduction of Thermal Conductivity in Cd and Sn Co-Doped Cu$_3$SbSe$_4$-Based Composites with a Secondary-Phase Cdse. *J. Mater. Sci.* **2020**, *56*, 4727–4740. [CrossRef]
170. Sharma, S.D.; Bayikadi, K.; Raman, S.; Neeleshwar, S. Synergistic Optimization of Thermoelectric Performance in Earth-Abundant Cu$_2$ZnSnS$_4$ by Inclusion of Graphene Nanosheets. *Nanotechnology* **2020**, *31*, 365402. [CrossRef]
171. Zhao, L.J.; Yu, H.J.; Yang, J.; Wang, M.Y.; Shao, H.; Wang, J.L.; Shi, Z.Q.; Wan, N.; Hussain, S.; Qiao, G.J.; et al. Enhancing Thermoelectric and Mechanical Properties of P-Type Cu$_3$SbSe$_4$-Based Materials via Embedding Nanoscale Sb$_2$Se$_3$. *Mater. Chem. Phys.* **2022**, *292*, 126669. [CrossRef]
172. Hu, Z.Q.; Liang, X.L.; Dong, D.M.; Zhang, K.R.; Li, Z.; Song, J.M. To Improve the Thermoelectric Properties of Cu$_2$GeSe$_3$ via Gese Compensatory Compositing Strategy. *J. Alloys Compd.* **2022**, *921*, 166181. [CrossRef]
173. Sun, F.H.; Dong, J.F.; Tang, H.; Zhuang, H.L.; Li, J.F. ZnO-Nanoparticle-Dispersed Cu$_{11.5}$Ni$_{0.5}$Sb$_4$S$_{13-\delta}$ Tetrahedrite Composites with Enhanced Thermoelectric Performance. *J. Electron. Mater.* **2018**, *48*, 1840–1845. [CrossRef]

174. Sun, F.H.; Dong, J.F.; Tang, H.; Shang, P.P.; Zhuang, H.L.; Hu, H.; Wu, C.F.; Pan, Y.; Li, J.F. Enhanced Performance of Thermoelectric Nanocomposites Based on $Cu_{12}Sb_4S_{13}$ Tetrahedrite. *Nano Energy* **2019**, *57*, 835–841. [CrossRef]
175. Hu, H.H.; Sun, F.H.; Dong, J.F.; Zhuang, H.L.; Cai, B.; Pei, J.; Li, J.F. Nanostructure Engineering and Performance Enhancement in Fe_2O_3-Dispersed $Cu_{12}Sb_4S_{13}$ Thermoelectric Composites with Earth-Abundant Elements. *ACS Appl. Mater. Interfaces* **2020**, *12*, 17852–17860. [CrossRef] [PubMed]
176. Muchtar, A.R.; Srinivasan, B.; Tonquesse, S.L.; Singh, S.; Soelami, N.; Yuliarto, B.; Berthebaud, D.; Mori, T. Physical Insights on the Lattice Softening Driven Mid-Temperature Range Thermoelectrics of Ti/Zr-Inserted SnTe-an Outlook Beyond the Horizons of Conventional Phonon Scattering and Excavation of Heikes' Equation for Estimating Carrier Properties. *Adv. Energy Mater.* **2021**, *11*, 2101122. [CrossRef]
177. Bo, L.; Zhang, R.P.; Zhao, H.Y.; Hou, Y.B.; Wang, X.L.; Zhu, J.L.; Zhao, L.H.; Zuo, M.; Zhao, D.G. Achieving High Thermoelectric Properties of Cu_2Se via Lattice Softening and Phonon Scattering Mechanism. *ACS Appl. Energy Mater.* **2022**, *5*, 6453–6461. [CrossRef]
178. Tan, G.J.; Hao, S.Q.; Hanus, R.C.; Zhang, X.M.; Anand, S.; Bailey, T.P.; Rettie, A.J.E.; Su, X.; Uher, C.; Dravid, V.P.; et al. High Thermoelectric Performance in $SnTe-AgSbTe_2$ Alloys from Lattice Softening, Giant Phonon-Vacancy Scattering, and Valence Band Convergence. *ACS Energy Lett.* **2018**, *3*, 705–712. [CrossRef]
179. Hanus, R.; Agne, M.T.; Rettie, A.J.E.; Chen, Z.W.; Tan, G.J.; Chung, D.Y.; Kanatzidis, M.G.; Pei, Y.Z.; Voorhees, P.W.; Snyder, G.J. Lattice Softening Significantly Reduces Thermal Conductivity and Leads to High Thermoelectric Efficiency. *Adv. Mater.* **2019**, *31*, e1900108. [CrossRef]
180. Slade, T.J.; Anand, S.; Wood, M.; Male, J.P.; Imasato, K.; Cheikh, D.; Al Malki, M.M.; Agne, M.T.; Griffith, K.J.; Bux, S.K.; et al. Charge-Carrier-Mediated Lattice Softening Contributes to High zT in Thermoelectric Semiconductors. *Joule* **2021**, *5*, 1168–1182. [CrossRef]
181. Pöhls, J.H.; MacIver, M.; Chanakian, S.; Zevalkink, A.; Tseng, Y.C.; Mozharivskyj, Y. Enhanced Thermoelectric Efficiency through Li-Induced Phonon Softening in $CuGaTe_2$. *Chem. Mater.* **2022**, *34*, 8719–8728. [CrossRef]
182. Zhang, Z.X.; Zhao, K.P.; Chen, H.Y.; Ren, Q.Y.; Yue, Z.M.; Wei, T.R.; Qiu, P.F.; Chen, L.D.; Shi, X. Entropy Engineering Induced Exceptional Thermoelectric and Mechanical Performances in $Cu_{2-y}Ag_yTe_{1-2x}S_xSe$. *Acta Mater.* **2022**, *224*, 117512. [CrossRef]
183. Cai, J.F.; Yang, J.X.; Liu, G.Q.; Wang, H.X.; Shi, F.F.; Tan, X.J.; Ge, Z.H.; Jiang, J. Ultralow Thermal Conductivity and Improved Zt of $CuInTe_2$ by High-Entropy Structure Design. *Mater. Today Phys.* **2021**, *18*, 100394. [CrossRef]
184. Fan, Y.J.; Wang, G.Y.; Zhang, B.; Li, Z.; Wang, G.W.; Zhang, X.; Huang, Y.L.; Chen, K.S.; Gu, H.S.; Lu, X.; et al. Synergistic Effect of $CuInSe_2$ Alloying on Enhancing the Thermoelectric Performance of $CuSnSe_3$ Compounds. *J. Mater. Chem. A* **2020**, *8*, 21181–21188. [CrossRef]
185. Qiu, P.F.; Qin, Y.T.; Zhang, Q.H.; Li, R.X.; Yang, J.; Song, Q.F.; Tang, Y.S.; Bai, S.Q.; Shi, X.; Chen, L.D. Intrinsically High Thermoelectric Performance in $AgInSe_2$ N-Type Diamond-Like Compounds. *Adv. Sci.* **2018**, *5*, 1700727. [CrossRef]

Disclaimer/Publisher's Note: The statements, opinions and data contained in all publications are solely those of the individual author(s) and contributor(s) and not of MDPI and/or the editor(s). MDPI and/or the editor(s) disclaim responsibility for any injury to people or property resulting from any ideas, methods, instructions or products referred to in the content.

Article

Band Structure, Phonon Spectrum and Thermoelectric Properties of Ag$_3$CuS$_2$

Dmitry Pshenay-Severin [1,*], Satya Narayan Guin [2], Petr Konstantinov [1], Sergey Novikov [1], Ekashmi Rathore [2], Kanishka Biswas [2] and Alexander Burkov [1]

[1] Ioffe Institute, St Petersburg 194021, Russia
[2] Jawaharlal Nehru Centre for Advanced Scientific Research, Bengaluru 560064, India
* Correspondence: d.pshenay@mail.ru

Abstract: Sulfides and selenides of copper and silver have been intensively studied, particularly as potentially efficient thermoelectrics. Ag$_3$CuS$_2$ (jalpaite) is a related material. However very little is known about its physical properties. It has been found that the compound undergoes several structural phase transitions, having the tetrahedral structural modification I4$_1$/amd at room temperature. In this work, its band structure, phonon spectrum and thermoelectric properties were studied theoretically and experimentally. Seebeck coefficient, electrical conductivity and thermal conductivity were measured in a broad temperature range from room temperature to 600 K. These are the first experimental data on transport properties of jalpaite. Ab initio calculations of the band structure and Seebeck coefficient were carried out taking into account energy dependence of the relaxation time typical for the scattering of charge carriers by phonons. The results of the calculations qualitatively agree with the experiment and yield large values of the Seebeck coefficient characteristic for lightly doped semiconductor. The influence of intrinsic defects (vacancies) on the transport properties was studied. It was shown that the formation of silver vacancies is the most probable and leads to an increase of hole concentration. Using the temperature dependent effective potential method, the phonon spectrum and thermal conductivity at room temperature were calculated. The measurements yield low lattice thermal conductivity value of 0.5 W/(m K) at 300 K, which is associated with the complex crystal structure of the material. The calculated room temperature values of the lattice thermal conductivity were also small (0.14–0.2 W/(m K)).

Keywords: jalpaite; thermoelectric properties; Seebeck coefficient; electric conductivity; thermal conductivity

Citation: Pshenay-Severin, D.; Guin, S.N.; Konstantinov, P.; Novikov, S.; Rathore, E.; Biswas, K.; Burkov, A. Band Structure, Phonon Spectrum and Thermoelectric Properties of Ag$_3$CuS$_2$. *Materials* **2023**, *16*, 1130. https://doi.org/10.3390/ma16031130

Academic Editors: Bao-Tian Wang and Peng-Fei Liu

Received: 30 December 2022
Revised: 18 January 2023
Accepted: 22 January 2023
Published: 28 January 2023

Copyright: © 2023 by the authors. Licensee MDPI, Basel, Switzerland. This article is an open access article distributed under the terms and conditions of the Creative Commons Attribution (CC BY) license (https://creativecommons.org/licenses/by/4.0/).

1. Introduction

In the present work, the band structure, Seebeck coefficient and lattice properties of copper and silver sulfide—Ag$_3$CuS$_2$ (mineral jalpaite) were studied. Sulfides and selenides of copper and silver were intensively studied, particularly in thermoelectricity. Interest in the family of these materials arose in connection with the concept of "phonon glass, electron crystal" (PGEC), put forward by Slack [1]. In copper selenide Cu$_2$Se rather high values of thermoelectric figure of merit $ZT = S^2\sigma/\kappa$ of up to 0.4–1.6 in the temperature range 500–1000 K were obtained [2,3] (here, σ and κ are electrical and thermal conductivities, S is the Seebeck coefficient and T is the absolute temperature). In this temperature range, copper selenide is in a superionic cubic β phase with the antifluorite structure (transition temperature 413 K). The increased copper ion mobility leads to a strong disorder in the cationic sublattice and to a strong scattering of phonons, resulting in the deacrease of the thermal conductivity down to 0.5 W/(m K), which is one of the reasons for the high thermoelectric figure of merit of the material. This led to the extension of the PGEC concept and the emergence of a new paradigm—"phonon liquid, electron crystal" (PLEC) [2]. In addition to thermoelectric applications, copper selenide was studied as a material for solar cells [4,5]. The study of the electronic spectrum showed that it is a direct band gap semiconductor with a band gap of 1.1–1.73 eV [5–7].

High mobility of copper ions in the high-temperature phase was also observed in more complex silver and copper sulfides [8,9]. According to the literature data, Ag_3CuS_2 undergoes several structural phase transitions [10,11]. At room temperature, it has a tetragonal structure. With increasing temperature the material undergoes a transition to a body-centered (at 387 K) and then to a face-centered cubic phase (at 549 K). According to X-ray data, a strong disorder was observed in Cu-sublattice of both cubic phases, which led authors to the conclusion that the phases are superionic. At room temperature, Ag_3CuS_2 is a direct-gap semiconductor with band extrema at the Γ point of the Brillouin zone with a band gap of 1.05 eV [12], therefore, besides thermoelectric, it was also considered for photovoltaic applications [13,14].

At room temperature, Ag_3CuS_2 exists in a tetragonal modification (space group $I4_1/amd$, No.141). There are 4 formula units (24 atoms) in the unit cell, but often the structure is represented using a non-elementary body-centered tetragonal cell (8 formula units, 48 atoms; see the Figure 1, left panel). Ag atoms occupy two non-equivalent positions [10,11]: 8c positions (0, 0, 0) for Ag1 and 16g positions (−0.3127, −0.0627, 0.875) for Ag2. The environment of silver atoms can be represented as a distorted octahedra (2 + 4) for Ag1, in which 2 of the surrounding sulfur atoms are at a distance of 2.530 Å, and the remaining 4 are at a distance of 3.072 Å. For silver atoms in the Ag2 position, the environment is described as a distorted tetrahedral (2 + 2) with two closer sulfur atoms at a distance of 2.552 Å and two farther apart at a distance of 2.957 Å. The Shannon radius for sulfur is 1.7 Å; for silver in the 2-, 4-, and 6-coordination configurations, they are 0.81, 1.14, and 1.29 Å, respectively. The smallest of the distances to sulfur atoms in both silver configurations are in good agreement with the sum of the Shannon radii for sulfur and silver in the two-coordination configuration −2.51 Å. The sum of the Shannon radii for the 4- and 6-coordination environments of silver and sulfur are 2.84 and 2.99 Å, that are closer to the distances to farther S atoms in these configurations. The copper atoms are in positions 8e (0, 0.25, 0.5319). Their closest neighbors are 2 sulfur atoms located almost on the same straight line at a distance of 2.1822 Å [11]. For comparison, the sum of the Shannon radii of sulfur and copper in the 2-coordination configuration (0.6 Å) for this case is 2.3 Å. Below the room temperature, at 250 K, Ag_3CuS_2 transforms into another tetragonal phase $I4_1/a$ of similar atomic coordination but with reduced symmetry of interatomic distances.

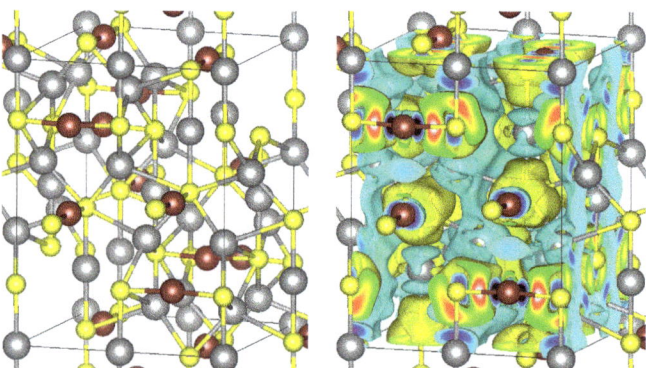

Figure 1. Body-centered tetragonal unit cell of Ag_3CuS_2 in jalpaite $I4_1/amd$ structure (**left** panel). Ag, Cu and S atoms are depicted using gray, brown and yellow spheres, respectively. Redistribution of electron density (**right** panel). The yellow contours correspond to an increase, and the cyan contours correspond to a decrease in the electron density compared to the electron density of atoms placed at the same positions.

To our knowledge, there is no information available in literature on electronic and thermal transport properties of Ag_3CuS_2.

2. Sample Preparation and Characterisation

Samples of Ag_3CuS_2 were prepared by melting of appropriate amounts of the constituting elements. The elemental silver (Ag, 99.999%, metal basis, Alfa Aesar), elemental copper (Cu, 99.9999%, metal basis, Alfa Aesar), and elemental sulfur (S, 99.999%, Alfa Aesar) were mixed in appropriate ratios in a quartz tube. The tubes were sealed under high vacuum ($\sim 10^{-5}$ Torr) and slowly heated to 773 K over 12 h, then heated to 1273 K in 5 h, soaked for 24 h, and subsequently slow cooled to room temperature. The structure of the synthesized compounds was characterized by X-ray powder diffraction.

Thermoelectric properties - Seebeck coefficient, S, electrical resistivity, ρ, and thermal conductivity, κ, - were measured using two home-made setups. The thermal conductivity was measured by steady-state classic procedure with setup, described in Ref. [15]. The setup is designed to make simultaneous measurement of thermal conductivity, electrical resistivity and Seebeck coefficient. However, the resistance of the Ag_3CuS_2 samples was too high for simultaneous resistivity and Seebeck coefficient measurements. Therefore, using this setup only the thermal conductivity was measured at temperatures from 100 K to 600 K. The resistivity and Seebeck coefficients were measured with another setup with higher input impedance and therefore capable to measure samples with higher resistance [16,17]. The standard uncertainty of the thermal conductivity measurements is below $\pm 10\%$. This uncertainty is mainly related to the uncertainty of determination of the heat energy density passing through sample crossection. Usually the samples have a regular, high quality shape, therefore the sample cross-section determination error is small. However in the present case the cylindrical samples of Ag_3CuS_2 have porous imperfections, that are difficult to take into account accurately when measuring the sample cross-section. Therefore we rise the error bar for the thermal conductivity in the present case to about 15%. The electrical resistivity 4-point DC measurement uncertainty is mainly determined by the shape factor, i.e., uncertainty of determination of the sample cross-section, and in the present case we estimate it as $\pm 5\%$. The Seebeck coefficient measurement uncertainty was estimated as $\pm(5\% + 0.5\ \mu V/K)$ [16]. However, to comply with this uncertainty, the sample resistance should not exceed the value of order 100 kOhm.

3. Experimental Results

The powder XRD results, shown in Figure 2, confirm the tetragonal crystal structure without significant amount of impurity phases.

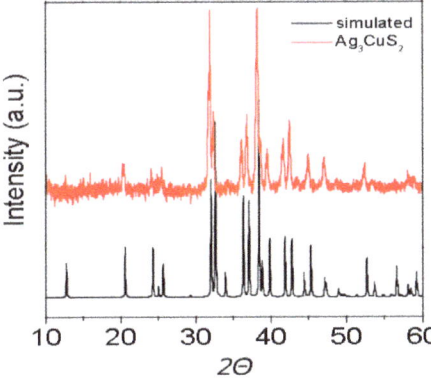

Figure 2. X-ray experimental diffraction pattern of Ag_3CuS_2 and simulated X-ray diffraction of $I4_1/amd$ structure.

The thermal conductivity was measured from 100 K to 600 K. To cover this temperature range we use two setups. In both cases the same steady-state method of the measurements was used, however at lower temperatures, from 100 K to 350 K–380 K, the radiation heat losses from the sample are comparatively small and are taken into account by introducing calibrated corrections [17], whereas at higher temperatures special sample holder room with active thermal guard shielding is used to suppress the radiation heat losses [15]. The measurements results are presented in Figure 3.

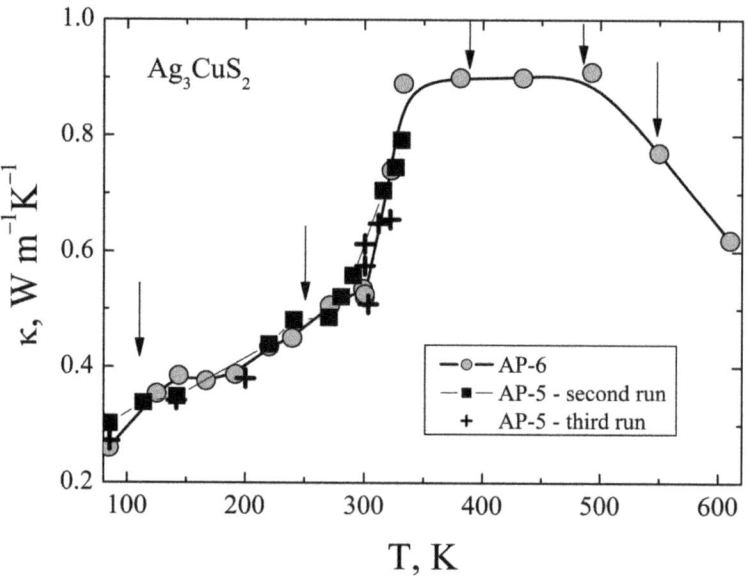

Figure 3. Thermal conductivity of Ag_3CuS_2. The arrows indicate temperatures of the structural phase transitions according to Ref. [11].

The thermal conductivity of the compound is very low and shows an unusual temperature dependence. It attains the maximum value of 0.7 $WK^{-1}m^{-1}$ above 300 K, is almost independent of temperature up to 500 K, and decreases linearly at 500 K to 600 K. Below 300 K the thermal conductivity is roughly a linear function of temperature. At 100 K it has value of only 0.3 $WK^{-1}m^{-1}$. The sample resistivity is typical for undoped semiconductors and is very high, therefore the electronic contribution to the thermal conductivity is negligible. As it was mentioned in Introduction, Ag_3CuS_2 undergoes a series of structural phase transitions: $P - based$ <–110 K–> $I4_1/a$ <–250 K–> $I4_1/amd$ <–387 K–> $Im3m$ <–483 K–> $Im3M + Fm3m$ <–549 K–>$Fm3m$ [11]. There is a clear correspondence between these phase transition temperatures and peculiarities on the temperature dependence of the thermal conductivity at 483 K, and possibly at 110 K and 250 K. However, the sharp increase of the thermal conductivity at 300 K apparently does not correspond to a phase transition. On the other hand, the phase transition $I4_1/amd$ <—> $Im3m$ at 387 K has no clear signature in the temperature dependence of the thermal conductivity. According to Ref. [11] this structural transition is accompanied by appearance of superionic phase with large structural disorder. Therefore one can expect a decrease of the lattice thermal conductivity at this temperature, which, however is not observed in our results. More detailed investigations of the thermal conductivity with thermal cycling across the transition at 387 K are necessary to resolve this apparent contradiction.

The electrical resistivity of Ag_3CuS_2 is shown in the Figure 4.

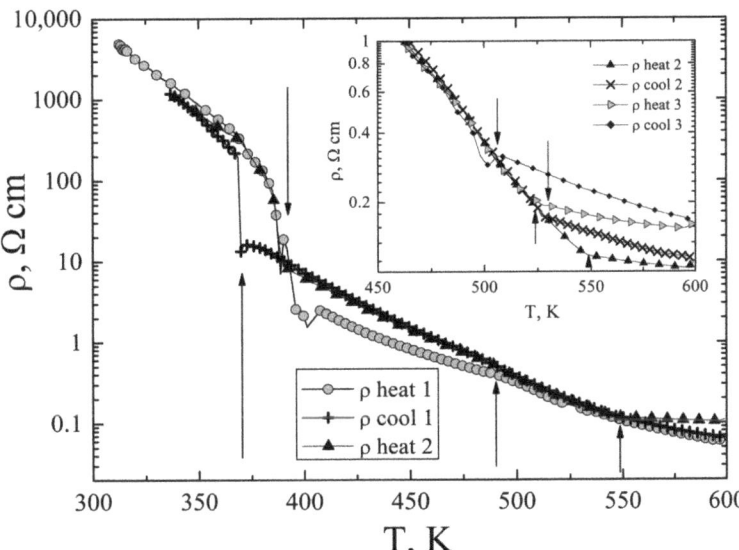

Figure 4. Electrical resistivity of Ag_3CuS_2. The arrows indicate temperatures at which the resistivity has peculiarities, these temperatures well correlate with the structural phase transitions according to Ref. [11]. The inset shows the evolution of the peculiarity near to 500 K with thermal cycling. The upward pointing arrows indicate the on-heating temperature of the transition, while the downward pointing arrows show the on-cooling transition temperature according to the resistivity data.

The resistivity was measured in dynamical temperature regime, i.e., with controlled continuous temperature variation during the measurements. The temperature variation rate in these measurements was about 5 K/min. The resistivity is large and has typical for undoped semiconductors temperature variation. The Figure 4 presents the resistivity temperature dependence from 300 K to 600 K, measured in three heating-cooling cycles of one Ag_3CuS_2 sample. On heating above 370 K a sharp drop of the resistivity is observed with pronounced temperature hysteresis on cooling. This behaviour of the resistivity is typical for a first-order phase transition, the temperature coincides with the temperature of the $I4_1/amd$ <—> $Im3m$ transformation [11]. The resistivity temperature dependence around this temperature is well reproducible on thermal cycling. The transition temperature, according to the resistivity data is 390 K ± 2 K on heating, and 375 K ± 10 K on cooling. There are another, less pronounced, peculiarities on the resistivity temperature dependence: change of the slope of $\ln \rho$ vs T dependence near to 480 K and 550 K. These peculiarities are more clearly visible on the temperature dependence of the resistivity temperature derivative $d\rho(T)/dT$, Figure 5. The temperatures of these peculiarities roughly correspond to the temperatures of the phase transitions $Im3m$ <–483 K–> $Im3M + Fm3m$ and $Im3M + Fm3m$ <–549 K–> $Fm3m$ [11]. However, as it is demonstrated in the inset to the Figure 4, the temperatures and even the magnitude of the peculiarities are dependent on the thermal history of the samples.

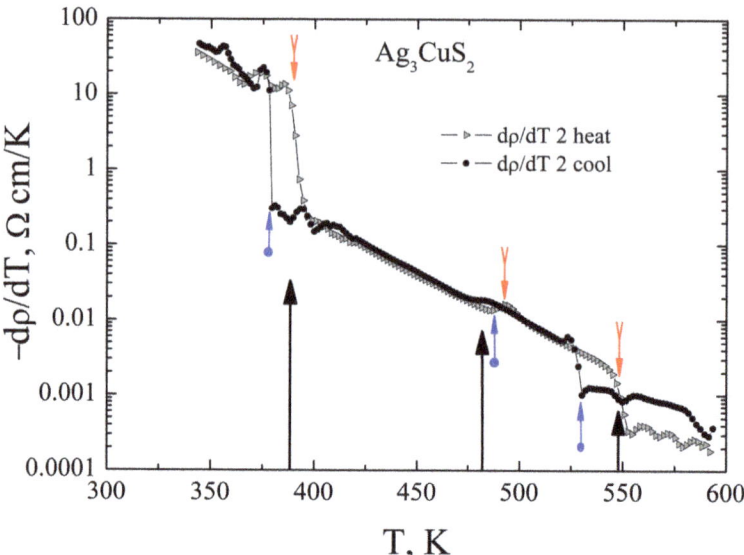

Figure 5. Temperature derivative of the resistivity of Ag_3CuS_2 on the second heating-cooling cycle. The large arrows indicate temperatures of the structural phase transitions according to Ref. [11]. The small downward pointing arrows indicate the on-heating temperature of the transition, while the upward pointing arrows show the on-cooling transition temperature according to the resistivity data.

The Seebeck coefficient was the most difficult to measure because of the high resistivity of the samples. Therefore sufficiently reliable data on the Seebeck coefficient were obtained only at temperatures above $I4_1/amd \longleftrightarrow Im3m$ transition. The results, together with simultaneously measured resistivity, are presented in Figure 6.

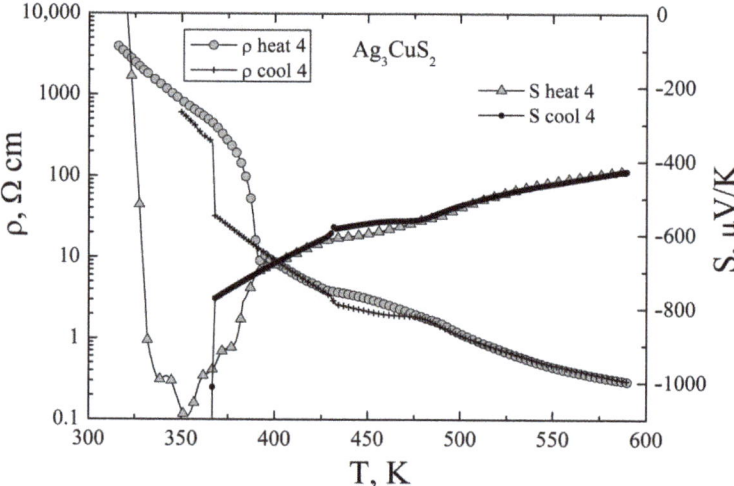

Figure 6. Temperature dependencies of the Seebeck coefficient and of the resistivity of Ag_3CuS_2 sample on the fourth heating-cooling cycle. The inset shows the thermoelectric power factor S^2/ρ.

The Seebeck coefficient at high temperatures is large in magnitude and has negative sign, the magnitude and the temperature variation are consistent with n-type semiconductor in intrinsic conductivity regime. Immediately below $I4_1/amd \longleftrightarrow Im3m$ transition the

Seebeck coefficient is negative. However we can not reliably measure its temperature variation and even the sign with further decrease of temperature.

In spite of the large Seebeck coefficient, the thermoelectric power factor S^2/ρ of the compound, shown in the inset of the Figure 6, is very small with maximum value of about 0.6 µW/cm K^2 at 600 K. This is due to the large electrical resistivity of the undoped compound. Further studies of the compouns are neccesary to find suitable ways for charge carrier optimisation.

Using the resistivity temperature dependence we estimate the band gap of the compound, Figure 7.

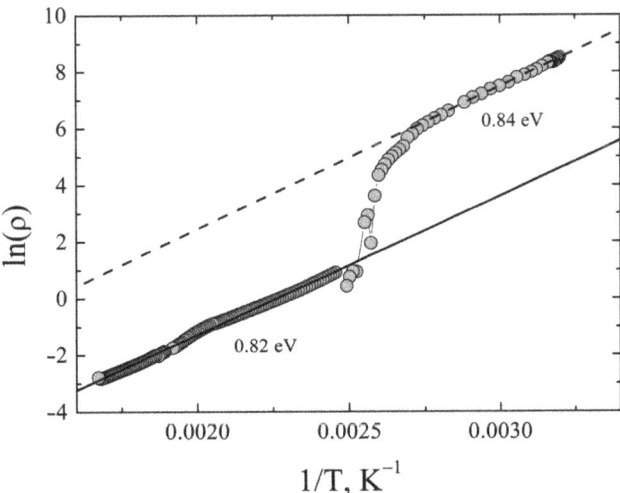

Figure 7. The dependence of the $\ln \rho(T)$ vs $1/T$ for Ag$_3$CuS$_2$ sample.

The estimate gives roughly the same band gap of $\approx$0.8 eV for the room-temperature phase $I4_1/amd$ and for all high-temperature modifications. This value of the band gap is considerably higher than theoretical one obtained using gradient-corrected local density functional approximation, discussed below. However, it is in quite good agreement with the value of 1.05 eV, obtained from optical reflectance data and hybrid functional calculations [12].

4. Ab Initio Calculation of Band Structure and Transport Properties

Ab initio calculations of lattice parameters and band structure were performed using density functional theory as implemented in VASP program [18,19]. We used generalized gradient PBE approximation for the density functional, plane wave energy cutoff of 400 eV and Monkhorst-Pack grid of $4 \times 4 \times 4$. The equilibrium lattice parameters were obtained through the calculation of equation of states with the unit cell shape and atomic force relaxation better than 1 meV/Å. The obtained parameters $a = 8.8974$ Å, $c = 11.6566$ Å can be compared with the experimental values $a = 8.6476$ Å, $c = 11.7883$ Å from Ref. [11] and $a = 8.6705$ Å, $c = 11.7573$ Å from Ref. [10]. The averaged deviation from experimental values for a and c are 2.8% and -1.0%, respectively. The calculations by Savory [12] gave larger deviations in the case of PBEsol density functional (-4.9% and 3.9%), but the use of hybrid functional allowed them to improve the agreement with experiment, e.g., in HSE06 approximation the deviation was 2.2% and 0.1% [12]. Here, we used the values obtained in PBE approximation that provided a balance of accuracy and computational efficiency. The atomic positions, obtained in this approximation, are given in the Table 1 and demonstrate good agreement with the experiment [11].

To investigate bonding character of Ag$_3$CuS$_2$, we calculated electron density. The redistribution of the electron density in the crystal compared to the density in atoms at the same positions is shown in the Figure 1 (right panel). One can see an increase in the electron density around the sulfur atoms and a decrease around silver and copper. Numerically, this redistribution can be characterized by the Bader ionic charges, which amounted to 0.38 (Ag1), 0.32 (Ag2), 0.3 (Cu), and −0.67 (S). For comparison, in PbTe the effective ion charges are ±0.64, and in NaCl they are ±0.86. Considering that there are 4 metal cations per 2 sulfur ions, the degree of ionicity is apparently close to that of PbTe. Another way to estimate the degree of ionicity f of a compound is using the electronegativity difference ΔX according to the formula $f = 1 - \exp(-0.67 \Delta X^2)$ [20]. Using the electronegativity values from [20] X_{Pb} = 2.62 and X_{Te} = 3.14, one can get $f = 17\%$ for PbTe. Estimations using Pauling's electronegativities gave ionicity for PbTe about 20% [21]. For silver and copper sulfide, taking into account the close electronegativity values of the metals X_{Ag} = 2.88 and X_{Cu} = 2.86, and also X_S = 3.44, a similar value $f = 19\%$ was obtained. Thus, the bonding character in Ag$_3$CuS$_2$ is of covalent-ionic type with a degree of ionicity of about 19%.

Recently, another approach to the estimation of bonding character was proposed in [22]. It is based on a charge-transfer index c, calculated from Bader charges Q_i and nominal oxidation state $Q_{ox,i}$ of the i-th ion: $c = (1/N) \sum_{i=1}^{N} Q_i / Q_{ox,i}$, where N is a total number of atoms. Using obtained Bader charges and nominal oxidation states of Ag^{1+}, Cu^{1+} and S^{2-}, one can obtain charge-transfer index for jalpaite $c = 0.33$. For comparison, the calculations for PbTe gave $c = 0.32$. According to [22], small values of c correspond to covalent bonding, $c = 0.3$–0.6 is typical for polar III-V compounds and nitrides, and the values close to unity correspond to ionic bonding. That is the use of charge-transfer index leads to qualitatively similar conclusion of mainly covalent type of bonding in jalpaite with a degree of ionicity close to that in PbTe.

Table 1. Atomic positions in Ag$_3$CuS$_2$ (I4$_1$/amd).

Trots et al. [11] (Expt.)				
Ag1	8c	0	0	0
Ag2	16g	−0.3127	−0.0627	0.875
Cu	8e	0	0.25	0.5319
S	16h	0	−0.0023	0.2146
PBE (This Work)				
Ag1	8c	0	0	0
Ag2	16g	−0.3134	−0.0634	0.875
Cu	8e	0	0.25	0.5281
S	16h	0	0.0052	0.2137

Figure 8 shows the results of band structure calculations for Ag$_3$CuS$_2$ (I4$_1$/amd) in the PBE approximation. On the whole, they agree with those given in [12] except much smaller band gap $\epsilon_g = 0.24$ eV. This is a well known deficiency of the PBE approximation which usually underestimates the band gaps. The larger value of $\epsilon_g = 1.05$ eV was obtained using HSE06 hybrid functional by Savory [12] and was confirmed in our calculations of the band energies at the Γ point in the same approximation. The value of 1.05 eV was also obtained by Savory [12] experimentally in optical reflectance measurements.

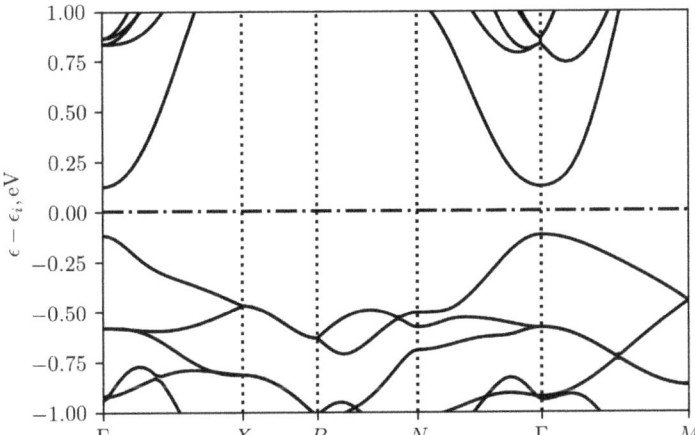

Figure 8. Band structure of Ag$_3$CuS$_2$ (I4$_1$/amd), calculated using generalized gradient PBE approximation. Energy is counted from the middle of the band gap ϵ_i. Special points in the Brillouin zone are denoted according to [23].

Ag$_3$CuS$_2$ turns out to be a direct band gap semiconductor with band extrema at the Γ point. Figure 9 shows the energy dependencies of the atomic orbitals projected density of states. It can be seen that the main contribution near the top of the valence band comes from the p-S, d-Ag, and d-Cu states with a small addition of the s-Ag contribution. In the depth of the valence band, the contribution of the d states of cations and the p states of sulfur dominates. The s states of sulfur form mainly a narrow band approximately 13 eV below the considered valence band (not shown in the Figure 9). The states near the bottom of the conduction band are formed by s-Ag, d-Cu, p-S, s-S, d-Ag in the order of decreasing contribution.

The Seebeck coefficient and the Hall concentration were calculated using band structure interpolation on 32 × 32 × 32 k-point grid in BoltzTraP [24] program. For calculation of the transport coefficients, the energies of the conduction band states were shifted upward by 0.8 eV to correct ϵ_g using the so-called scissors operator. For the relaxation time, the energy dependence $\tau(\epsilon) \sim 1/\text{DOS}(\epsilon)$ was used, which is typical for the scattering of charge carriers on acoustical phonons. The results of calculations at room temperature are shown in the Figure 10. The position of the chemical potential at 300 K for intrinsic conduction, when the concentrations of electrons and holes are the same, is slightly shifted from the middle of the band gap ϵ_i to the bottom of the conduction band (by about 7 meV). This agrees with the band structure plot and the density of states: the latter is higher near the top of the valence band than near the bottom of the conduction band. Estimates of the effective masses of the density of states near the band extrema also gave larger values for holes $m_p = 0.64 m_0$ than for electrons $m_n = 0.49 m_0$. If the proportionality constant in $\tau(\epsilon)$ dependence is the same for electrons and holes, the mobility ratio u_n/u_p for non-degenerate statistics turns out to be 2.4. The Seebeck coefficient in the case of intrinsic conduction turns out to be negative and equal to about −800 μV/K at room temperature. The maximum values of Seebeck coefficient depending on the concentration are in the range of ±1700 μV/K. By the order of magnitude, it agrees with the measured values of Seebeck coefficient above room temperature, which exceeds −1000 μV/K. At the same time, the experimental resistivity is greater than 1 kOhm·cm, i.e., apparently, the carrier concentration in the sample corresponds to the intrinsic conduction, which at room temperature, taking into account the magnitude of the band gap, should be small. The experimental data indicate a possible change of the Seebeck coefficient sign from positive to negative near room temperature that should be connected with the phase transition. A similar behavior was observed in undoped AgCuS [25,26] samples, in which a double p-n-p conductivity sign reversal associated with two phase transitions at 361 K and 439 K was found. In that case, the range of Seebeck coefficient variation exceeded 1700 µV/K. Calculations showed [25] that the change in sign of the Seebeck coefficient in AgCuS was due to passing of the system through an intermediate semimetallic state in the course of the phase transition at 361 K, when the chemical potential was shifted from the top of the valence band to the bottom of the conduction band and back.

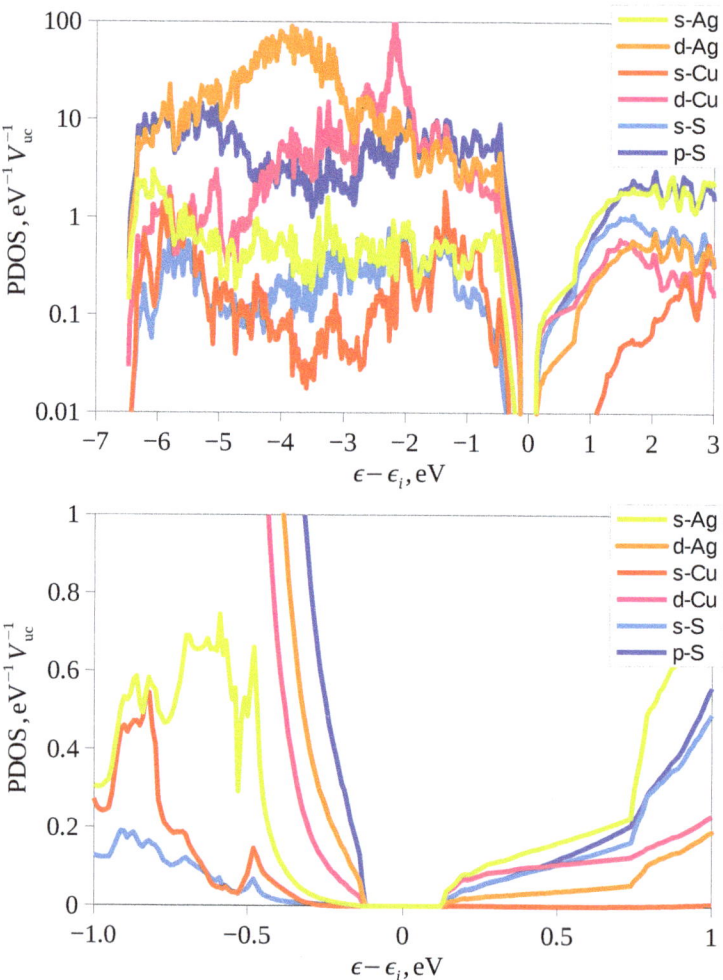

Figure 9. Atomic state projected density of electron states (PDOS) in Ag_3CuS_2 ($I4_1/amd$). Upper panel shows PDOS in logarithmic scale in the wide energy range, while lower panel shows PDOS around the band gap. Energy is counted from the middle of the band gap ϵ_i.

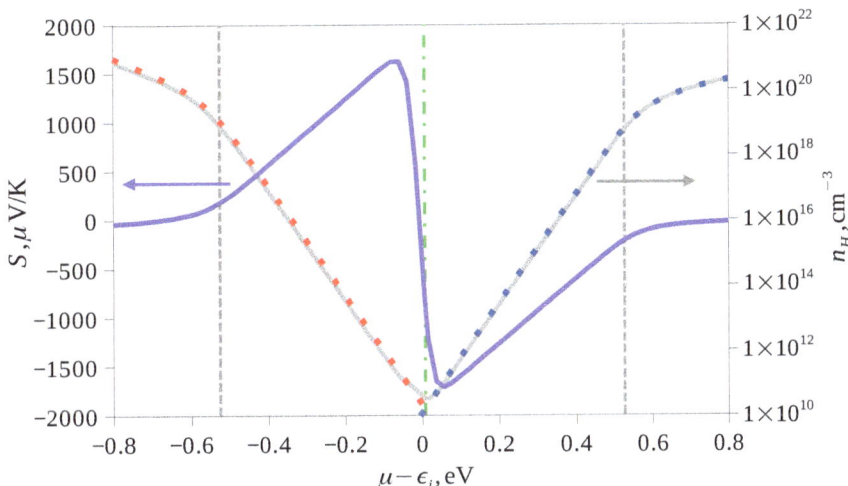

Figure 10. The dependence of the Seebeck coefficient (blue curve, left axis) and the Hall concentration (gray curve, right axis) on the chemical potential at 300 K. The vertical gray dashed lines show the boundaries of the band gap, and the green dash-dotted line shows the position of the chemical potential at equal concentrations of electrons and holes. Red (blue) dots on the Hall concentration plot show the concentrations of holes (electrons) separately. The chemical potential is counted from the middle of the band gap ϵ_i.

To optimize Ag_3CuS_2 for thermoelectric applications, it is necessary to find methods of its alloying. As a first step, we performed supercell calculations for a crystal in the presence of native point defects—copper, silver, and sulfur vacancies. We used a supercell with 96 atoms containing 16 formula units. In the supercell, one of the selected atoms was removed, and the positions of atoms in the presence of a defect were optimized up to residual interatomic forces less than 0.01 eV/Å while the cell shape and the volume were kept unchanged to better model the limit of low impurity concentration [27]. For each of the considered defects, the density of electronic states was calculated on a $9 \times 9 \times 9$ k-point grid and the anticipated type of doping was determined from the shift of electronic chemical potential. The defect formation energies were calculated following the method described in Ref. [27]. The obtained formation energies appeared to be negative: $\epsilon_f(V_{Ag}) = -0.58$ eV, $\epsilon_f(V_{Cu}) = -0.43$ eV and $\epsilon_f(V_S) = -0.53$ eV. This is connected to the fact that considered structural modification of jalpaite occurs at room temperature, and the total energy calculations were carried out for zero temperature. Thus, obtained results requires further corrections due to lattice contribution and were used here only for qualitative conclusions on the relative probability of defects formation.

The calculations showed that the formation of a silver vacancy is most probable, followed by that of sulfur and copper. Figure 11 shows the energy dependences of the density of states near the band gap in comparison with pure Ag_3CuS_2. Calculations showed that in the presence of copper or silver vacancies the chemical potential is shifted down into the valence band, which suggests that these defects should lead to the p-type doping. In the case of sulfur vacancies the valence band remains completely filled and doping does not occur. In all considered cases, simultaneously with doping, the band gap increases, which is especially strong in the presence of sulfur vacancies, which is obviously related to the large contribution of the p states of sulfur to the density of states near the band gap boundaries. The comparison of DOS before and after ionic relaxation for the case of Ag vacancies (see gray dotted and solid curves in the Figure 11) showed that an increase in the band gap occurs mainly during the process of atomic relaxation. In all considered cases, the density of states in the valence band in the presence of vacancies increases with the hole energy steeper than in pure Ag_3CuS_2. Thus, the formation of vacancies should be accompanied by an increase of hole effective mass. The presence of cation vacancies and the effect p-type doping was also experimentally observed in the related compound AgCuS [26].

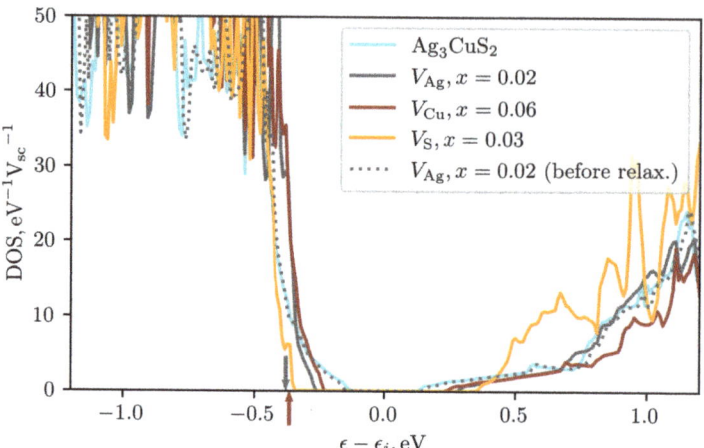

Figure 11. The density of states around the band gap of pure Ag_3CuS_2 and in the presence of Ag, Cu, and S vacancies with the atomic fraction x. The arrows show the positions of Fermi energy for Ag and Cu vacancies. Gray dotted curve shows DOS in the presence of Ag vacancy before atomic relaxation. Energy is counted from the middle of the band gap ϵ_i.

5. Ab Initio Calculation of Phonon Spectrum and Lattice Thermal Conductivity

As was mentioned above, the crystal structure at zero temperature differs from the structure at room temperature. The available data on the phonon spectrum demonstrate imaginary phonon frequencies (see the Figure 12), indicating an instability of the tetragonal modification at zero temperature [28]. Therefore, in the present work, we took into account the stabilization of phonon modes due to anharmonicity. The calculation was carried out using the temperature dependent effective potential approach (TDEP) [29,30]. In this method, the potential energy of atoms in a crystal is expanded into a series in terms of atomic displacements with expansion coefficients that depend on temperature. To determine these coefficients, we performed molecular dynamics modeling of the crystal at room temperature. The effective force constants of the 2nd and 3rd orders were determined using the least squares method from the data on atomic coordinates and interatomic forces obtained from molecular dynamics calculations. Using these effective force constants, the phonon spectrum was calculated by the lattice dynamics method.

Molecular dynamics calculations were performed using the VASP program in the NVT ensemble. The calculation was carried out at the experimental values of the lattice parameters for 300 K from [11] for a supercell of 96 atoms. The cutoff energy was 400 eV, and the Monkhorst-Pack grid in the Brillouin zone of the supercell was $2 \times 2 \times 2$. The step of integrating the equations of motion was equal to 1fs. The system was equilibrated for 1000 steps, and then data were collected from 2000 simulation steps. With the obtained data, the force constants of the 2nd and 3rd orders, which best describe the surface of constant energy, were calculated in the TDEP [29,30] program. The resulting phonon spectrum in Ag_3CuS_2 ($I4_1/amd$) is depicted in the Figure 13. It demonstrates the absence of imaginary modes, i.e., the stabilization of the structure. The Figure 14 shows the phonon density of states and the contribution to the latter from vibrations of atoms of each type. Acoustic modes occupy a small frequency range from 0 to 1 THz. From 0 to 2 THz, the main contribution to vibrations comes from silver atoms. In the range from 2 to 3 THz, there is a narrow peak associated with copper ions vibrations with an additional contribution from silver ions. Then, there are three bands of optical phonons. The frequency range from 3.5 to 6.5 THz is associated mainly with vibrations of sulfur atoms with a small addition from vibrations of silver atoms. Modes in the range 6.5–9 THz are only of sulfur atoms, and a narrow band near 10 THz contains approximately equal contributions from vibrations of sulfur and copper.

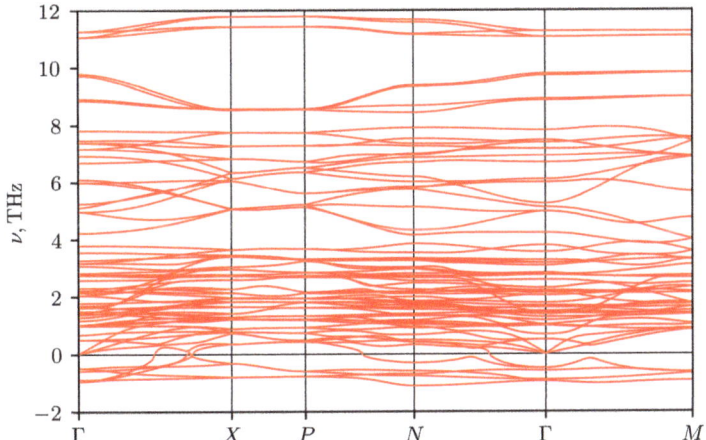

Figure 12. Phonon spectrum of Ag_3CuS_2 ($I4_1/amd$), calculated at zero temperature, using equilibrium lattice parameters and PBEsol density functional approximation [28].

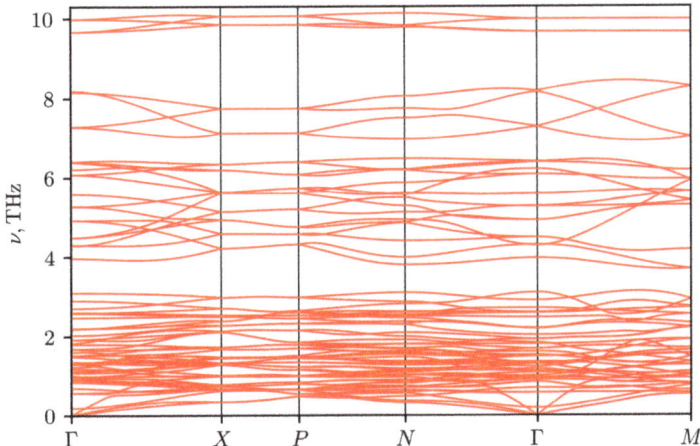

Figure 13. Phonon spectrum of Ag_3CuS_2 ($I4_1/amd$) calculated at room temperature.

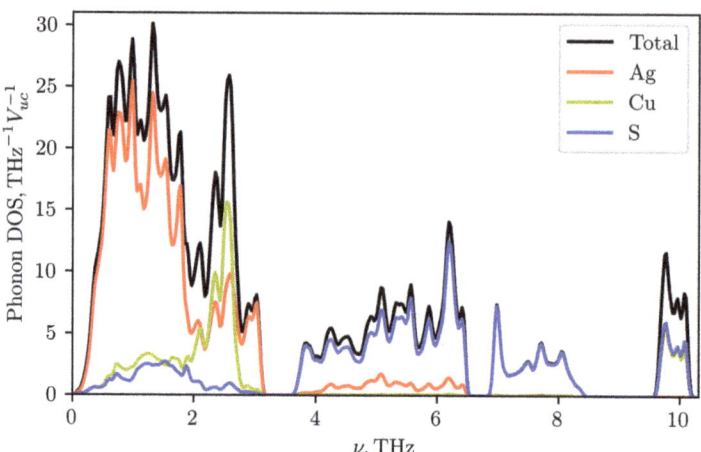

Figure 14. Total and atom-projected phonon density of states in Ag_3CuS_2 ($I4_1/amd$) calculated at room temperature.

The lattice thermal conductivity of Ag_3CuS_2 at room temperature was calculated using the TDEP program with the obtained force constants of the 2nd and 3rd order. Due to the large number of optical modes with low group velocity and the strong anharmonicity of vibrations, the calculation gave very low thermal conductivity values: 0.14 and 0.2 W/(m K) in the a-b plane and along the c axis, respectively. The Figure 15 shows the frequency-resolved phonon contributions to the thermal conductivity averaged over directions (red curve). In addition, the cumulative contribution to thermal conductivity from phonons with different frequencies are also shown (cyan curve). It can be seen that approximately half of the contribution comes from acoustic and low-frequency optical modes with frequencies up to 3 THz, where the contributions of silver and copper vibrations are predominant. Almost the entire remaining contribution comes from optical modes associated with vibrations of sulfur atoms in the range from 3.5 to 6.5 THz. Experimental data also give a low value for thermal conductivity—about 0.5 W/(m K), but it is noticeably higher than the calculated value. The reason for the discrepancy between the estimates and experiment requires further investigation. One possible reason for the discrepancy may lie in the description of the phonon spectrum. Experimental data on the phonon frequencies of jalpaite are not available in the literature, but from a comparison of the Figures 12 and 13 it can be seen that the spectrum obtained by the TDEP method is shifted to lower frequencies: the frequency range has decreased from 12 to 10 THz. The difference in the the range of phonon frequencies can be due to different approximations for density functional used in [28] (PBEsol) and in the present work (PBE). In addition, an account of anharmonicity by effective harmonic potential, obtained in TDEP approach, can result not only in the stabilization of phonon modes but also in a decrease of maximum frequencies in the spectrum depending on the type and sign of anharmonicity (see, e.g., [29]). In addition, the cumulative contribution to thermal conductivity from the phonons with different mean free paths l_p showed that up to 80% of thermal conductivity is due to phonons with $l_p < 1.5$ nm (see blue curve in the Figure 15). For such small mean free paths the usual approach based on the kinetic equation for the description of thermal conductivity may becomes questionable. It would be interesting in the future to compare the obtained spectrum with the results of a direct calculation of the quasiparticle spectrum of phonons using data on molecular dynamic trajectories [31]. The comparison of thermal conductivity estimates with direct calculations using the molecular dynamics would also be desirable.

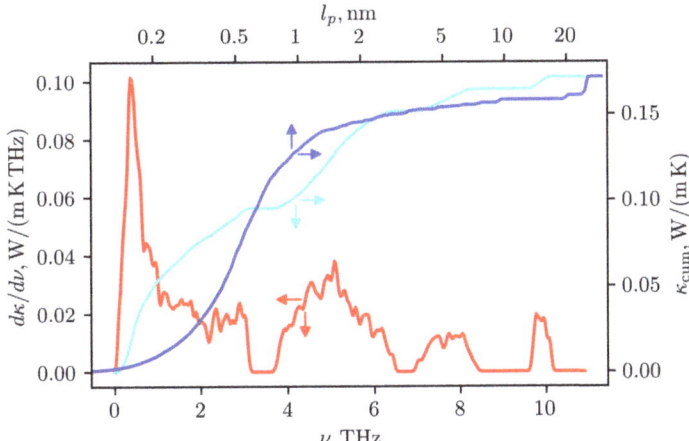

Figure 15. Spectral (red curve) and cumulative (cyan curve) contributions of phonons into lattice thermal conductivity at 300 K as a functions of phonon frequency. Blue curve shows cumulative thermal conductivity as a function of phonon mean free path l_p. Arrows are directed to the axes used for plotting corresponding curves.

6. Conclusions

In this work we presented experimental and theoretical study of band structure and thermelectric properties of jalpaite (Ag_3CuS_2). The Seebeck coefficient, electrical conductivity and thermal conductivity were measured in a broad temperature range. The samples have intrinsic conduction with high resistivity above 1 kΩcm at room temperature, that decreases down to 0.1 Ωcm at 600 K. The temperature dependence of resistivity demonstrate several peculiarities close to the phase transition temperatures given in [11]. The Seebeck coefficient is negative and has a value down to −1000 μV/K near room temperature and around −400 μV/K at 600 K. Close to room temperature and below the Seebeck coefficient has a tendency of a sign change that can be caused by approaching the low-temperature phase transition. The thermal conductivity demonstrates unusual temperature dependence and the value of 0.5 W/(m K) at the room temperature. Low values of thermal conductivity and high Seebeck coefficient can be preferential for thermoelectric applications under proper doping conditions.

Theoretically, electronic band structure, Seebeck and Hall coefficients were studied at the room temperature. The calculations showed that the material tends to demonstrate negative Seebeck coefficient in intrinsic doping region with the magnitude close to experimental values. The investigation of point defects using supercell approach showed that the copper and silver vacancies can lead to the p-type doping, while sulfur vacancies should lead only to decrease of hole mobility. The anaharmonisity stabilised phonon spectrum at room temperatures was obtained using temperature dependent effective potential approach. Both, the theoretical estimates and the experiment, yield a low lattice thermal conductivity due to complex crystal structure.

Author Contributions: Conceptualization, K.B. and A.B.; Investigation, D.P.-S., S.N.G., P.K., S.N., E.R., K.B. and A.B.; Methodology, K.B. and A.B.; Software, D.P.-S.; Writing—original draft, D.P.-S. and A.B.; Writing—review & editing, S.N.G., P.K., S.N., E.R. and K.B. All authors have read and agreed to the published version of the manuscript.

Funding: This research received no external funding.

Institutional Review Board Statement: Not applicable.

Informed Consent Statement: Not applicable.

Data Availability Statement: The data that support the findings of this study are available from the corresponding author upon reasonable request.

Conflicts of Interest: The authors declare no conflict of interest.

References

1. Slack, G.A. *New Materials and Performance Limits for Thermoelectric Cooling*; CRC Handbook of Thermoelectrics; CRC Press: Boca Raton, FL, USA, 1995; Chapter 34, pp. 407–440.
2. Liu, H.; Shi, X.; Xu, F.; Zhang, L.; Zhang, W.; Chen, L.; Li, Q.; Uher, C.; Day, T.; Snyder, G.J. Copper ion liquid-like thermoelectrics. *Nat. Mater.* **2012**, *11*, 422–425. [CrossRef] [PubMed]
3. Yu, B.; Liu, W.; Chen, S.; Wang, H.; Wang, H.; Chen, G.; Ren, Z. Thermoelectric properties of copper selenide with ordered selenium layer and disordered copper layer. *Nano Energy* **2012**, *1*, 472–478. [CrossRef]
4. Okimura, H.; Matsumae, T.; Makabe, R. Electrical properties of $Cu_{2-x}Se$ thin films and their application for solar cells. *Thin Solid Films* **1980**, *71*, 53–59. [CrossRef]
5. Mane, R.; Kajve, S.; Lokhande, C.; Han, S.H. Studies on p-type copper (I) selenide crystalline thin films for hetero-junction solar cells. *Vacuum* **2006**, *80*, 631–635. [CrossRef]
6. Sorokin, G.; Papshev, Y.; Oush, P. Photoconductivity of Cu_2S, Cu_2Se and Cu_2Te. *Sov. Phys. Solid State* **1965**, *7*, 2244.
7. Terekhov, V.; Gorbachev, E.; Domashevskaya, E.; Kashkarov, V.; Teterin, Y. Electronic structure of copper chalcogenides using X-ray spectral and X-ray photoelectron data. *Sov. Phys. Solid State* **1983**, *25*, 2482.
8. Kadrgulov, R.; Yakshibaev, R.; Khasanov, M. Phase relations, ionic transport and diffusion in the alloys of Cu_2S-Ag_2S mixed conductors. *Ionics* **2001**, *7*, 156–160. [CrossRef]
9. Akmanova, G.; Yakshibaev, R.; Davletshina, A.; Bikkulova, N. Changes in the nature of chemical bonding in solid solutions between copper and silver chalcogenides. *Inorg. Mater.* **2020**, *56*, 1–6. [CrossRef]
10. Baker, C.; Lincoln, F.; Johnson, A. Crystal structure determination of Ag_3CuS_2 from powder X-ray diffraction data. *Aust. J. Chem.* **1992**, *45*, 1441–1449. [CrossRef]
11. Trots, D.; Senyshyn, A.; Mikhailova, D.; Vad, T.; Fuess, H. Phase transitions in jalpaite, Ag_3CuS_2. *J. Phys. Condens. Matter* **2008**, *20*, 455204. [CrossRef]
12. Savory, C.N.; Ganose, A.M.; Travis, W.; Atri, R.S.; Palgrave, R.G.; Scanlon, D.O. An assessment of silver copper sulfides for photovoltaic applications: Theoretical and experimental insights. *J. Mater. Chem. A* **2016**, *4*, 12648–12657. [CrossRef] [PubMed]
13. Lei, Y.; Yang, X.; Gu, L.; Jia, H.; Ge, S.; Xiao, P.; Fan, X.; Zheng, Z. Room-temperature preparation of trisilver-copper-sulfide/polymer based heterojunction thin film for solar cell application. *J. Power Sources* **2015**, *280*, 313–319. [CrossRef]
14. Liu, Z.; Han, J.; Guo, K.; Zhang, X.; Hong, T. Jalpaite Ag_3CuS_2: A novel promising ternary sulfide absorber material for solar cells. *Chem. Commun.* **2015**, *51*, 2597–2600. [CrossRef] [PubMed]
15. Vedernikov, M.; Konstantinov, P.; Burkov, A. Development of automated techniques of measuring of temperature dependences of the transport properties of thermoelectric materials. In *Proceedings of the Eigth International Conference on Thermoelectric Energy Conversion*; Scherer, H., Ed.; INPL: Nancy, France, 1989; pp. 45–48.
16. Burkov, A.; Heinrich, A.; Konstantinov, P.; Nakama, T.; Yagasaki, K. Experimental set-up for thermopower and resistivity measurements at 100–1300 K. *Meas. Sci. Technol.* **2001**, *12*, 264–272. [CrossRef]
17. Burkov, A.T.; Fedotov, A.I.; Novikov, S.V. Chapter Methods and apparatus for measuring thermopower and electrical conductivity of thermoelectric materials at high temperatures. In *Thermoelectrics for Power Generation—A Look at Trends in the Technology*; INTECH: Vienna, Austria, 2016; pp. 353–389. [CrossRef]
18. Kresse, G.; Furthmüller, J. Efficient iterative schemes for ab initio total-energy calculations using a plane-wave basis set. *Phys. Rev. B* **1996**, *54*, 11169. [CrossRef]
19. Kresse, G.; Joubert, D. From ultrasoft pseudopotentials to the projector augmented-wave method. *Phys. Rev. B* **1999**, *59*, 1758. [CrossRef]
20. Tantardini, C.; Oganov, A.R. Thermochemical electronegativities of the elements. *Nat. Commun.* **2021**, *12*, 2087. [CrossRef]
21. Ravich, Y.I.; Efimova, B.A.; Smirnov, I.A. *Semiconducting Lead Chalcogenides*; Springer: Boston, MA, USA, 1970. 10.1007/978-1-4684-8607-0_2. [CrossRef]
22. Mori-Sánchez, P.; Pendás, A.M.; Luaña, V. A Classification of Covalent, Ionic, and Metallic Solids Based on the Electron Density. *J. Am. Chem. Soc.* **2002**, *124*, 14721–14723. [CrossRef]
23. Hinuma, Y.; Pizzi, G.; Kumagai, Y.; Oba, F.; Tanaka, I. Band structure diagram paths based on crystallography. *Comput. Mater. Sci.* **2017**, *128*, 140–184. [CrossRef]
24. Madsen, G.K.; Singh, D.J. BoltzTraP. A code for calculating band-structure dependent quantities. *Comput. Phys. Commun.* **2006**, *175*, 67–71. [CrossRef]
25. Guin, S.N.; Pan, J.; Bhowmik, A.; Sanyal, D.; Waghmare, U.V.; Biswas, K. Temperature Dependent Reversible p–n–p Type Conduction Switching with Colossal Change in Thermopower of Semiconducting AgCuS. *J. Am. Chem. Soc.* **2014**, *136*, 12712–12720. [CrossRef] [PubMed]

26. Dutta, M.; Sanyal, D.; Biswas, K. Tuning of p–n–p-Type Conduction in AgCuS through Cation Vacancy: Thermopower and Positron Annihilation Spectroscopy Investigations. *Inorg. Chem.* **2018**, *57*, 7481–7489. . [CrossRef]
27. Van de Walle, C.G.; Neugebauer, J. First-principles calculations for defects and impurities: Applications to III-nitrides. *J. Appl. Phys.* **2004**, *95*, 3851–3879. [CrossRef]
28. Togo, A. Phonon Database. 2018. Available online: http://phonondb.mtl.kyoto-u.ac.jp/ph20180417/d005/mp-5725.html (accessed on 17 December 2022).
29. Hellman, O.; Abrikosov, I.A.; Simak, S.I. Lattice dynamics of anharmonic solids from first principles. *Phys. Rev. B* **2011**, *84*, 180301. [CrossRef]
30. Hellman, O.; Abrikosov, I.A. Temperature-dependent effective third-order interatomic force constants from first principles. *Phys. Rev. B* **2013**, *88*, 144301. [CrossRef]
31. Carreras, A.; Togo, A.; Tanaka, I. DynaPhoPy: A code for extracting phonon quasiparticles from molecular dynamics simulations. *Comput. Phys. Commun.* **2017**, *221*, 221–234. [CrossRef]

Disclaimer/Publisher's Note: The statements, opinions and data contained in all publications are solely those of the individual author(s) and contributor(s) and not of MDPI and/or the editor(s). MDPI and/or the editor(s) disclaim responsibility for any injury to people or property resulting from any ideas, methods, instructions or products referred to in the content.

Article

Effect of Ni Substitution on Thermoelectric Properties of Bulk β-Fe$_{1-x}$Ni$_x$Si$_2$ ($0 \leq x \leq 0.03$)

Sopheap Sam [1], Soma Odagawa [1], Hiroshi Nakatsugawa [1,*] and Yoichi Okamoto [2]

[1] Yokohama National University, Yokohama 240-8501, Japan
[2] National Defense Academy, Yokosuka 239-8686, Japan
* Correspondence: nakatsugawa-hiroshi-dx@ynu.ac.jp

Abstract: A thermoelectric generator, as a solid-state device, is considered a potential candidate for recovering waste heat directly as electrical energy without any moving parts. However, thermoelectric materials limit the application of thermoelectric devices due to their high costs. Therefore, in this work, we attempt to improve the thermoelectric properties of a low-cost material, iron silicide, by optimizing the Ni doping level. The influence of Ni substitution on the structure and electrical and thermoelectric characteristics of bulk β-Fe$_x$Ni$_{1-x}$Si$_2$ ($0 \leq x \leq 0.03$) prepared by the conventional arc-melting method is investigated. The thermoelectric properties are reported over the temperature range of 80–800 K. At high temperatures, the Seebeck coefficients of Ni-substituted materials are higher and more uniform than that of the pristine material as a result of the reduced bipolar effect. The electrical resistivity decreases with increasing x owing to the increases in metallic ε-phase and carrier density. The ε-phase increases with Ni substitution, and solid solution limits of Ni in β-FeSi$_2$ can be lower than 1%. The highest power factor of 200 μWm^{-1}K^{-2} at 600 K is obtained for x = 0.001, resulting in the enhanced ZT value of 0.019 at 600 K.

Keywords: iron silicide; bipolar effect; Ni doping; thermoelectric properties; ZT values

Citation: Sam, S.; Odagawa, S.; Nakatsugawa, H.; Okamoto, Y. Effect of Ni Substitution on Thermoelectric Properties of Bulk β-Fe$_{1-x}$Ni$_x$Si$_2$ ($0 \leq x \leq 0.03$). *Materials* **2023**, *16*, 927. https://doi.org/10.3390/ma16030927

Academic Editors: Bao-Tian Wang and Peng-Fei Liu

Received: 13 December 2022
Revised: 16 January 2023
Accepted: 17 January 2023
Published: 18 January 2023

Copyright: © 2023 by the authors. Licensee MDPI, Basel, Switzerland. This article is an open access article distributed under the terms and conditions of the Creative Commons Attribution (CC BY) license (https://creativecommons.org/licenses/by/4.0/).

1. Introduction

Thermoelectricity has been considered a potential technique to recover waste heat into electrical energy through the Seebeck effect without exhaust gas pollution to the environment, with no moving parts, and with no necessary maintenance required. To achieve highly efficient thermoelectric (TE) devices, finding promising semiconducting materials is the main challenge. Traditional TE materials such as PbTe and Bi$_2$Te$_3$ are high-priced and toxic; therefore, researchers have been trying to develop abundant and non-toxic materials, such as binary copper chalcogenide [1], copper sulfide compound [2,3], iron silicide [4–8], and other materials, in order to replace those traditional ones. Iron silicide compound is an abundant and non-toxic material having three different kinds of phases, such as the cubic ε-phase with space group $P2_13$ [4,5], the tetragonal α-phase with space group $P4/mmm$ [6,7], and the orthorhombic β-phase with space group $Cmce$ [8]. According to Piton and Fay diagram [9], the semiconducting β-phase can be formed at a temperature below 1259 K and depends on the kind and amount of external impurity doping, whereas the metallic ε and α-phases are grown at a higher temperature. It is noticed that its ε and α-phases are a metal that is not suitable for TE applications due to the deterioration of the Seebeck coefficient ($S = -\Delta V/\Delta T$, where ΔV and ΔT are the TE voltage and temperature difference across the material, respectively). However, its β-phase, a semiconducting material with a small band gap of around 0.7 eV [10], is suitable in TE applications. In addition, compared to other traditional TE materials (PbTe and Bi$_2$Te$_3$), β-FeSi$_2$ can work at high temperatures due to strong oxidation resistance, good thermal stability, and low cost [11–15]. However, due to its narrow band gap and low carrier concentration (n_H) of around 10^{16} cm^{-3}, the bipolar effect usually occurs in a non-doped β-FeSi$_2$, especially in high-temperature regions, resulting in a decline of the |S|. The Seebeck effect is generated by two types of carriers having opposite signs. With the increased temperature and low n_H, the total Seebeck effect is cancelled out, which is unfavorable for TE applications [16–18]. Therefore, as temperature increases, the TE performance is defined by $ZT = S^2\rho^{-1}\kappa^{-1}T$, where S, ρ, κ, and T are Seebeck coefficient, electrical resistivity, total thermal conductivity, and temperature, respectively. The Seebeck coefficient worsens due to the decrease in |S| caused by the bipolar effect. The pristine β-FeSi$_2$ has a low value of ZT of only round 2×10^{-4} [19]. To solve this issue, doping with impurities having a large valence electron in either Fe or Si sites is considered an effective technique for increasing the n_H, resulting in an improvement in the stability of |S| [17]. In addition, the ρ is inversely proportional to the n_H; therefore, it can be simultaneously decreased owing to the increase in n_H. As a result, the ZT can be significantly improved due to the monotonicity in |S| and the decrease in ρ. Theoretically, it was reported that the optimum n_H to improve the TE performance of β-FeSi$_2$ is approximately within the range of 1×10^{20} to 2×10^{21} cm^{-3} [15]. In fact, we prepared β-Fe$_{0.97}$Co$_{0.03}$Si$_2$ with the arc melting method and found a ZT value of 0.099 at 800 K [20]. Furthermore, many previous works [21–30] attempted to enhance the n_H of β-FeSi$_2$ via doping with various impurities.

Ito et al. reported the TE characteristics of β-FeSi$_{2-x}$P$_x$ fabricated by mechanical alloyed (MA) and hot-pressed (HP) method. By doping P at the Si site, the S was negative, indicating the n-type material with the optimum concentration of $x = 0.02$, and the ρ slightly decreased with a considerable increase in P content due to the increase in n_H. As a result, the highest ZT of about 0.033 at 672 K was obtained with $x = 0.02$, which was more than 11 times higher than that of the non-doped sample [21]. In addition, Tani and Kido found that the ZT of β-FeSi$_2$ can be enhanced up to 0.14 at 847 K by doping with Pt as an impurity [22]. Ohtaki et al. investigated various impurities for doping, such as Cu, Zn, Nb, Ag, Sb, and Mn, by analyzing the microstructural changes and TE performance of β-FeSi$_2$. They reported that the microstructures were remarkably changed by those impurities. The highest ZT value of about 0.026 was obtained by 3% Mn doping at 873 K [23]. In addition, Chen et al. investigated the thermoelectric characteristics of Co addition on β-FeSi$_2$ fabricated by rapid solidification and followed the HP method. It was reported that the optimum doping to achieve maximum $ZT = 0.25$ was obtained in Fe$_{0.94}$Co$_{0.06}$Si$_2$ samples due to the enhancement in S and a significant reduction in ρ [24]. Furthermore, Du et al. attempted to improve the TE performance of the previous Fe$_{0.94}$Co$_{0.06}$Si$_2$ by doping with an additional impurity element named Ru. It was found that Ru doping significantly decreases the thermal conductivity because the strain field and mass oscillation scatter the phonons, resulting in the improvement of $ZT = 0.33$ at 900 K [25]. Moreover, Dabrowski et al. investigated the effects of several dopants, namely, Mn, Co, Al, and P, on the TE properties of β-FeSi$_2$. They reported that compared to other dopants, P was not effective at improving the ZT due to only a slight decrease in ρ, where the n_H of the P-doped sample was probably lower than that of other impurity-doped samples; however, the highest ZT was obtained in a Co-doped sample, probably due to the high n_H [26]. Qiu et al. have recently reported that by doping 16% Ir into the Fe site of β-FeSi$_2$, the ZT can be greatly improved to 0.6 at 1000 K due to the significant reductions in ρ and κ, resulting from high n_H and phonon–electron scattering, respectively [27]. Based on a series of previous reports, it is worth noticing that doping with elements having large valence electrons to either Fe or Si sites of β-FeSi$_2$-based materials is remarkably effective at improving the n_H and the TE performance.

Since Ni has two valence electrons more than Fe, the n_H can be possibly increased by substituting Ni into the Fe site of β-FeSi$_2$. Komabayashi et al. reported that the S, ρ, and power factor ($PF = S^2\rho^{-1}$) at room temperature of Fe$_{0.94}$Ni$_{0.06}$Si$_{2.05}$ thin film fabricated by the RF sputtering method were -113 μVK^{-1}, 0.076 Ωcm, and 17 μWm^{-1}K^{-2}, respectively [28]. In addition, Nagai et al. investigated the effect of Ni addition on the PF of β-FeSi$_2$ fabricated by mechanical alloying and hot-pressing techniques. The highest |S| was obtained after 1% Ni doping—240 μVK^{-1} at 600 K—and ρ significantly decreased with Ni addition. This indicates that both S and ρ can be simultaneously improved by Ni addition, contributing to the enhancement of PF [29]. Furthermore, Tani and Kido reported that the ρ of bulk β-Fe$_{1-x}$Ni$_x$Si$_2$ decreased with the substitution of Ni owing to the increase in n_H [30]; thus, a reduction in the bipolar effect should be achieved. However, there are only a few reports regarding the effect of Ni doping of β-FeSi$_2$, and an investigation on the thermal conductivity (κ) and the ZT values has not been reported. Moreover, the optimum Ni doping concentration needed to improve the TE performance of β-Fe$_{1-x}$Ni$_x$Si$_2$ also has not been investigated yet.

In this work, we attempted to improve the electrical and thermoelectric properties of the bulk of binary β-FeSi$_2$ by Ni substitution into the Fe site prepared by the facile arc-melting techniques and directly followed by a heat treatment and annealing process. For the β-Fe$_{1-x}$Ni$_x$Si$_2$ system ($0 \leq x \leq 0.03$), a detailed investigation of the optimum doping level of Ni to enhance TE performance is reported for the temperature range of 80–800 K.

2. Materials and Methods

2.1. Sample Fabrication

The raw materials of Fe grain (99.9% up, 3Nup, High Purity Chemicals, Japan), Si grain (99.999%, 5N, High Purity Chemicals, Japan), and Ni grain (99.9%, 3N, High Purity Chemicals, Japan) were prepared following the composition of Fe$_x$Ni$_{1-x}$Si$_2$, where $0 \leq x \leq 0.03$. The melting process was performed by using the arc-melting method under a vacuum of about $2-5 \times 10^{-5}$ torr in an argon (Ar) atmosphere to prevent oxidation during melting. In addition, titanium (Ti) 10 g was set and initially melted before the main materials to remove the residual oxygen inside the melting chamber. To get an ingot with a homogeneous material distribution, it was flipped and remelted three times. The numerical control (NC) wire-cutting machine (EC-3025, Makino) was then used to slice the ingots into small pieces (sample's size W × L × T = 7 × 7 × 1.5 mm) to facilitate the characterization of TE properties. The pieces were then polished in order to remove the oxidized surface before the heat-treatment process. The metallic ε and α-phases were formed during the arc-melting process. In order to transform into the β-phase, the heat-treatment process at 1423 K for 3 h, and consequently, the annealing process at 1113 K for 20 h, were applied for all samples in vacuumed silica quartz ampule. The first step of heat treatment was to additionally homogenize the material distribution, and the second step was to transform it into a single β-phase. The heat treatment and annealing process followed that of reference [23], where the optimum condition was reported.

2.2. Sample Characterization

The CuKα high-resolution X-ray diffractometer (SmartLab, Rigaku, Tokyo, Japan) was used for the powder X-ray diffraction (XRD) measurements. With Rietveld analysis utilizing the RIETAN-FP software, calculation of the crystal structure parameters and phase identification were carried out by using the measured XRD data. A scanning electron microscope (VE-8800, KEYENCE, Osaka, Japan) apparatus was then used to observe the surface structure of each of the fabricated materials. The elemental analysis was performed with a scanning electron microscope (SU8010, Hitachi High-Technologies, Tokyo, Japan) equipped with a Bruker EDS XFlash5060FQ detector. The Archimedes method was performed to measure the relative density with a gravity measurement kit (SMK-401, SHIMADZU Co., Kyoto, Japan). ResiTest8300 (TOYO Co., Aichi-ken, Japan) apparatus was used to measure mobility (μ_H) and carrier density (n_H) at room temperature. In addition, the electrical resistivity (ρ) and Seebeck coefficient (S) were also measured by using the ResiTest8300 at temperatures of 80–395 K and by homemade apparatus under an Ar atmosphere at temperatures of 400–800 K. The thermal conductivity (κ_{total}) was measured by using a power efficiency measurement (PEM-2, ULVAC, Inc., Kanagawa, Japan) system and the ZT can be calculated by $ZT = S^2T/(\rho \kappa_{total})$.

3. Results and Discussions

Figure 1 shows the X-ray diffraction (XRD) peaks of $Fe_xNi_{1-x}Si_2$ ($0 \leq x \leq 0.03$) at 300 K within the angles of $20° \leq 2\theta \leq 90°$. Mainly, the β-phase was achieved for $0 \leq x \leq 0.03$; however, a trace of the ε-phase still remained at $2\theta \approx 45.2°$ on the right of the indexed peak (421), as zoomed in on in the inset of Figure 1. The intensity of this ε-phase peak increases with increasing Ni concentration from 0 to 0.03; the low intensity occurred at $x \leq 0.005$. The XRD peaks of our $0 \leq x \leq 0.005$ samples are similar to those of the study of Dąbrowski et al., who reported that a single β-phase was obtained by doping with other impurities, such as aluminum (Al) and phosphorus (P) [26]. Therefore, it is considered that the $0 \leq x \leq 0.005$ samples had very small amounts of the ε-phase.

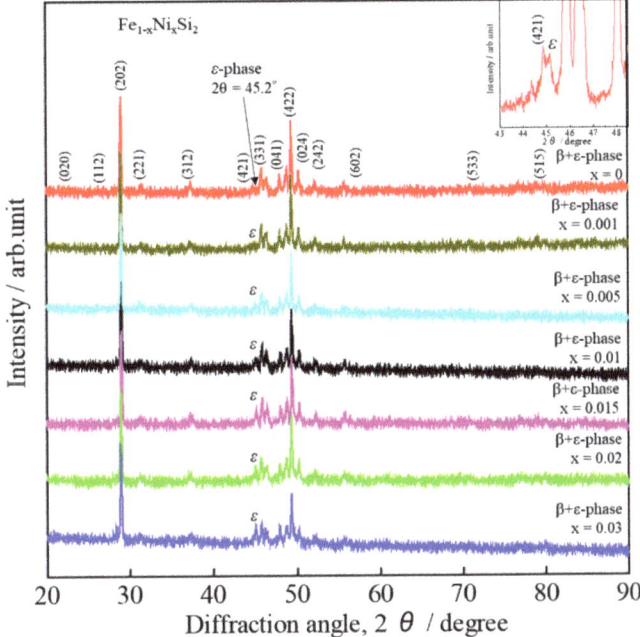

Figure 1. X-ray diffraction patterns of β-$Fe_{1-x}Ni_xSi_2$ ($0 \leq x \leq 0.03$) at room temperature.

Figure 2 shows the SEM images of $Fe_{1-x}Ni_xSi_2$ ($0 \leq x \leq 0.03$) captured at room temperature. The identification of phase transition by using the SEM micrograph can also be found in the previous reports of Dąbrowski et al. [26,31]. Figure 2a shows that before heat treatment, the ε and α-phases were formed at x = 0 (the bright grain represents the ε-phase and the dark grain represents α-phase). The white dots are not the microstructures but merely dust contaminated by the polished substrate. After heat treatment, for $0 \leq x \leq 0.005$, the samples were grown in a single β-phase, as shown in Figure 2b–d, and for $0.01 \leq x \leq 0.03$, the samples were grown with the majority of the β-phase and the minority of the ε-phase, as shown in Figure 2e–g. It is observed that the area or amount of the ε-phase increases with increasing Ni addition (x).

To observe the Ni distribution in each phase, the SEM-EDS measurement for elemental analysis was performed for the $0.005 \leq x \leq 0.03$ sample after the heat treatment. As shown in the color mapping of Figure 3, Ni was homogenously distributed for x = 0.005 due to the formation of a single β-phase, whereas the Ni-richness was distributed in the area of the ε-phase for $0.01 \leq x \leq 0.03$, as can be seen in the green. This tendency indicates that the semiconducting phase is moderately transformed into the metallic ε-phase by increasing Ni substitution. Furthermore, a portion of the Ni concentration is accumulated in the grain boundaries between ε-phase and β-phase, probably due to the large particle size of raw material. This issue can probably be solved by ball milling, followed by fast-sintering techniques. By utilizing the ball-milling method, the particle size of Ni can be significantly reduced, and the fast-sintering techniques could help to reduce grain growth. As the particle size reduces, the Ni might more homogenously distribute, leading to simultaneously eliminating the accumulation of Ni and grain growth. In addition, Table 1 also shows the quantitative analysis of the $0.005 \leq x \leq 0.03$ sample. In the area of the β-phase for all samples, the atomic concentration of Fe was approximately 1/3, and that of Si was approximately 2/3. This indicates that Fe:Si ratio is 1:2, corresponding to β-FeSi$_2$. On the other hand, in the area of the ε-phase, the atomic concentration of Fe was approximately $\frac{1}{2}$, and that of Si was also 1/2. This indicates that the Fe:Si ratio is about 1:1, corresponding to ε-FeSi. In the β-phase area of the $0.005 \leq x \leq 0.03$ sample, the actual Ni composition ranged from 0.003(1) to 0.010(4), indicating that the solid solution limit of Ni for β-FeSi$_2$ is lower than $x = 0.01$. When the value is higher than 0.01, it facilitates the formation of the ε-phase. As a result, a single β-phase could be obtained in the $0 \leq x \leq 0.005$ samples, as verified with the SEM image in Figure 2b–d. Moreover, as shown in Table 1, Ni in both β and ε-phases linearly increases with x, but the slope of the ε-phase is around six times that of the β-phase.

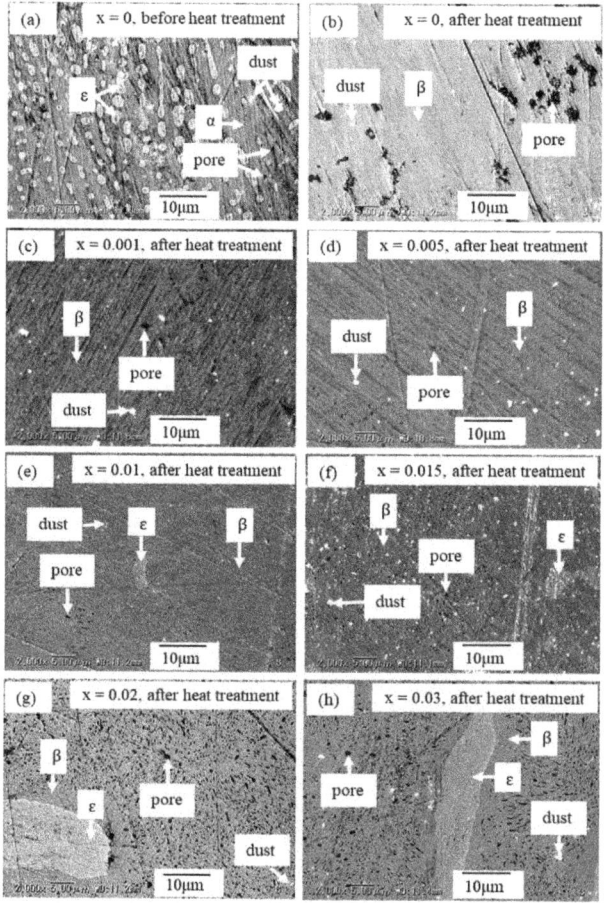

Figure 2. SEM image of β-Fe$_{1-x}$Ni$_x$Si$_2$: (**a**) before heat treatment for $x = 0$, (**b–h**) after heat treatment for $0 \leq x \leq 0.03$.

Moreover, Figure 2 shows that pore size after heat treatment is larger than that before heat treatment. This enlargement of pore size happens when the volume β-FeSi$_2$ occupies varies with the volumes of metallic ε and α-phases during the heat-treatment process (ε-FeSi + α-Fe$_2$Si$_5$ → β-FeSi$_2$). However, the relative densities range from 95.6(1)–98.7(1)%, as shown in Table 2. These values are as high as for a sample prepared by hot-pressing (HP) techniques [32], but are relatively higher than those for samples prepared by pulse plasma sintering (PPS) [28] or spark plasma sintering (SPS) [33]. This result suggests that the proposed arc-melting and direct-heat-treatment method is efficient at fabricating a high-relative-density sample that contributes positively to the decrease in electrical resistivity, which is good for TE application. The three dimensions of the orthorhombic crystal structure of β-FeSi$_2$ were provided by our previous report [20].

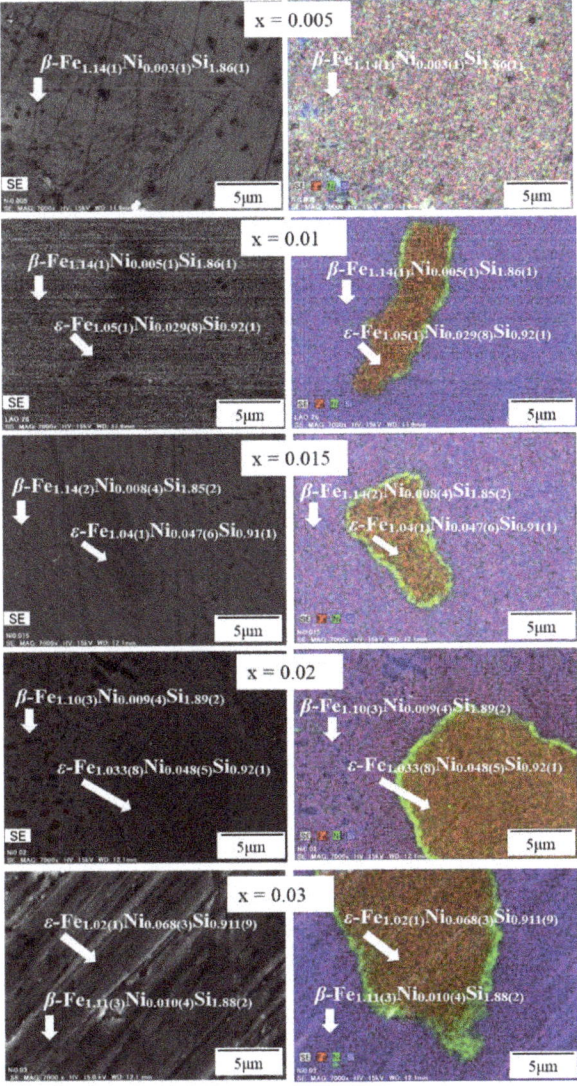

Figure 3. SEM-EDS mapping of β-Fe$_{1-x}$Ni$_x$Si$_2$ (0.005 ≤ x ≤ 0.03). Fe, Ni, and Si are mapped with red, green, and blue, respectively.

Table 1. Elemental composition of β-Fe$_{1-x}$Ni$_x$Si$_2$ (0.005 ≤ x ≤ 0.03) quantified by SEM-EDS analysis.

x	Area	Element	Atomic %	Composition Ratio	Symbol
0.005	β	Fe	38.1(4)	1.14(1)	β-Fe$_{1.14(1)}$Ni$_{0.003(1)}$Si$_{1.86(1)}$
		Ni	0.10(5)	0.003(1)	
		Si	61.8(4)	1.86(1)	
0.01	β	Fe	37.9(5)	1.14(1)	β-Fe$_{1.14(1)}$Ni$_{0.005(1)}$Si$_{1.86(1)}$
		Ni	0.17(4)	0.005(1)	
		Si	62.9(5)	1.86(1)	
	ε	Fe	52.7(4)	1.05(1)	ε-Fe$_{1.05(1)}$Ni$_{0.029(8)}$Si$_{0.92(1)}$
		Ni	1.5(4)	0.029(8)	
		Si	45.8(6)	0.92(1)	
0.015	β	Fe	38.2(6)	1.14(2)	β-Fe$_{1.14(2)}$Ni$_{0.008(4)}$Si$_{1.85(2)}$
		Ni	0.3(1)	0.008(4)	
		Si	61.5(6)	1.85(2)	
	ε	Fe	52.0(6)	1.04(1)	ε-Fe$_{1.04(1)}$Ni$_{0.047(6)}$Si$_{0.91(1)}$
		Ni	2.3(3)	0.047(6)	
		Si	45.7(7)	0.91(1)	
0.02	β	Fe	36.8(9)	1.10(3)	β-Fe$_{1.10(3)}$Ni$_{0.009(4)}$Si$_{1.89(2)}$
		Ni	0.3(1)	0.009(4)	
		Si	62.9(8)	1.89(2)	
	ε	Fe	51.7(7)	1.033(8)	ε-Fe$_{1.033(8)}$Ni$_{0.048(5)}$Si$_{0.92(1)}$
		Ni	2.4(2)	0.048(5)	
		Si	45.9(6)	0.92(1)	
0.03	β	Fe	37.1(9)	1.11(3)	β-Fe$_{1.11(3)}$Ni$_{0.010(4)}$Si$_{1.88(2)}$
		Ni	0.3(1)	0.010(4)	
		Si	62.5(8)	1.88(2)	
	ε	Fe	51.1(5)	1.02(1)	ε-Fe$_{1.02(1)}$Ni$_{0.068(3)}$Si$_{0.911(9)}$
		Ni	3.4(2)	0.068(3)	
		Si	45.5(4)	0.911(9)	

Table 2. Summary of thermoelectric properties of β-Fe$_{1-x}$Ni$_x$Si$_2$ (0 ≤ x ≤ 0.03) at 300 K, where L_O, r = −1/2, n_H, μ_H, S, ρ, and κ are Lorenz number, scattering factor (for acoustic phonon scattering), carrier density, mobility, Seebeck coefficient, electrical resistivity, and thermal conductivity, respectively.

x	L_O [V^{2}K^{-2}]	r	n_H [cm^{-3}]	μ_H [cm^{2}V^{-1}s^{-1}]	|S| [μVK^{-1}]	ρ [Ωcm]	κ [Wm^{-1}K^{-1}]	Relative Density [%]
0	1.792 × 10^{-8}	−1/2	2.3(2) × 10^{16}	37(4)	127	7.10	7.16	98.0(1)
0.001	1.624 × 10^{-8}	−1/2	1.2(4) × 10^{17}	35(7)	393	1.39	8.25	98.3(1)
0.005	1.656 × 10^{-8}	−1/2	2.6(2) × 10^{17}	34(3)	194	0.69	8.57	96.24(8)
0.01	1.648 × 10^{-8}	−1/2	3.2(7) × 10^{17}	27(5)	205	0.69	8.44	97.5(3)
0.015	1.674 × 10^{-8}	−1/2	4.8(4) × 10^{17}	20(1)	176	0.63	7.27	98.7(1)
0.02	1.766 × 10^{-8}	−1/2	1.2(4) × 10^{18}	11(3)	135	0.45	7.12	95.6(2)
0.03	2.139 × 10^{-8}	−1/2	2.3(2) × 10^{18}	10(1)	62	0.26	6.18	95.6(1)

The bonds of the Fe1 and sites are formed geometrically in eight coordinates, four each to Si1 and Si2, and the bonds of Fe–Si vary in length from 2.361(5) to 2.402(6) Å and from 2.282(5) to 2.415(4) Å, respectively.

The Rietveld analysis of β-Fe$_{0.995}$Ni$_{0.005}$Si$_2$ after heat treatment is shown in Figure A1 (Appendix A). The calculated data, experimental data, and difference between the data, are represented by green, red, and blue lines, respectively. According to the analysis, it is considered that after the process of heat treatment, the sample is successfully grown in the β-phase with a trace of the metallic ε-phase.

Therefore, the result of the Rietveld analysis agrees with that of the SEM image. The orthorhombic structure (*Cmce* space group) was chosen for Rietveld analysis. As Ni was partially substituted into the Fe sites, 1−x was assigned as the occupied rate of Fe1 and Fe2, and x was assigned as the occupied rate of Ni1 and Ni2. In addition, both the Fe site and Si site were assigned with the isotropic atomic displacement B with the value of 1.0 Å^2. A split pseudo-Voigt function was used to fit the Bragg peak profiles. A summary of the structural parameters, which were calculated by the Rietveld analysis, is reported in Table A1 (Appendix B). The lattice constants (a, b, c), interact atomic distances (Si-Fe), interacting atomic angles (Fe-Si-Fe), and reliability factor for weight diffraction patterns (R_{wp}) with x dependences are plotted in Figure 4. In Figure 4a, the variations in lattice constants a, b, and c with x are negligible. In addition, as shown in Table A1, the change in unit-cell volume (V) is almost within the error range. The effect of Ni substitutions is probably not significant for the lattice constants due to the low solubility limit of Ni. Figure 4b shows that the atomic distances of Si1-Fe1 and Si1-Fe2 tend to slightly rise with increasing x, though there is no significant change in Si2-Fe1 or Si2-Fe2 with x. In addition, the interactive atomic angles of both Fe1-Si1-Fe1 and Fe2-Si2-Fe2 slightly rise with x, but those of both Fe1-Si2-Fe1 and Fe1-Si2-Fe1 slightly decline as x increases, as shown in Figure 4c. It is considered that both Fe1 and Fe2 are slightly changed with Ni addition. Therefore, the Ni population should equally occupy both Fe1 and Fe2 sites, which is similar to a previous study wherein Co was doped into the β-FeSi$_2$ system [20,34]. Figure 4d shows the reliability factor R_{wp} with x dependences. The R_{wp} value for x = 0 is about 3.316%, indicating a good fit between the observed and computed intensities. However, as x increases, the R_{wp} moderately increases, probably due to the increasing amount of the ε-phase, which is verified with XRD patterns in Figure 1 and the SEM image in Figure 2.

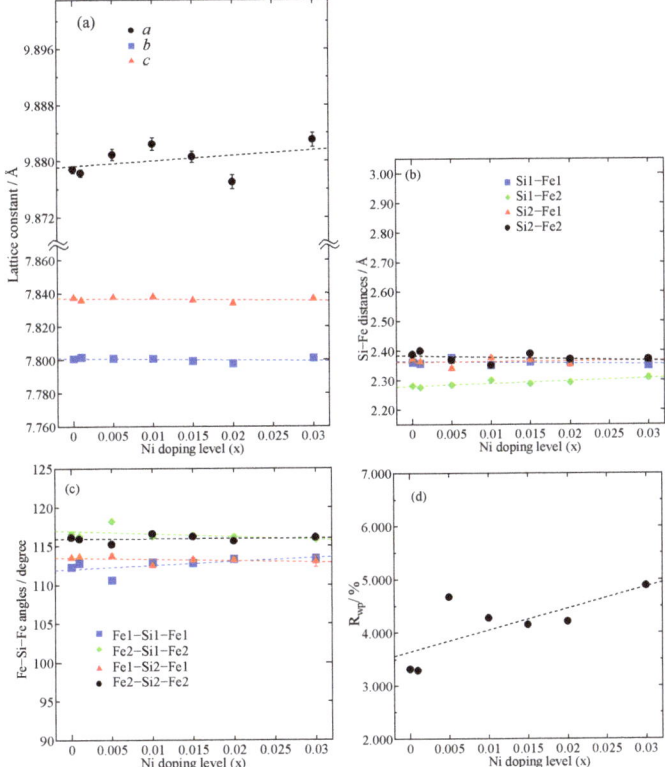

Figure 4. (a) Lattice constants a, b, c; (b) interactive atomic distances of Fe-Si, (c) interactive atomic angles of Fe-Si-Fe, and (d) reliability factor R_{wp} for β-Fe$_{1-x}$Ni$_x$Si$_2$ ($0 \leq x \leq 0.03$) at room temperature.

The electrical resistivity (ρ) with respect to the temperature dependence of $Fe_{1-x}Ni_xSi_2$ is shown in Figure 5. The ρ significantly decreases as Ni content increases from x = 0 to x = 0.03. The decrease in ρ is mainly caused by the increases in ε-phase and carrier concentration (n_H), as shown in the inset of Figure 5. This tendency can be explained by Drude's theory in Equation (1):

$$\rho = n_H^{-1} |e|^{-1} \mu_H^{-1} \quad (1)$$

where e and μ_H are elementary charge and carrier mobility, respectively [35]. Equation (1) expresses that ρ is inversely proportional to n_H. Therefore, as n_H increases, ρ can be effectively obtained. In Table 2, the ρ of the non-doped sample is 7.10 Ωcm with the n_H of only around $1.3(2) \times 10^{16} cm^{-3}$. As x increases from 0.001 to 0.03, the ρ remarkably decreases from 1.39 to 0.26 Ωcm due to the increase in n_H from $1.2(4) \times 10^{17}$ to $2.3(2) \times 10^{18} cm^{-3}$. Furthermore, the increase in the ε-phase with Ni substitution, as discussed for the microstructures above, should also contribute to the reduction in ρ. For x = 0.01, the ρ value of our sample was almost similar to that of the one Tani and Kido prepared by pressure-sintering techniques [30]. However, for x = 0.03, the ρ of our sample was about two times lower due to the higher μ_H. The μ_H of our sample was $10(1)$ $cm^2V^{-1}s^{-1}$, and that of their sample was only around 0.27 $cm^2V^{-1}s^{-1}$. If we compare another dopant, cobalt (Co), at the same doping level, the Ni-substituted material has a much higher value of ρ than the Co-substituted materials. This is because Co has a higher solid solution limit in β-$FeSi_2$; its value is up to 0.116, as reported by Kojima et al. [36]. In addition, Nagai et al. reported that the ρ of the x = 0.06 thin film was 0.076 Ωcm. Such a low value of ρ in a thin-film sample should be mainly affected by the large μ_H. It is considered that the ρ of the bulk sample prepared by the arc-melting method should drastically decrease if the doping amount is up to x = 0.06 due to the increase in n_H; however, the thermoelectric power will be deteriorated due to the effect of the metallic ε-phase.

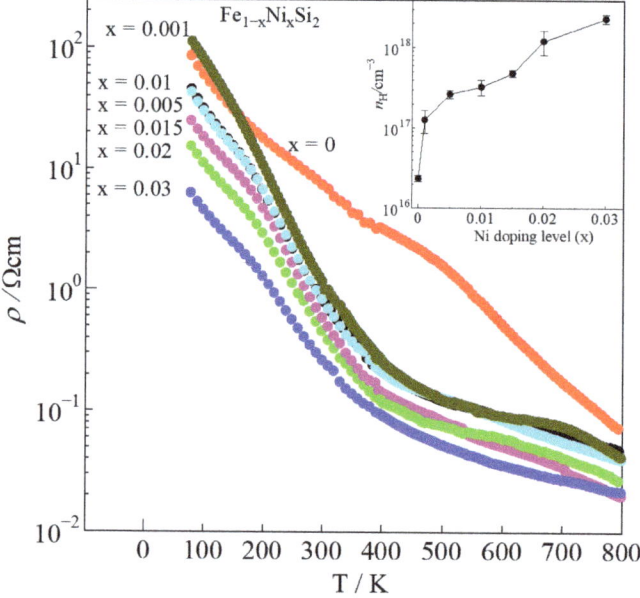

Figure 5. Electrical resistivity of β-$Fe_{1-x}Ni_xSi_2$ ($0 \leq x \leq 0.03$) with respect to temperature, where the carrier concentration, n_H, at 300 K is plotted in the inset.

The Seebeck coefficient (S) with temperature dependence is shown in Figure 6. The $|S|$ of the non-doped sample (x = 0) remarkably decreases from about 290 μVK^{-1} to approximately 0 µVK^{-1} as the temperature increases from 420 to 800 K. It is considered that the bipolar effect dominates at high-temperature regions in pristine β-FeSi$_2$ due to low carrier density. For the Ni-doped β-FeSi$_2$ system, the S is positive at temperatures 80–115 K and 80–195 K for x = 0.001 and 0.03, respectively, indicating the p-type materials. At higher temperatures, the S becomes negative, indicating an n-type conduction material. This result is consistent with that of Tani and Kido [30], who also reported that the sign of the Hall coefficient (R_H) changes from positive to negative at 160 K. It is considered that conduction is dominated by both holes and electrons, and its ratio varies depending on temperature. In addition, as x increases, the |S| becomes more stable from room temperature to 800 K. This tendency suggests that the bipolar effect is remarkably reduced with Ni substitution due to the increase in n_H. The bipolar effect was reduced by Ni doping; however, at high temperatures, it was not completely eliminated. This is probably due to the much lower actual Ni doping concentration in the β-phase. The increase in n_H contributes to the reduction in |S|. The relationship between |S| and n_H can be expressed by Mott's formula:

$$S = \frac{k_B^2 T}{3|e|\hbar^2} m^* \left(\frac{\pi}{3n_H}\right)^{2/3} \quad (2)$$

where k_B, T, e, $\hbar$, m^*, and n_H are Boltzmann constant, temperature, elementary charge, Planck constant, effective mass, and carrier concentration, respectively [37]. Equation (2) indicates that the |S| is inversely proportional to n_H; therefore, as can be seen in Figure 6, for 0.001 ≤ x ≤ 0.03, the |S| of the Ni-doped samples decreases with x. Furthermore, the inset of Figure 6 shows that μ_H decreases with x, probably owing to the difference in the effective mass between the electron and the hole. This tendency can be expressed by Equation (3):

$$m^* = \frac{e\tau}{\mu_H} \quad (3)$$

where e, τ, and μ_H are elementary charge, scattering time, and mobility, respectively [35]. When the effective mass of the electron is larger than that of the hole, the mobility of the electron is lower. As shown in Figure A2 (Appendix A), for 0.001 ≤ x ≤ 0.03, the |S| decreases with n_H, and the tendency of the experimental values of |S| fits with that of the calculated values (solid black curve, in the case of m^* = 0.1 m_e) using the Mott's formula in Equation (2). It is confirmed that Mott's theory implies for 0.001 ≤ x ≤ 0.03. For 0 ≤ x < 0.001, the experimental value of |S| is out of the fitting curve; therefore, this might be possibly described by a two-carrier model [38]. The highest value of |S| was obtained for the x = 0.001 sample with the value of 450 µVK^{-1} at 450 K.

Figure 7 shows the power factor (PF) with temperature dependence. The PF is calculated by $PF = S^2 \rho^{-1}$. The improvement in PF contributes positively to enhancing TE performance (ZT). The PF of the non-doped sample exhibited the highest value of around 3.5 µWm^{-1}K^{-2} at around 450 K, as shown in the inset of Figure 7. By doping with Ni, the PF can be significantly improved; the maximum value was around 200 µWm^{-1}K^{-2} at 600 K, achieved by the x = 0.001 sample. The enhancement in PF is caused not only by the remarkable increase in, S but also by the reduction in ρ. Compared to previous work reported by Komabayashi et al., the PF of thin-film x = 0.06 was 17 µWm^{-1}K^{-2} at 300 K [28]. This value is similar to that of our bulk x = 0.001 sample with the PF of about 13 µWm^{-1}K^{-2} at 300 K. The thin-film sample usually had a much lower ρ than that of the bulk sample, which provided a better PF. However, the high value of |S| for our bulk sample also increased the PF, which is comparable to that of the thin-film sample. In addition, Nagai et al. reported that the highest PF of the bulk x = 0.01 samples prepared by mechanical milling and hot pressing was about 50 µWm^{-1}K^{-2} at 650 K [29], whereas that of our x = 0.01 sample prepared by arc-melting was about 130 µWm^{-1}K^{-2} at 750 K. The higher PF in our sample is owed to the larger |S|. It is considered that our sample had a lower ε-phase amount as a result of the heat-treatment process, as that was not applied to their sample. This might be a reason why the |S| of their sample was lower. Therefore, heat treatment is necessary for the fabrication of β-FeSi$_2$ for TE applications.

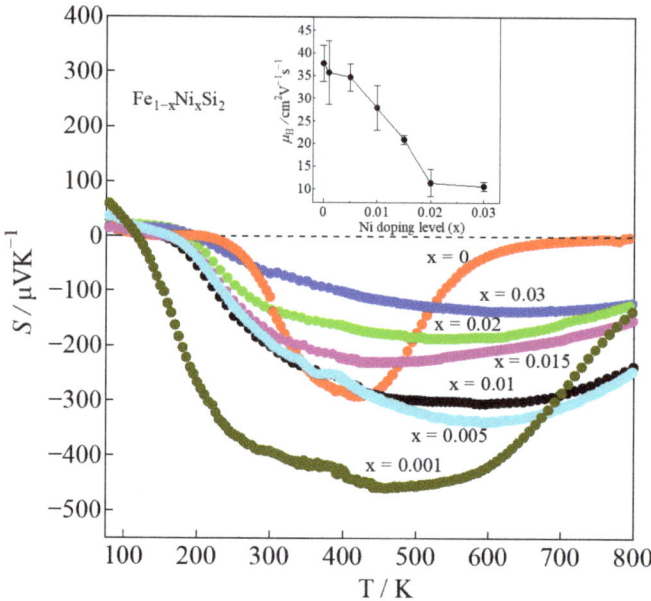

Figure 6. Seebeck coefficient of β-Fe$_{1-x}$Ni$_x$Si$_2$ ($0 \leq x \leq 0.03$) with respect to temperature, where the carrier mobility, μ_H, at 300 K is plotted in the inset.

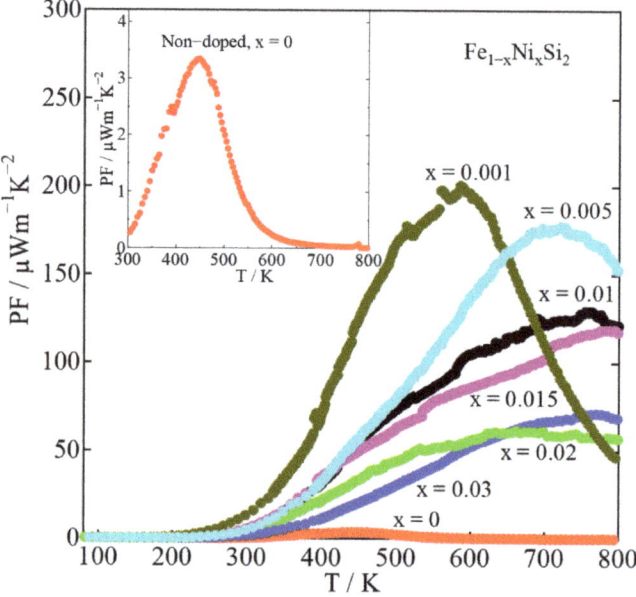

Figure 7. Power factor (PF) of β-Fe$_{1-x}$Ni$_x$Si$_2$ ($0 \leq x \leq 0.03$) with respect to temperature, where the inset plots the data of x = 0.

The total thermal conductivity ($\kappa_{total} = \kappa_l + \kappa_e$, where κ_l and κ_e are the lattice and electronic thermal conductivity, respectively) of all samples is plotted in Figure 8. It is shown that the κ_{total} of Ni-doped samples is slightly higher than that of the on-doped one. This is probably because of the increase in the metallic ε-phase with increasing Ni content, as can be proved by XRD patterns and SEM-EDS analysis. As shown in the inset of Figure 8, the electronic thermal conductivity (κ_e) increased with x due to the decrease in ρ. The κ_e was calculated by the Wiedemann–Franz law ($\kappa_e = L_O T/\rho$, where L_O is the Lorenz number). The L_O was calculated by the measured Seebeck coefficient $|S|$ in the case of acoustic phonon scattering ($r = -1/2$). Equation (4) explains the relationship between L_O and r:

$$L_O = \left(\frac{k_B}{e}\right)^2 \left[\frac{(r+\frac{7}{2})F_{r+\frac{5}{2}}(\eta)}{(r+\frac{3}{2})F_{r+\frac{1}{2}}(\eta)} - \left\{\frac{(r+\frac{5}{2})F_{r+\frac{3}{2}}(\eta)}{(r+\frac{3}{2})F_{r+\frac{1}{2}}(\eta)}\right\}^2\right] \quad (4)$$

where the function is given as: $F_n(\eta) = \int_0^\infty \frac{\chi^n}{1+e^{\chi-\eta}}d\chi$, $\chi = \frac{E}{k_B T}$, $\eta = \frac{E_F}{k_B T}$ and E_F is Fermi energy [39]. Table 2 shows that the values of L_O increase with x for $0.001 \leq x \leq 0.03$. This tendency shows that the β-phase moderately transforms into the ε-phase as the level of Ni doping increases. The increase in L_O also contributes to the high κ_e because of its proportionality.

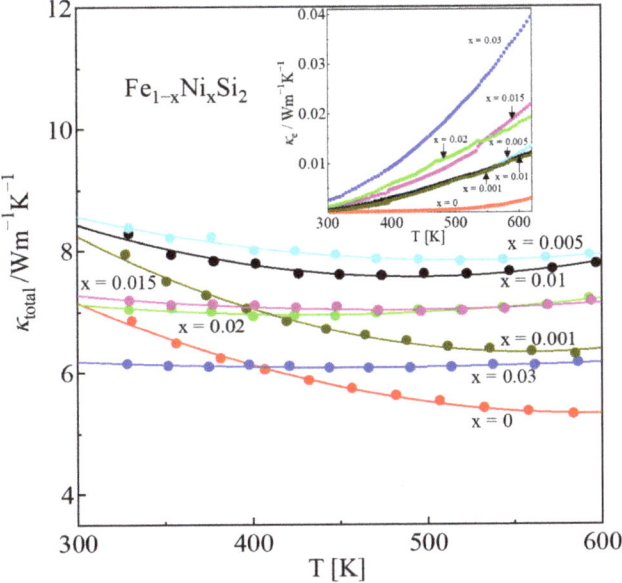

Figure 8. Thermal conductivity ($\kappa_{total} = \kappa_L + \kappa_e$) of β-Fe$_{1-x}Ni_xSi_2$ ($0 \leq x \leq 0.03$) with respect to temperature, where the inset plots the electronic thermal conductivity ($\kappa_e = L_O T/\rho$).

The ZT value with temperature dependences is plotted in Figure 9. In addition, the inset of Figure 9 shows that the $x = 0$ sample had the highest ZT value of 2.6×10^{-4} at the temperature of 450 K. If we compare the ZT value of pristine material to Ni-doped materials, its value is very low. In the Ni-doped system, the maximum ZT of around 0.019 at 600 K was obtained in $x = 0.001$ owing to the enhancement of the power factor. When x is higher than 0.001, the ZT is decreased due to the reduction in $|S|$ caused by the increased amount of metallic ε-phase. Therefore, it is considered that for Ni-doped β-FeSi$_2$, the low doping amount, below 1%, is more effective at improving the TE properties.

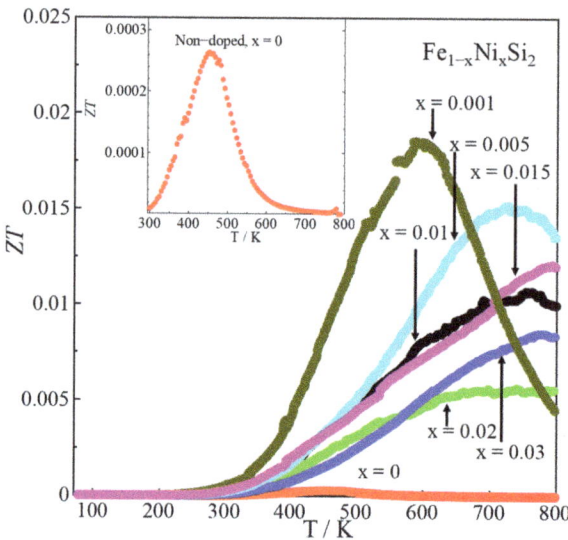

Figure 9. ZT of β-Fe$_{1-x}$Ni$_x$Si$_2$ ($0 \leq x \leq 0.03$) with respect to temperature, where the inset plots the data for x = 0.

4. Conclusions

Thermoelectric (TE) materials β-Fe$_{1-x}$Ni$_x$Si$_2$ were fabricated for $0 \leq x \leq 0.03$ by the conventional arc-melting method, followed by a heat treatment and annealing process. Traces of the ε-phase were formed for all samples; the lowest amount of it was obtained at $x \leq 0.005$. The solid solution limit of Ni in the β-phase is below x = 0.01, and the ε-phase increases with increasing Ni concentration. As x increases, the electrical resistivity (ρ) and Seebeck coefficient $|S|$ decrease, owing to the increases in ε-phase and carrier density. As a result, the optimum doping amount to achieve a maximum power factor (PF) of around 200 μWm^{-1}K^{-2} was obtained in the x = 0.001 sample due to significant enhancement in $|S|$. The PF value of this sample was comparable to that of the thin-film sample reported by Komabayashi et al. [28]. However, this value is higher than that of the hot-pressed sample reported by Nagai et al. [29], resulting from the improvement in $|S|$. The improvement in PF led to obtaining the ZT of 0.019 at 600 K in the same x = 0.001 sample. It would be worth investigating a method to increase Ni's solubility in β-FeSi$_2$. As Ni solubility increases, the ε-phase can be reduced, resulting in an improvement in S and a decrease in thermal conductivity (κ). Therefore, ZT can be more significantly enhanced, making the material suitable for industrial waste heat recovery in mid–high temperature applications.

Author Contributions: Conceptualization, S.S., H.N. and Y.O.; methodology, S.S. and S.O.; formal analysis, S.S., S.O., H.N. and Y.O.; investigation, S.S. and S.O.; resources, H.N., and Y.O.; data curation, S.S. and S.O., writing—original draft preparation, S.S.; writing—review and editing, S.S. and H.N. All authors have read and agreed to the published version of the manuscript.

Funding: This research received no external funding.

Institutional Review Board Statement: Not applicable.

Informed Consent Statement: Not applicable.

Data Availability Statement: Not applicable.

Acknowledgments: The XRD measurements and SEM-EDS analysis were carried out using the Yokohama National University Instrumental Analysis and Evaluation Center equipment.

Conflicts of Interest: The authors declare no conflict of interest.

Appendix A

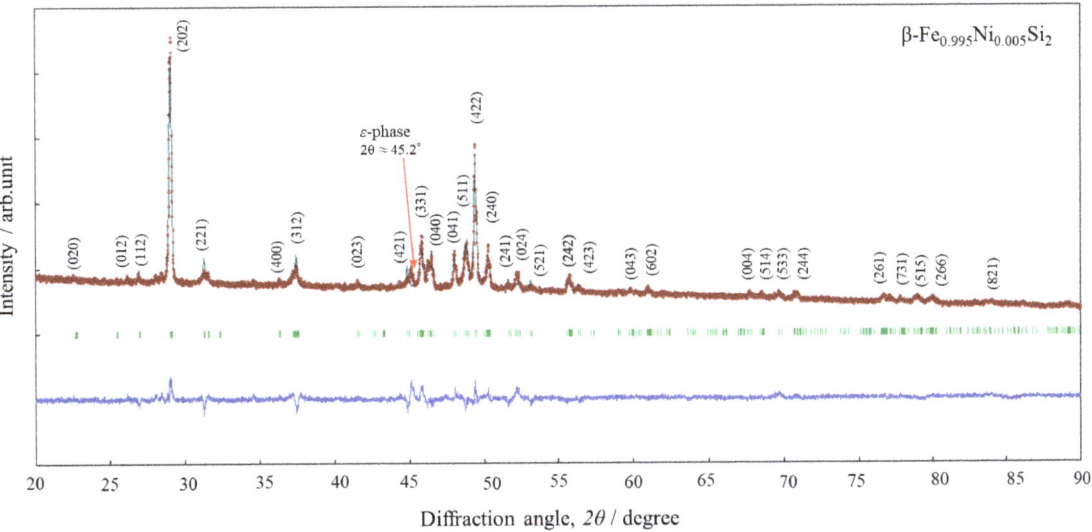

Figure A1. Rietveld analysis of β-Fe$_{0.995}$Ni$_{0.005}$Si$_2$.

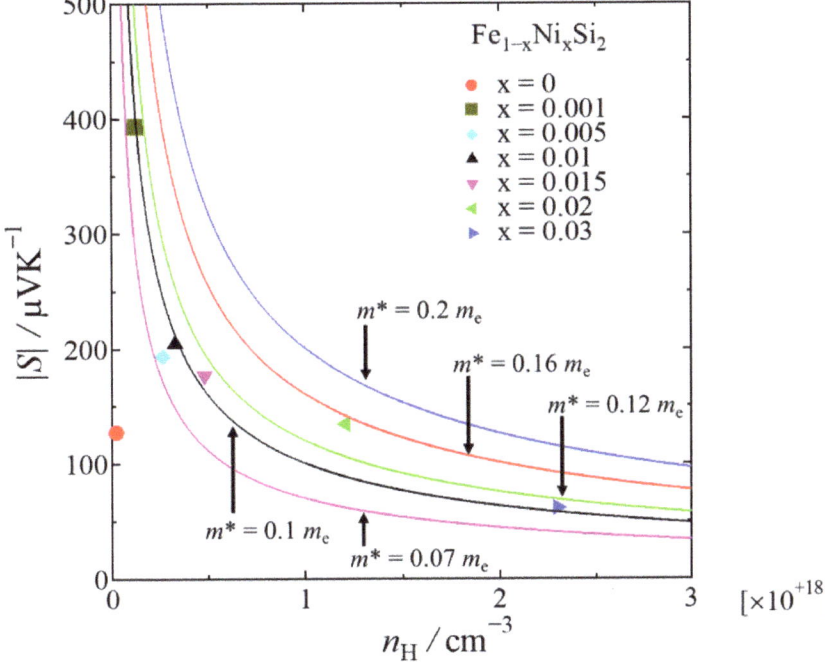

Figure A2. Absolute Seebeck coefficient with respect to carrier concentration at room temperature, where the solid curves represent the calculated data estimated by using Mott's formula at various effective masses ($m^* = x \, m_e$, where x is variable and m_e is the static mass of electron, i.e., 9.10938×10^{-31} kg).

Appendix B

Table A1. Crystal-structure parameters of β-Fe$_{1-x}$Ni$_x$Si$_2$ ($0 \leq x \leq 0.03$) at room temperature.

Samples			Fe$_{1-x}$Ni$_x$Si$_2$						
Composition, x			0	0.001	0.005	0.01	0.015	0.02	0.03
Space Group			Cmce	Cmce	Cmce	Cmce	Cmce	Cmce	Cmce
a (Å)			9.8788(5)	9.8783(5)	9.8809(8)	9.8824(9)	9.8806(8)	9.877(1)	9.883(1)
b (Å)			7.8008(4)	7.8016(4)	7.8008(7)	7.8008(7)	7.7993(6)	7.7978(8)	7.8011(9)
c (Å)			7.8372(4)	7.8357(4)	7.8375(7)	7.8378(7)	7.8356(6)	7.8338(8)	7.8365(9)
V (Å^3)			603.96(5)	603.88(5)	604.12(9)	604.23(9)	603.83(8)	603.4(1)	604.1(1)
Fe1		x	0.2160(2)	0.2163(3)	0.2178(4)	0.2173(3)	0.2170(4)	0.2170(4)	0.2179(5)
		y	0	0	0	0	0	0	0
		z	0	0	0	0	0	0	0
		B (Å^2)	0.1	0.1	0.1	0.1	0.1	0.1	0.1
		g	1.000	0.999	0.995	0.990	0.985	0.980	0.970
Ni1		x	N/A	0.2163(3)	0.2178(4)	0.2173(3)	0.2170(4)	0.2170(4)	0.2179(5)
		y	N/A	0	0	0	0	0	0
		z	N/A	0	0	0	0	0	0
		B (Å^2)	N/A	0.1	0.1	0.1	0.1	0.1	0.1
		g	N/A	0.001	0.005	0.010	0.015	0.020	0.030
Fe2		x	1/2	1/2	1/2	1/2	1/2	1/2	1/2
		y	0.3014(4)	0.3019(4)	0.2987(6)	0.3015(5)	0.3034(5)	0.3031(6)	0.3036(7)
		z	0.1940(4)	0.1943(4)	0.1967(5)	0.1970(5)	0.1944(5)	0.1962(5)	0.1965(6)
		B (Å^2)	0.1	0.1	0.1	0.1	0.1	0.1	0.1
		g	1.000	0.999	0.995	0.990	0.985	0.980	0.970
Ni2		x	N/A	1/2	1/2	1/2	1/2	1/2	1/2
		y	N/A	0.3019(4)	0.2987(6)	0.3015(5)	0.3034(5)	0.3031(6)	0.3036(7)
		z	N/A	0.1943(4)	0.1967(5)	0.1970(5)	0.1944(5)	0.1962(5)	0.1965(6)
		B (Å^2)	N/A	0.1	0.1	0.1	0.1	0.1	0.1
		g	N/A	0.001	0.005	0.010	0.015	0.020	0.030
Si1		x	0.1217(5)	0.1227(6)	0.1178(8)	0.1231(8)	0.1228(8)	0.1236(9)	0.124(1)
		y	0.2811(7)	0.2801(7)	0.281(1)	0.2795(9)	0.2784(9)	0.276(1)	0.277(1)
		z	0.0394(4)	0.0377(4)	0.0410(6)	0.0388(6)	0.0394(6)	0.0385(6)	0.0404(8)
		B (Å^2)	0.3	0.3	0.3	0.3	0.3	0.3	0.3
		g	1.0	1.0	1.0	1.0	1.0	1.0	1.0
Si2		x	0.3761(5)	0.3758(6)	0.3742(8)	0.3775(7)	0.3764(8)	0.3755(8)	0.3762(9)
		y	0.0399(5)	0.0387(6)	0.0434(8)	0.0442(7)	0.0414(7)	0.0442(8)	0.0445(9)
		z	0.2220(6)	0.2219(7)	0.2200(9)	0.2217(9)	0.2222(9)	0.2200(9)	0.223(1)
		B (Å^2)	0.3	0.3	0.3	0.3	0.3	0.3	0.3
		g	1.0	1.0	1.0	1.0	1.0	1.0	1.0
R_{wp} (%)			3.316	3.292	4.669	4.272	4.148	4.209	4.881
R_P (%)			2.108	2.107	3.397	3.012	2.773	2.786	3.163
R_R (%)			29.041	30.762	38.361	35.726	36.636	36.753	40.663
R_e (%)			0.792	1.429	1.222	1.069	0.922	0.883	1.029
R_B (%)			8.543	9.144	10.839	10.024	10.650	9.684	11.392
R_F (%)			8.603	8.788	8.217	8.323	9.418	7.582	8.542
$S = R_{wp}/R_e$			4.187	2.305	3.821	3.996	4.499	4.767	4.743
Si1—Fe1/Ni1 (Å)			2.361(5)	2.357(6)	2.377(8)	2.352(7)	2.363(7)	2.365(9)	2.351(9)
Si1—Fe1/Ni1 (Å)			2.402(6)	2.391(6)	2.426(9)	2.390(8)	2.382(8)	2.366(8)	2.37(1)
Si1—Fe2/Ni2 (Å)			2.282(5)	2.277(6)	2.284(8)	2.301(7)	2.289(8)	2.294(8)	2.31(1)
Si1—Fe2/Ni2 (Å)			2.415(4)	2.430(5)	2.366(6)	2.407(6)	2.420(6)	2.418(7)	2.410(9)
Fe1/Ni1—Si1—Fe1/Ni1 (deg.)			112.3(2)	112.8(2)	110.6(3)	112.9(3)	112.8(3)	113.3(3)	113.4(4)
Fe2/Ni2—Si1—Fe2/Ni2 (deg.)			116.6(2)	116.3(2)	118.2(3)	116.2(3)	116.4(3)	116.2(3)	115.8(2)
Si2—Fe1/Ni1 (Å)			2.372(5)	2.363(6)	2.340(7)	2.376(7)	2.370(8)	2.354(8)	2.372(9)
Si2—Fe1/Ni1 (Å)			2.381(5)	2.386(6)	2.399(8)	2.399(7)	2.386(7)	2.400(8)	2.385(9)
Si2—Fe2/Ni2 (Å)			2.322(6)	2.314(7)	2.355(8)	2.335(9)	2.315(9)	2.34(1)	2.32(1)
Si2—Fe2/Ni2 (Å)			2.388(5)	2.400(5)	2.368(9)	2.351(7)	2.390(7)	2.371(8)	2.371(9)
Fe1/Ni1—Si2—Fe1/Ni1 (deg.)			113.5(2)	113.6(2)	113.7(3)	112.5(2)	113.2(3)	113.2(3)	113.0(7)
Fe2/Ni2—Si2—Fe2/Ni2 (deg.)			116.1(2)	115.9(2)	115.2(3)	116.6(2)	116.2(3)	115.6(3)	116.1(3)

References

1. Chen, X.Q.; Yang, J.P.; Wu, T.; Li, L.; Luo, W.; Jiang, W.; Wang, L.J. Nanostructured Binary Copper Chalcogenides: Synthesis Strategies and Common Applications. *Nanoscale* **2018**, *10*, 15130–15163. [CrossRef] [PubMed]
2. Chen, X.Q.; Zhang, H.; Zhao, Y.Y.; Liu, W.D.; Dai, W.; Lu, X.F.; Wu, C.; Luo, W.; Fan, Y.C.; Wang, L.J.; et al. Carbon-Encapsulated Copper Sulfide Leading to Enhanced Thermoelectric Properties. *ACS Appl. Mater. Interfaces* **2019**, *11*, 22457–22463. [CrossRef] [PubMed]
3. Chen, X.Q.; Fan, S.J.; Han, C.; Wu, T.; Wang, L.J.; Jian, W.; Dai, W.; Yang, J.P. Multiscale architectures boosting thermoelectric performance of copper sulfide compound. *Rare Met.* **2021**, *40*, 2017–2025. [CrossRef] [PubMed]
4. Wever, F.; Moller, H. The crystal structure of iron silicide, FeSi. *Z. Kristallogr.* **1930**, *75*, 362–365.
5. Pauling, L.; Soldate, A.M. The nature of the bonds in the iron silicide, FeSi, and related crystals. *Acta Cryst.* **1948**, *1*, 212–216. [CrossRef]
6. Sidorenko, F.A.; Gel'd, P.V.; Shumilov, M.A. Investigation of α-leboite transition. *Fiz. Met. Metalloved.* **1960**, *6*, 861.
7. Sakata, T.; Sakai, Y.; Yoshino, H.; Fujii, H.; Nishida, I. Studies on the formation of $FeSi_2$ from the $FeSi$-Fe_2Si_5 eutectic. *J. Less Common Met.* **1978**, *61*, 301–308. [CrossRef]
8. Dusausoy, Y.; Protas, J.; Wandji, R.; Roques, B. Structure cristalline du disiliciure de fer, $FeSi_2 β$. *Acta Cryst.* **1971**, *B27*, 1209–1218. [CrossRef]
9. Piton, J.P.; Fay, M.F. Phase changes in iron–silicon alloys around the composition $FeSi_2$. *C. R. Acad. Sci.* **1968**, *C266*, 514–516.
10. Clark, S.J.; Al-Allak, H.M.; Brand, S.; Abram, R.A. Structure and electronic properties of $FeSi_2$. *Phys. Rev. B* **1998**, *58*, 10389–10393. [CrossRef]
11. Watanabe, T.; Hasaka, M.; Morimura, T.; Nakashima, H. Thermoelectric properties of the Co-doped β-$FeSi_2$ mixed with Ag. *J. Alloys Compd.* **2006**, *417*, 241–244. [CrossRef]
12. Arushanov, E.; Lisunov, K.G. Transport properties of β-$FeSi_2$. *Jpn. J. Appl. Phys.* **2015**, *54*, 07JA02. [CrossRef]
13. Nozariasbmarz, A.; Agarwal, A.; Coutant, Z.A.; Hall, M.J.; Liu, J.; Liu, R.; Malhotra, A.; Norouzzadeh, P.; Öztürk, M.C.; Ramesh, V.P.; et al. Thermoelectric silicides: A review. *Jpn. J. Appl. Phys.* **2017**, *56*, 05DA04. [CrossRef]
14. Isoda, Y.; Udono, H. Preparation and Thermoelectric Properties of Iron Disilicide. In *Handbook of Thermoelectrics and Its Energy Harvesting, Preparation, and Characterization in Thermoelectrics*; Rowe, D.M., Ed.; CRC Press: Boca Raton, FL, USA; Taylor & Francis Group: Abingdon, UK, 2017; Chapter 18. [CrossRef]
15. Burkov, A.T. Silicide Thermoelectrics: Materials for Energy Harvesting. *Phys. Status Solidi A* **2018**, *215*, 1800105. [CrossRef]
16. Bahk, J.H.; Shakouri, A. Enhancing the thermoelectric figure of merit through the reduction of bipolar thermal conductivity with heterostructure barriers. *Appl. Phys. Lett.* **2014**, *105*, 052106. [CrossRef]
17. Gong, J.J.; Hong, A.J.; Shuai, J.; Li, L.; Yan, Z.B.; Ren, Z.F.; Liu, J.-M. Investigation of the bipolar effect in the thermoelectric material $CaMg2Bi2$ using a first-principles study. *Phys. Chem. Chem. Phys.* **2016**, *18*, 16566–16574. [CrossRef]
18. Chen, Z.W.; Zhang, X.; Ren, J.; Zeng, Z.; Chen, Y.; He, J.; Chen, L.; Pei, Y. Leveraging bipolar effect to enhance transverse thermoelectricity in semimetal Mg_2Pb for cryogenic heat pumping. *Nat. Commun.* **2021**, *12*, 3837. [CrossRef]
19. Du, X.L.; Qiu, P.F.; Chai, J.; Mao, T.; Hu, P.; Yang, J.; Sun, Y.Y.; Shi, X.; Chen, L.D. Doubled Thermoelectric Figure of Merit in p-Type β-$FeSi_2$ via Synergistically Optimizing Electrical and Thermal Transports. *ACS Appl. Mater. Interfaces* **2020**, *12*, 12901–12909. [CrossRef]
20. Sam, S.; Nakatsugawa, H.; Okamoto, Y. Optimization of Co additive amount to improve thermoelectric properties of β-$FeSi_2$. *Jpn. J. Appl. Phys.* **2022**, *61*, 111002. [CrossRef]
21. Ito, M.; Nagai, H.; Oda, E.; Katsuyama, S.; Majima, K. Effects of P doping on the thermoelectric properties of β-$FeSi_2$. *J. Appl. Phys.* **2002**, *91*, 2138. [CrossRef]
22. Tani, J.; Kido, H. Thermoelectric properties of Pt-doped β-$FeSi_2$. *J. Appl. Phys.* **2000**, *88*, 5810. [CrossRef]
23. Ohtaki, M.; Ogura, D.; Eguchi, K.; Arai, H. Thermoelectric Properties of Sintered $FeSi_2$ with Microstructural Modification. *Chem. Lett.* **1993**, *22*, 1067–1070. [CrossRef]
24. Chen, H.Y.; Zhao, X.B.; Stiewe, C.; Platzek, D.; Mueller, E. Microstructures and thermoelectric properties of Co-doped iron disilicides prepared by rapid solidification and hot pressing. *J. Alloys Compd.* **2007**, *433*, 338–344. [CrossRef]
25. Du, X.L.; Hu, P.; Mao, T.; Song, Q.F.; Qiu, P.F.; Shi, X.; Chen, L.D. Ru Alloying Induced Enhanced Thermoelectric Performance in $FeSi_2$-Based Compounds. *ACS Appl. Mater. Interfaces* **2019**, *11*, 32151–32158. [CrossRef] [PubMed]
26. Dabrowski, F.; Ciupinski, L.; Zdunek, J.; Krsuzewski, J.; Zybala, R.; Michalski, A.; Kurzydloski, J. Microstructure and thermoelectric properties of p and n type doped β-$FeSi_2$ fabricated by mechanical alloying and pulse plasma sintering. *Mater. Today-Proc.* **2019**, *8*, 531–539. [CrossRef]
27. Qiu, P.F.; Cheng, J.; Chai, J.; Du, X.L.; Xia, X.G.; Ming, C.; Zhu, C.X.; Yang, J.; Sun, Y.Y.; Xu, F.F.; et al. Exceptionally Heavy Doping Boosts the Performance of Iron Silicide for Refractory Thermoelectrics. *Adv. Energy Mater.* **2022**, *12*, 2200247. [CrossRef]
28. Komabayashi, M.; Hijikata, K.; Ido, S. Effects of Some Additives on Thermoelectric Properties of $FeSi_2$ Thin Films. *Jpn. J. Appl. Phys.* **1991**, *30*, 331–334. [CrossRef]
29. Nagai, H.; Maeda, I.; Katsuyama, S.; Majima, K. The Effects of Co and Ni Doping on the Thermoelectric Properties of Sintered β-$FeSi_2$. *J. Jpn. Soc. Powder Powder Metall.* **1994**, *41*, 560–564. [CrossRef]
30. Tani, J.; Kido, H. Electrical properties of Co-doped and Ni-doped β-$FeSi_2$. *J. Appl. Phys.* **1998**, *84*, 1408. [CrossRef]

31. Dąbrowski, F.; Ciupiński, Ł.; Zdunek, J.; Chromiński, W.; Kruszewski, M.; Zybała, R.; Michalski, A.; Kurzydłowski, K.J. Microstructure and Thermoelectric Properties of Doped FeSi$_2$ with Addition of B4C Nanoparticles. *Arch. Metall. Mater.* **2021**, *66*, 1157–1162. [CrossRef]
32. Qu, X.; Lü, S.; Hu, J.; Meng, Q. Microstructure and thermoelectric properties of β-FeSi$_2$ ceramics fabricated by hot-pressing and spark plasma sintering. *J. Alloys Compd.* **2011**, *509*, 10217–10221. [CrossRef]
33. Nogi, K.; Kita, T. Rapid production of β-FeSi$_2$ by spark-plasma sintering. *J. Mater. Sci.* **2000**, *35*, 5845–5849. [CrossRef]
34. Dezsi, I.; Fetzer, C.; Kiss, M.; Degroote, S.; Vantomme, A. Site location of Co in β-FeSi$_2$. *J. Appl. Phys.* **2005**, *98*, 073523. [CrossRef]
35. Takeuchi, T. New Thermoelectric Materials with Precisely Determined Electronic Structure and Phonon Dispersion. In *Handbook of Thermoelectrics and Its Energy Harvesting, Preparation, and Characterization in Thermoelectrics*; Rowe, D.M., Ed.; CRC Press: Boca Raton, FL, USA; Taylor & Francis Group: Abingdon, UK, 2017; Chapter 7.
36. Kojima, T.; Masumoto, K.; Okamoto, M.A.; Nishida, I. Formation of β-FeSi$_2$ from the sintered eutectic alloy FeSi-Fe$_2$Si$_5$ doped with cobalt. *J. Less Common Met.* **1990**, *159*, 299–305. [CrossRef]
37. Mott, N.F. Conduction in glasses containing transition metal ions. *J. Non-Cryst. Solids* **1968**, *1*, 1–17. [CrossRef]
38. Zhang, Q.; Liao, B.L.; Lan, Y.C.; Lukas, K.; Liu, W.S.; Esfarjani, K.; Opeil, C.; Broido, D.; Chen, G.; Ren, Z.F. High thermoelectric performance by resonant dopant indium in nanostructured SnTe. *Proc. Natl. Acad. Sci. USA* **2013**, *110*, 13261–13266. [CrossRef]
39. Zhao, L.D.; Lo, S.H.; He, J.Q.; Li, H.; Biswas, K.; Androulakis, J.; Wu, C.I.; Hogan, T.P.; Chung, D.Y.; Dravid, V.P.; et al. High Performance Thermoelectrics from Earth-Abundant Materials: Enhanced Figure of Merit in PbS by Second Phase Nanostructures. *ACS J. Am. Chem. Soc.* **2011**, *133*, 20476–20487. [CrossRef]

Disclaimer/Publisher's Note: The statements, opinions and data contained in all publications are solely those of the individual author(s) and contributor(s) and not of MDPI and/or the editor(s). MDPI and/or the editor(s) disclaim responsibility for any injury to people or property resulting from any ideas, methods, instructions or products referred to in the content.

Article

An Electrical Contacts Study for Tetrahedrite-Based Thermoelectric Generators

Rodrigo Coelho [1], Yassine De Abreu [2], Francisco Carvalho [3], Elsa Branco Lopes [1] and António Pereira Gonçalves [1,*]

[1] C2TN, DECN, Instituto Superior Técnico, Universidade de Lisboa, Campus Tecnológico e Nuclear, 2695-066 Bobadela, Portugal
[2] CESI, Campus D'enseignement Supérieur et de Formation Professionnelle, 15C Av. Albert Einstein, 69100 Villeurbanne, France
[3] DEEC, Instituto Superior Técnico, Universidade de Lisboa, 1049-001 Lisboa, Portugal
* Correspondence: apg@ctn.tecnico.ulisboa.pt; Tel.: +351-219946182

Abstract: High electrical and thermal contact resistances can ruin a thermoelectric device's performance, and thus, the use of effective diffusion barriers and optimization of joining methods are crucial to implement them. In this work, the use of carbon as a $Cu_{11}Mn_1Sb_4S_{13}$ tetrahedrite diffusion barrier, and the effectiveness of different fixation techniques for the preparation of tetrahedrite/copper electrical contacts were investigated. Contacts were prepared using as jointing materials Ni and Ag conductive paints and resins, and a Zn-5wt% Al solder. Manual, cold- and hot-pressing fixation techniques were explored. The contact resistance was measured using a custom-made system based on the three points pulsed-current method. The legs interfaces (Cu/graphite/tetrahedrite) were investigated by optical and scanning electron microscopies, complemented with energy-dispersive X-ray spectroscopy, and X-ray diffraction. No interfacial phases were formed between the graphite and the tetrahedrite or Cu, pointing to graphite as a good diffusion barrier. Ag water-based paint was the best jointing material, but the use of hot pressing without jointing materials proves to be the most reliable technique, presenting the lowest contact resistance values. Computer simulations using the *COMSOL* software were performed to complement this study, indicating that high contact resistances strongly reduce the power output of thermoelectric devices.

Keywords: electrical contacts; tetrahedrite; diffusion barrier; contact resistances; computer simulations

Citation: Coelho, R.; De Abreu, Y.; Carvalho, F.; Branco Lopes, E.; Gonçalves, A.P. An Electrical Contacts Study for Tetrahedrite-Based Thermoelectric Generators. *Materials* **2022**, *15*, 6698. https://doi.org/10.3390/ma15196698

Academic Editor: Bao-Tian Wang

Received: 30 August 2022
Accepted: 20 September 2022
Published: 27 September 2022

Publisher's Note: MDPI stays neutral with regard to jurisdictional claims in published maps and institutional affiliations.

Copyright: © 2022 by the authors. Licensee MDPI, Basel, Switzerland. This article is an open access article distributed under the terms and conditions of the Creative Commons Attribution (CC BY) license (https://creativecommons.org/licenses/by/4.0/).

1. Introduction

Climate change and global warming have pushed mankind to novel attitudes towards energy production, with it being important that industries and cities to transit from fossil energy sources to renewable ones. Therefore, greener, further efficient, and smarter energetic systems have become more popular and widespread every year, boosting the need to search for new devices and materials. In this context, thermoelectric (TE) materials are quite attractive, since they can directly convert waste heat into usable electricity through the Seebeck effect [1,2]. Thermoelectric generators (TEGs) are eco-friendly devices based on TE materials that do not emit greenhouse gases and have no moving parts, allowing them to work for a long time with little or practically no maintenance. They are typically made from arrays of *n*- and *p*-type semiconductors (*n*- and *p*-type legs) that are connected electrically in series and thermally in parallel using electrodes (usually made of copper) to form the electrical circuits [3,4]. On the top and bottom of the connected legs and electrodes, there is usually a coverage, normally made by alumina or polymers, to electrically insulate them [4]. TEGs can have several geometries and sizes depending on the required applications, being devices with a high modularity and quite easy to install. These devices can be used in many industries for waste heat recovery, good examples with great potential being the cement, steel, ceramic, and glass ones [5–7]. However, TEGs can also be used for other applications, such as in biology, remote monitoring stations, sensors, personal devices (e.g., fitness bands), or electronics for the Internet of Things (IOT) [8–10]. Another important field of application is aerospace, especially for energy generation, either from radioisotope generators (to power exploration rovers on remote locations and for outer solar system missions) or from concentrated solar light (to power satellites on space [11,12]). Regardless of all of the mentioned applications, TEGs are not yet implemented on a large scale, mainly due to the cost of the technology (which uses expensive and rare elements), their low conversion efficiencies comparatively to other systems, such as the organic Rankine Cycle [13–16], and the toxicity of the constituents. Indeed, the most commercially used TE materials are based on bismuth telluride (Bi_2Te_3), silicon germanium (SiGe), and lead telluride (PbTe) [4,14], which are not attractive for widespread or large-scale applications [16]. In this context, and given the present TEGs market needs, new, alternative, and cheaper materials are being explored and studied. Among these, tetrahedrites, which belong to the copper antimony sulfosalts family, are seen as having good potential for TE applications. They are abundant on the Earth's crust, present low toxicity, and are highly available (even if synthetized), which makes them much cheaper (~7 USD/Kg) and more ecological when compared to the commercial ones [14,17].

The performance of a thermoelectric material can be evaluated through the calculation of its figure of merit (zT). This dimensionless parameter is given by $zT = (S^2 \bullet \sigma \bullet T)/\kappa$, where S is the Seebeck coefficient, κ and σ are the electrical and thermal conductivities and T is the absolute temperature [18]. Materials with zTs close to 1, though providing TEGs with low efficiencies, are already considered worthy for many TE applications. Tetrahedrites are p-type semiconductors that crystallize in a cubic unit cell (space group $I\bar{4}3m$) and can have several chemical compositions that give origin to different TE performances, with it being possible to achieve zTs close to unit at temperatures of 623 K [17,19–22]. However, to build a tetrahedrite-based TEG, it is just not enough to have TE materials with good performance, it is also fundamental that the materials do not deteriorate or react in the device at the working conditions. Consequently, it is important to study their stability and how to properly connect them to the TEG electrodes, especially because high electrical and thermal resistivity can considerably reduce the device's performance.

High resistivity can arise from reactions between the TE materials (legs) and the electrodes that connect them, which give rise to interfacial phases with different electrical and thermal properties. At the same time, the mentioned interfacial phases can have distinct coefficients of thermal expansion (CTE) that can damage the devices by detaching or breaking the legs, with most of the TE materials needing diffusion and/or buffer barriers to be exposed to working temperatures without being damaged. The majority of the diffusion barriers used in conventional TEGs are based in very thin metallic layers, such as Ni, Ag, Ti88-Al12 or Fe, which are specially selected due to their high electrical and thermal conductivities and low reactivity with their respective TE legs and electrodes [23–26]. Yet, those metals and alloys are not suitable to be used with tetrahedrites, since they easily react and form phases with the elements present in the matrix, such as S and Sb [2,27]. Therefore, our group decided to investigate carbon and gold as a diffusion barrier for tetrahedrites, with the preliminary results being presented in an international conference [28]. Taking into account the referred work and considering that flexible graphite is a material with high resistance against oxidation and thermal shock, good mechanical and thermal stability, and good electrical and thermal conductivity [29], it was selected to be tested as a diffusion barrier in this study.

Nevertheless, to setup a tetrahedrite-based TEG, it is not enough just to select the correct diffusion barriers, it is also necessary to know how to properly connect them to the legs and respective electrodes. Depending on the jointing approach, several techniques, or additional materials, such as solders or paints, can be required. In most of the commercial devices, the copper electrodes are fixed to the Bi_2Te_3 legs just by brazing or soldering [3,30]. Other efficient methods include the use of hot pressing, spark plasma sintering or the preparation of the contacts using thermal spray [3,31,32]. For a certain composition of the

TE legs, the correct jointing materials and methodologies must be developed in order to obtain the lowest contact resistances and the highest leg quality. Since most of the contact fabrication methodologies are unpublished or patented [33], it is not clear how the electrical contacts are prepared in the majority of commercial devices. Simultaneously, there are not many reported studies concerning the characterization of electrical contact resistances on TEGs, and manufacturers do not give information about them in most of the commercial devices (the TEGs datasheets).

One of the first studies devoted to the use of diffusion barriers and the measurement of contact resistances in TE materials was the work of O. J. Mengali and M. R. Seiler [34]. In this study, the contact resistances between Bi_2Te_3, Ag_2Te, and Ag_2Se (metalized with Ni and Sn layers) and copper blocks (soldered to the TE materials) were measured. A homemade contact measuring system, consisting of a potential probe apparatus with an alternative current (AC) potential and a moving contact probe made of tungsten, was used in this study. In the measurements, it was observed that some data deviated from the linearity due to localized variations of the resistivity and irregularities on the cross-sectional area of the materials. Among several hypotheses, the authors pointed the possibility of compound dissociation and oxidation (during the interlayers depositions) as the main causes for the observed deviations. Despite the reported issues, the measured contact resistances for Bi_2Te_3 were low, between 0.0074 mΩ.mm^2 and 3.7 mΩ.mm^2. At the same time, the authors noticed that the different techniques used for the leg's metallization and the overall state of the samples affected the resistance jumps and the quality of the contacts. They also observed that the contact resistance of the other TE materials (Ag_2Te and Ag_2Se) displayed similar values when metalized with Ni or Sn layers and fixed to copper blocks using the same conditions.

To measure the contact resistances in TE materials, Y. Kim et al. [35] also developed a custom-made apparatus based on the AC pulsed current method. The materials measured were SnSe and Bi_2Te_3 legs, both fixed to copper electrodes by hot pressing. On the measurements of the Cu/SnSe/Cu legs, resistance jumps between 379 $\mu\Omega$ and 15 mΩ were observed, the specific contact resistance of the legs was not mentioned in the publication. On the Bi_2Te_3 legs, contact resistances of 0.7 mΩ.mm^2 were measured, with these values lying in the ranges reported in the O. J. Mengali and M. R. Seiler work [34]. To check if the measurements were correct and to discard errors caused by sample heating during the current injection, the authors conducted some tests using a direct current (DC) technique. They used currents up to 500 mA and checked if the samples heated up due to the Peltier effect created by parasitic voltages. They observed that the legs heated up very easily, at least 2.3 °C degrees. Taking this into account, and to minimize the Peltier effects, the authors installed heat-dissipating blocks on their measuring system and decided to adopt the AC technique as the main measuring methodology.

Another study devoted to diffusion barriers and contact resistances of TE materials was conducted by Yohann Thimont and his team [36]. On their experimental apparatus, a tungsten scanning probe was used. To keep the samples in place during the measurements, two contact springs with a constant force of 4.528 N/mm were attached to copper blocks. The measured materials were TE legs made of magnesium silicide and silicon-germanium prepared with Ni diffusion barriers. The authors noted that the Ni barriers were deposited by a metallization process without specifying the technique or the conditions used. Nevertheless, a contact resistance of 0.45 mΩ.mm^2 between the Ni layer and the $Mg_2Si_{0.98}Bi_{0.02}$ leg was observed, while on the $MnSi_{1.75}Ge_{0.02}$ leg, a contact resistance of 4.1 mΩ.mm^2 was measured. The authors also observed that the contact resistance increased with the interface layers' thickness and with the time used in the metallization processes. During these experiments, the contact resistance between the Ni layers and the Cu blocks (used to fix the legs) was also measured, with values of 12 mΩ.mm^2 being obtained when pressures up to 0.25 MPa were applied. Higher contact resistances (around 30–40 mΩ.mm^2) were noted when low-to-no pressures were used. Therefore, the authors concluded that the use of high pressures improved the electrical contacts quality, but only to some extent. After a specific point, increasing the pressure did not significantly reduce the contact resistance. This phenomenon was explained as a consequence of the increase in the number of contacts between the TE materials and the electrodes (when pressure is applied/increased). Up to a point and while introducing more pressure, the number of contacts does not increase anymore, and the surface area starts to rise due to plastic deformation.

While some works exist on the investigation of diffusion barriers and electrical contacts of TE legs based on old and novel materials, there are almost no published studies conducted on copper sulfosalts. One of the few studies, performed on colusites, was conducted by Chetty et al. [37]. In the referred work, the authors measured the contact resistance of a $Cu_{26}Nb_2Ge_6S_{32}$ TE leg fixed to copper by hot pressing, obtaining values of 0.5 mΩ.mm^2. It is important to notice that gold layers were used as diffusion barriers between the colusite and the copper, with the objective of preventing the formation of interfacial phases during the leg's exposition to high temperatures and give rise to a good electrical contact. The authors reported the measured values as being in the range of other TE materials in development, such as skutterudites, half-Heusler, and MgAgSb-based compounds. However, and despite the low values obtained, a significant solubility of Nb in Au was detected, and no details/information were given about the type of setup or conditions used to measure the contact resistances. Since most of the jointing procedures for commercial TE materials are patented, and new approaches may be required to produce devices based on new materials, it is quite important to understand the techniques to manufacture good electrical contacts, especially on emerging TE materials such as the tetrahedrites or other materials from the copper sulfosalts family.

In this work, investigations on the use of carbon as tetrahedrites diffusion barrier, together with the exploration of different fixation techniques for the preparation of good electrical contacts between the tetrahedrite and copper, are presented. The objective is to evaluate carbon as a suitable diffusion barrier and find the best jointing materials and fabrication techniques necessary to build a tetrahedrite-based device. Since the development of tetrahedrite-based TEGs is still in its early stages and high electrical and thermal contact resistances can be critical for the operation of such devices, it is crucial to identify the most suitable materials and fabrication methods to produce commercially competitive TEGs. Computer simulations, to understand how the measured contact resistances affect the performance of a tetrahedrite-based thermocouple, were also performed, and the results confirm the importance of producing good electrical contacts.

2. Materials and Methods

Manganese doped tetrahedrites with $Cu_{11}Mn_1Sb_4S_{13}$ composition [27] were synthesized by solid state reaction from pure elements, Cu 99.9999%, Sb 99.9999%, Mn 99.9%, and S 99.5%, all from Alfa Aesar, Haverhill, MA, USA. The mixtures (~2 g/batch) were vacuum sealed (10^{-3} Pa) on quartz ampoules and melted at 1191 K on vertical furnaces. After the melting process, the materials where ground to powders, cold pressed at 512 MPa, sealed under vacuum, and annealed at 713 K for 5 days. The thermal treated materials were then manually crushed into fine powders and sintered by hot pressing. High density carbon dies with 10 mm internal diameter holes were used in the sintering procedure. In the majority of the samples, flexible graphite disks (thickness 0.5 mm, 99.8% purity, Sigma-Aldrich, St. Louis, MO, USA), with 10 mm diameter, were also inserted below and above the tetrahedrite powders to act as diffusion barriers. However, in one of them (sample A), the copper electrode was directly hot pressed with the powders and used to check if interfacial phases were formed. The densification was made by hot pressing at 848 K and applying a pressure of 60 MPa for 90 min. Pellets with ~10 mm diameter and ~3.5 mm thickness, with a relative density $\geq$ 88% (see Figure S1 and Table S1 in Supplementary Data File) and containing thin graphite disks in both top and bottom surfaces, were obtained. The pellets were cut into square prisms with ~7 $\times$ 7 $\times$ 3.5 mm^3 dimensions and linked to copper contacts using different jointing materials and procedures.

Nickel (CW2000, Chemtronics®, Cobb Center Drive, Kennesaw, GA, USA) and silver water-based (EM-Tec Ag46, Labtech International Ltd., East Sussex, UK) electrically conductive paints, silver (Pleco® 16047, Ted Pella Inc., Redding, CA, USA), and nickel (Pleco® 16059-10, Ted Pella Inc., Redding, CA, USA) water-based resins and the Zn-5wt% Al solder were used as jointing materials. Together with them, different fixation procedures were applied to the legs with the graphite layers, such as cold pressing (CP), hot pressing (HP), and a manual method (where the pressure was manually applied, ~2.5 MPa). A fixation method consisting of the direct hot press of copper into the legs, without the use of paints or solders (called Root HP) was also explored as an alternative way. The objective of using different jointing materials and procedures is to evaluate the most suitable combination for contact fixation/fabrication of the tetrahedrite legs. The effect of using different fixation pressures (in the contacts preparation) was also studied, whereby the pressures were increased up to the samples' breaking point. We must keep in mind that tetrahedrites operate at medium temperatures (between 293 and 623 K), which implies that common solders (used in commercial devices) cannot be applied. Moreover, all the fixation materials (paints, solders, etc.) must support continuous work at medium temperatures (up to 623 K). A summary of the jointing materials, techniques, and conditions used to prepare the TE legs studied in this work are presented in Table 1.

Table 1. Summary of the jointing materials, conditions and techniques used in the preparation of the TE legs studied in this work.

Sample	Jointing Material	Fixation Technique	Conditions
A	No paints or solders No graphite layer	HP	56 MPa, 1 h 30 min at 848 K
M1	Ni conductive paint	Manual	~2.5 MPa, ~5 min
M2	Water-based Ag Paint	Manual	~2.5 MPa, ~5 min
M3	Ni Resin	Manual	~2.5 MPa, ~5 min
M4	Ag Resin	Manual	~2.5 MPa, ~5 min
CP1	Ni conductive paint	CP	41 MPa, 6 h
CP2	Water-based Ag Paint	CP	16 MPa, 6 h
CP3	Ni Resin	CP	28 MPa, 4 h
CP4	Ag Resin	CP	32 MPa, 4 h
HP1	Ni conductive paint	HP	22 MPa, 1 h at 493 K
HP2	Water-based Ag Paint	HP	37 MPa, 1 h at 493 K
HP3	Water-based Ag Paint	HP	23 MPa, 1 h at 493 K
HP4	Water-based Ag Paint	HP	22 MPa, 1 h at 493 K
HP5	Ag Resin	HP	20 MPa, 2 h at 493 K
HP6	Ni Resin	HP	20 MPa, 2 h at 403 K
HP7	Ni Resin	HP	15 MPa, 2 h at 403 K
HP8	Zn-5wt% Al solder	HP	22 MPa, 25 min at 732 K
HP9	No paints or solders	HP	56 MPa, 1 h 30 min at 848 K
HP10	No paints or solders	HP	56 MPa, 1 h 30 min at 848 K

CP = Cold Pressing; HP = Hot Pressing.

To check the interface and porosity of the prepared samples, optical and scanning electron microscopy (SEM) observations, complemented with Energy-dispersive X-ray spectroscopy (EDS) analysis, were performed. For all the observations, the surface of the samples was dry polished using SiC sandpaper (P2500 Grit). Optical micrographs were acquired by a digital microscope (HIGH CLOUD, 500X-1500X, Beijing, China), while for SEM, two electron microscopes were used: (a) one JEOL JSM7001F SEM equipped with a field emission gun and an Oxford Instruments EDS system (both from Tokyo, Japan), and (b) one Phenom ProX Desktop SEM equipped with an EDS system (both from Waltham, Massachusetts, USA). The EDS analysis were performed with an accelerating voltage of 20 kV. The porosity of the materials was evaluated by using the *ImageJ* software version 1.5a to analyze the SEM micrographs taken at four distinct zones of the tetrahedrite pellets.

X-ray diffraction measurements were carried in a Bruker D2 PHASEER diffractometer (from Billerica, MA, USA) using a Bragg–Brentano geometry and Cu Kα radiation source (wavelength of 1.54060 nm). The current and tension were set to 30 kV and 10 mA, respectively. The tetrahedrite green pellets and their respective graphite layers were ground to powders and analyzed in low-noise Si single-crystal sample holders. All data was collected from 10° to 65° with a step of 0.02 and acquisition time of 0.85 s per step. To check the possibility of reaction between the copper contacts and tetrahedrite, the surface of the Sample A was dry polished using sandpaper (grits P600 and P1200), placed on a custom-made sample holder and scanned from 15° to 90°, with a step of 0.02° and an acquisition time of 8 s per step.

The acquired diffractograms were treated using *OriginPro* software version 9.0 and phase identification performed by comparison of the observed data with cards from the Crystallographic Open Database (COD) using *DIFRAC.EVA* software (version 5.1). The electrical contact resistance was measured using a custom-made scanning probe system, using a method previously described [28,38]. All samples were dry polished using SiC sandpapers with P600, P1000 and P1200 grits and cleaned with ethanol (95%). After cleaning, the legs were glued to microscope glass slides and mounted on the system. The measurements start by positioning the scanning tip on the top of the thermoelectric legs in a defined/initial zone (D = 0 μm), followed by the injection of 1 mA positive and negative electrical pulses of 1 ms duration. After each pulse, the difference in voltage and resistance were calculated using the equations:

$$\Delta V = (V_1 - V_2)/2 \tag{1}$$

$$R = \Delta V / I \tag{2}$$

where ΔV is the voltage variation, V_1 and V_2 are the voltage readings (in the two directions), R is the resistance, and I is the pulsed current.

The entire process was repeated with steps of 100 μm across the sample surface. The objective was to scan the TE legs to obtain plots of resistance versus distance that allow the identification of the resistance jumps between the copper contacts and the tetrahedrite bulk material. After the identification of such differences, the specific contact resistance was calculated by multiplying the jumps (in mΩ) by the specific contact area of the legs (in mm^2). The area of the TE legs was measured with the help of a common ruler (1 mm scale) and the error associated with the experimental set up was taken from the standard deviation of the fitted data "ROOT-MSE (SD)" tool [39,40], using the *OriginPro* software version 9.0. In each leg, the measurements were made along the two directions (first from left to right, L, and after from right to left, R), and often performed in more than one zone. A scheme of the measurement system is presented in Figure 1.

Computer simulations were made with the *COMSOL Multiphysics* software v5.5 to study the devices performance expected for the measured contact resistances. The simulations were based on the Finite Element Analysis (FEA) theory, where several equations are applied to specific points of a defined mesh. At the same time, several boundary conditions were applied to an optimized 3D model [38], consisting of a thermoelectric pair made by a tetrahedrite leg (*p*-type element) and a magnesium silicide material (*n*-type element). Copper electrodes with 1 mm thickness connect the two legs; they were covered with 2 mm thick alumina plates for electrical insulation. The tetrahedrite elements have a square shape area of 7×7 mm^2, while the Mg$_2$Si-based legs have an area of 4×4 mm^2, with both legs having a height of 3 mm. The space between the *n* and *p* elements was set to 1 mm, the alumina plates were rectangular with 13×8 mm^2. To perform the simulations, several materials properties, such as the electrical conductivity, thermal conductivity, Seebeck coefficient, and others, are added to the 3D CAD model presented below (Figure 2). All the material properties used were retrieved from the literature and from previous studies, including *COMSOL* materials database [22,38,41].

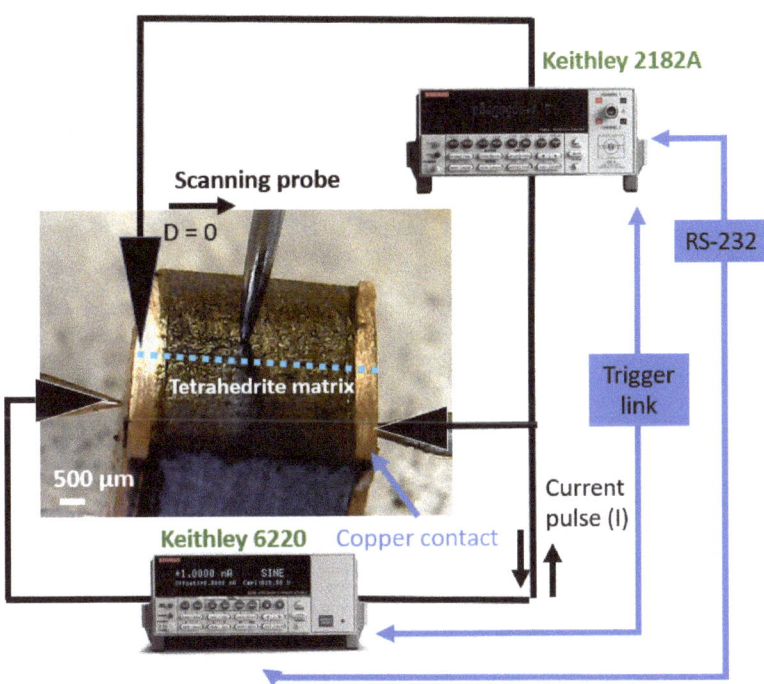

Figure 1. Scheme of the measuring system, with the TE leg mounted.

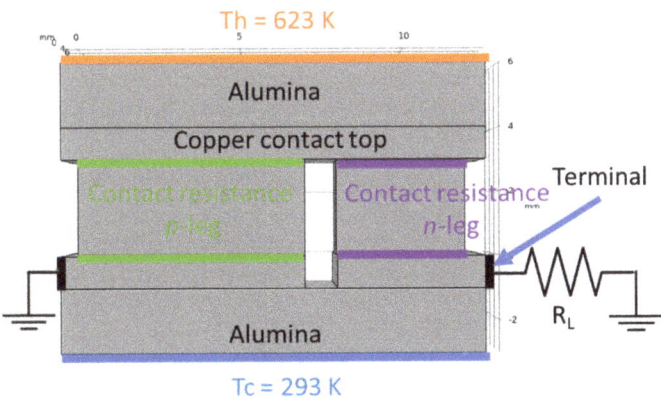

Figure 2. Electrical and thermal boundary conditions of the 3D model simulations. T_h and T_c represent the hot and cold side temperatures, respectively, and R_L is the load resistance.

For the simulations, the "thermoelectric effect" physics was selected and applied to the 3D model. The differential equations used on the model were based on Fourier's law and can be written as [42,43]:

$$\rho\, C_p\, u.\nabla T + \nabla.q = Q + Q_{ted} \quad (3)$$

$$q = -k\, \nabla T \quad (4)$$

where ρ is the density, C_p is the heat capacity at constant pressure, q is the heat flux by conduction, k is the thermal conductivity, Q_{ted} is the thermoelastic damping, Q is an additional heat source, and u is a velocity field vector (only used when parts of the model are moving on the materials frame). A hot surface on the top of the model, with the temperature of 623 K, and a convective heat flux on the bottom were defined as boundary conditions for the simulations. The convective heat flux, q_0, was determined according to the equation:

$$q_0 = h \cdot (T_{ext} - T) \tag{5}$$

where h is the heat transfer coefficient, defined as 2000 [38], T_{ext} is an external temperature (defined by the user), and T is the reference temperature. For the TE simulations of the pair, the temperature of the hot side was set as 623 K, while the cold side (defined as T_{ext}) was set to 293 K, with all the geometry components being thermally insulated.

To account for the contact resistance of the legs, two nodes named "contact impedance" were added to the simulation. In these nodes, it is possible to manually define a surface resistance for the p and n legs that can take into account the experimental values. For the calculation of the contact resistance nodes, the following equations were used:

$$n \cdot J_1 = 1/\rho_1 \, (V_1 - V_2) \tag{6}$$

$$n \cdot J_2 = 1/\rho_2 \, (V_1 - V_2) \tag{7}$$

where ρ is the surface resistance, V is the voltage, J is the current density, and n is a surface normal, with the numbers 1 and 2 referring to the two sides of each boundary (top and bottom) of the contact interface. A summary of all the boundary conditions used in the computer simulations can be observed in the model presented on Figure 2.

With the thermal and electrical boundary conditions defined and the proper regions selected, a normal mesh was built. On this mesh, all the described equations (from 3–7) were applied and solved. To obtain the typical current-voltage (IV) and current-power (IP) curves of a TE device, it is necessary to adjust R_L. In this study, R_L was changed by performing a parametric study with the model being in a stationary state. Four simulations were performed, one using contact resistances equivalent to the ones found on commercial devices and the other three using contact resistances (for the p leg) in the range of 50–700 mΩ.mm².

3. Results

The observation of sample A (Figure 3a,b), where copper was directly hot pressed to the tetrahedrite powders, without any graphite layer in between, shows no traces of the copper disk. X-ray diffraction and EDS analysis (see Supplementary Materials File, Figures S2 and S4) suggest the total reaction of copper with tetrahedrite and the formation of Cu_2S, Cu_3SbS_3 and Sb.

In contrast with these results, the samples covered with a graphite layer present no interfacial phases and a continuous coating, with good adhesion to the tetrahedrite (Figure 3c,d). Similarly, this is usually also the case for the samples with the copper electrodes connected to the graphite layers with the help of a jointing material, such as the nickel resin or the nickel paint, where no additional phases are observable, but some solubility between Ni and Cu is seen (Figures 4 and S5 Supplementary Materials File). The micrographs of the $Cu_{11}Mn_1Sb_4S_{13}$ tetrahedrite sample hot pressed using a Zn-5Al wt% solder as jointing material are presented on Figure 5. An interlayer in between the Cu disks and the graphite diffusion barrier (Figure 5b) can be observed. This interlayer was ascribed to the Zn-Al solder, since no reaction between the solder and the Cu plate was observed on the SEM-EDS analysis, performed after HP (Figure S6 and Table S5, Supplementary Materials File).

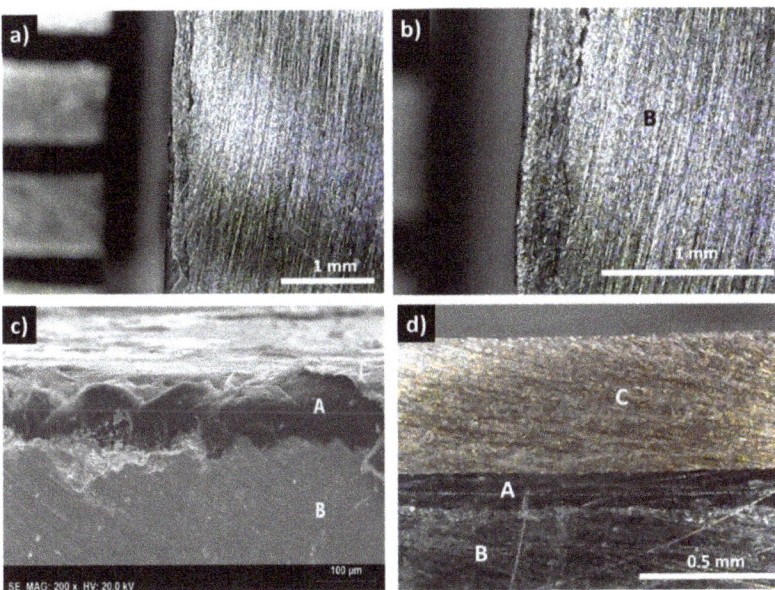

Figure 3. Optical and SEM observations of hot-pressed (HP) samples: (**a,b**) sample A, HP using $Cu_{11}Mn_1Sb_4S_{13}$ powder directly in contact with a copper plate; (**c**) $Cu_{11}Mn_1Sb_4S_{13}$ powder directly HP with a graphite layer; (**d**) sample HP with a graphite diffusion barrier (A) between the $Cu_{11}Mn_1Sb_4S_{13}$ powder (B) and the Cu contact (C).

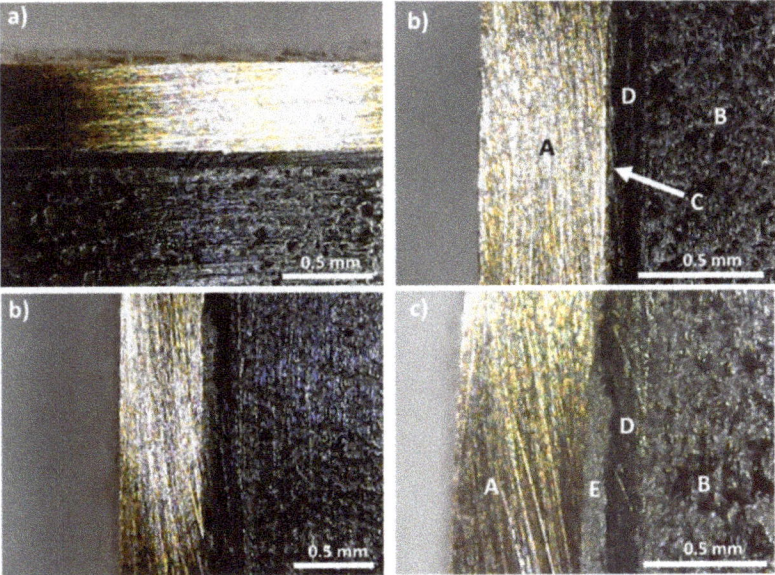

Figure 4. Examples of optical micrographs of $Cu_{11}Mn_1Sb_4S_{13}$ samples with copper electrodes connected to the graphite layers with the help of (**a,b**) nickel paint and (**c,d**) nickel resin jointing material; (A) copper contact; (B) tetrahedrite; (C) nickel paint; (D) graphite layer; (E) nickel resin.

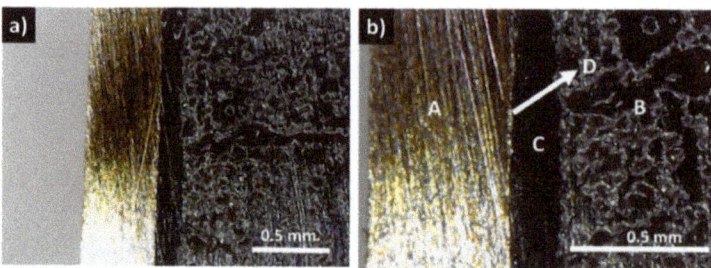

Figure 5. Optical micrographs of $Cu_{11}Mn_1Sb_4S_{13}$ sample (B), with a copper electrode (A), HP at 22 MPa (**a,b**) to the graphite layers (C), using a Zn-5Al wt% solder (D) as jointing material.

However, in the (CP and HP) pressed materials, visible cracks are often observed inside the tetrahedrite phase, with a good example being presented in Figure 6a,b, even for applied pressures as low as 22 MPa (the list of samples that present visible cracks can be consulted in the Supplementary Materials File, Table S2).

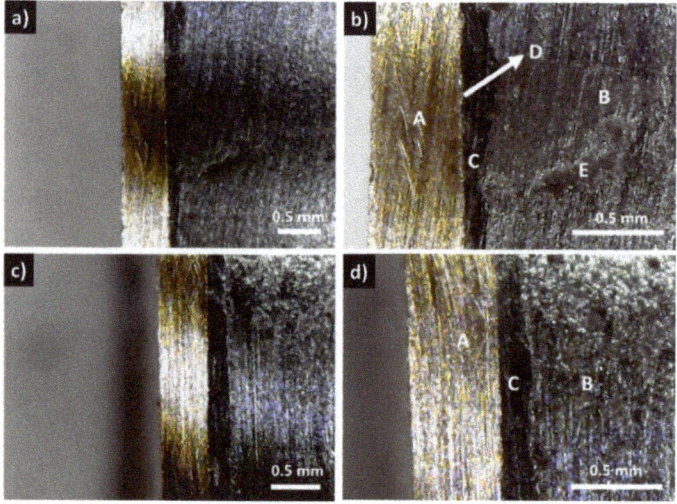

Figure 6. Optical micrographs of $Cu_{11}Mn_1Sb_4S_{13}$ samples (B) with copper electrodes (A) hot pressed at 22 MPa (**a,b**), and at 20 MPa (**c,d**), to the graphite layers (C), using silver conductive paint (D) and silver resin as jointing materials, where a crack is clearly seen (E).

Figure 7 presents typical curves resulting from the contact resistance measurements, with Figure 7a showing the scans for the TE legs prepared with Cu contacts and fixed using Ag water-based paint at different pressures, while in Figure 7b, the effects of using Ni resin and different preparation techniques and pressures are presented. In all graphs, the y-axis is defined by the resistance of each measured point versus the specific contact area of each sample (RA), allowing a direct comparison between the contact resistance jumps in all of the presented curves. While analyzing the plots, it is noticeable that RA increases with the distance for almost every scanned leg. However, jumps between the tetrahedrite and the copper electrodes, which are associated with the contact resistances, are observed. On TE legs presenting good electrical contacts, the RA values taken at the tetrahedrite material increases almost linearly with the distance, while the points taken on the Cu contacts follow a nearly flat tendency, due to the very low electrical resistivity of copper. It can also be seen that the measurements work as an indirect technique to evaluate the quality of the assembled TE legs: legs with low contact resistances present low jumps and an almost linear RA increase, while legs with low contact quality present high jumps and a flat linear tendency in the tetrahedrite phases.

The effect of applying different pressures on the preparation of the contacts with Ag water-based paint and Ni resin are presented in Figure 7a,b, respectively. Increasing the compressive forces during contacts preparation generally reduces the resistances. However, is not possible to indefinitely increase them, as the probability of inducing cracks substantially rises at high pressures. In fact, it was noticed that for almost every TE leg where the contacts were prepared by HP and CP (made on already sintered legs) cracks frequently appeared, even when low compressive forces were used. However, no evident correlation can be found between crack formation and the HP process, indicating that most of the visible cracks should be formed due to high pressures used for joining (>22 MPa) and not due to thermal expansion of the materials during the HP process.

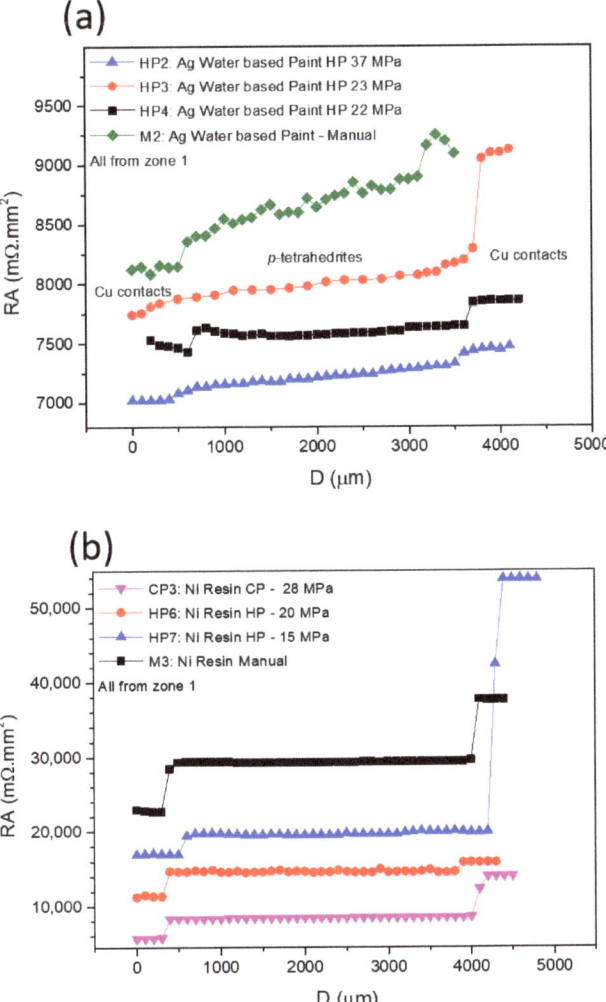

Figure 7. Contact resistance results for: (**a**) those fixed using Ag Paint and different pressures; and (**b**) those fixed using Ni Resin and using different techniques and pressures.

In Figure 7b (Ni resin), higher contact resistance jumps can be observed for the manual and CP preparation methods, while the HP tends to give smaller jumps (even when lower pressures are used). However, for the sample hot pressed at 15 MPa, the resistance jump on the right (corresponding to the measurement of the second contact) is much higher than on the left (first contact measurement). This behavior is also observed on pressed samples prepared using other jointing materials and fixation methods (e.g., Figure 7a, Ag water-based paint HP at 23 MPa) and can be ascribed to damage on the contact zones that are made by the scanning tip.

Figure 8 displays examples of measurements made on samples: (a) prepared by HP at 37 MPa using Ag paint (sample HP2), and (b) prepared by "root HP" (sample HP9). Both samples present low contact resistances (101–43 mΩ.mm^2), lower than those obtained by other the jointing materials and fixation methods/conditions.

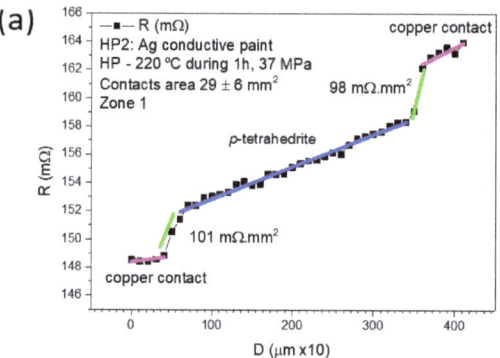

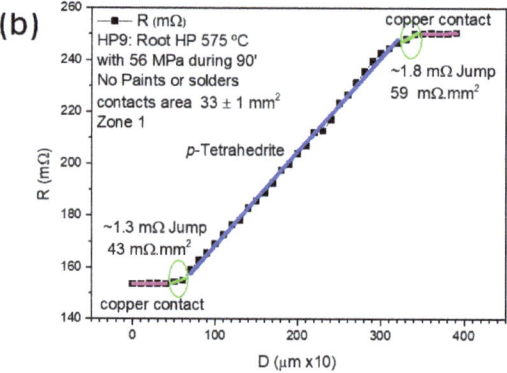

Figure 8. Contact resistance results for: (**a**) HP leg with Ag paint; and (**b**) "root HP" leg with no paints or solders.

A list of the contact resistance results is presented in Table 2. It is possible to see that in many cases, the resistances for each leg are similar (or in the same range) independently of the scanned zones or directions. However, in other cases, the second measurement (from right to left) shows much higher contact resistances and, consequently, were not considered for the analysis. This behavior can be ascribed to the degradation of the contact interface by the measuring tip, as the second measurement is performed through the same path as the first one (see below Figure 9 the example shown for Figure 10d, where the tip path is clearly visible in the tetrahedrite phase). Moreover, in several of the pressed samples, it was not possible to perform the measurements due to high resistances or noise. Contact resistances not considered or not possible to measure were labeled as NP.

Table 2. Contact resistance results.

Sample	Jointing	Area	Contact Resistance (m$\Omega \cdot$mm^2)	
			L	R
M1	Ni paint	56 ± 8 mm^2	1: 414 ± 134 2: 302 ± 136	1: NP 2: NP
M2	Ag paint	64 ± 8 mm^2	1: 211 ± 78 2: 608 ± 127	1: 371 ± 98 2: 307 ± 90
M3	Ni resin	46 ± 8 mm^2	1: 6825 ± 1168 2: 6552 ± 1127	1: 8008 ± 1361 2: 6325 ± 1090
M4	Ag resin	46 ± 8 mm^2	1: 319 ± 66 2: 273 ± 58	1: 182 ± 44 2: 228 ± 51
CP1	Ni paint	24 ± 5 mm^2	1: 300 ± 70 2: 293 ± 75	1: 314 ± 72 2: 353 ± 86
CP2	Ag paint	60 ± 8 mm^2	1: 174 ± 52 2: 420 ± 732 3: 150 ± 65	1: NP 2: NP 3: NP
CP3	Ni resin	35 ± 6 mm^2	1: 2520 ± 450 2: 2485 ± 615	1: 5425 ± 948 2: 5845 ± 1191
CP4	Ag resin	31 ± 6 mm^2	1: 78 ± 23 2: 156 ± 100	1: NP 2: NP
HP1	Ni paint	49 ± 7 mm^2	NP	NP
HP4	Ag paint	35 ± 6 mm^2	1: 180 ± 52 2: 598 ± 210	1: 202 ± 55 2: 1090 ± 293
HP3	Ag paint	38 ± 6 mm^2	1: 138 ± 45 2: 268 ± 89	1: NP 2: NP
HP2	Ag paint	29 ± 6 mm^2	1: 101 ± 25 2: NP	1: 98 ± 25 2: NP
HP8	Zn-5Al wt% solder	49 ± 7 mm^2	1: NP 2: NP	1: 289 ± 55 2: NP
HP6	Ni resin	49 ± 7 mm^2	1: 3332 ± 589 2: 3234 ± 624	1: 1225 ± 288 2: 1274 ± 344
HP7	Ni resin	65 ± 1 mm^2	1: 2535 ± 163 2: 2405 ± 284	1: NP 2: NP
HP5	Ag resin	45 ± 7 mm^2	1: 461 ± 87 2: 45 ± 315	1: 224 ± 51 2: 842 ± 435
HP9	No paint or solder	33 ± 1 mm^2	1: 43 ± 45 2: 124 ± 22 3: 65 ± 28	1: 59 ± 46 2: 46 ± 21 3: 59 ± 27
HP10	No paint or solder	28 ± 1 mm^2	1: 62 ± 72 2: NP	1: 139 ± 75 1: NP

R = right; L = left; NP = not considered/measured.

The jointing material that gave origin to the highest contact resistances (independently of the technique used) was Ni resin. Nevertheless, a tendency to decrease the resistance is seen when we pass from "Manual" to "Cold Pressing" and from "Cold Pressing" to "Hot Pressing" (Figure 7b). This tendency to decrease the contact resistance from "Manual" to "Cold Pressing" is also observed for the other jointing materials, while for "Manual" to "Hot pressing" a larger range of values is observed, with no evident trend. Despite these facts, it is possible to observe in Figure 7a a continuous reduction in the contact resistance with the increase in the pressure applied, especially if we exclude the right contact of the HP3 leg (pressed at 23 MPA). At the same time, while analyzing Table 2, it was also seen that the Ni conductive paint can lose its conductivity when exposed to high temperatures, and thus, it was not possible to perform the measurements when prepared by hot pressing.

The fixation technique that resulted in the highest values of contact resistances was the manual one. The higher reproducibility and lowest contact resistances were observed for the "root HP" legs, with contacts prepared by directly hot pressing the $Cu_{11}Mn_1Sb_4S_{13}$ tetrahedrite powder with the graphite layer and the copper contacts, without using any paint, resin, or solder.

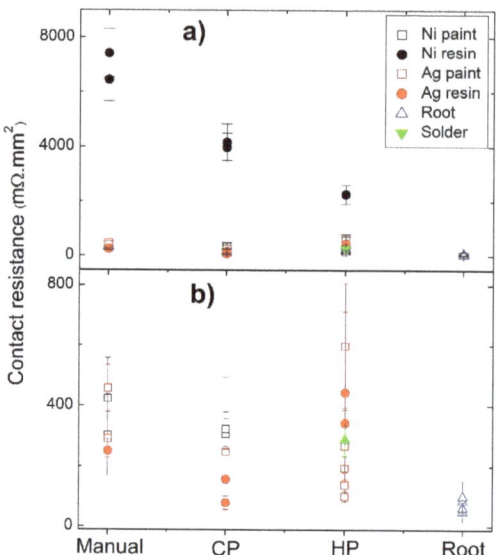

Figure 9. Contact resistance as a function of the jointing material and fixation method: (**a**) all samples; (**b**) samples with resistances lower than 800 mΩ.mm².

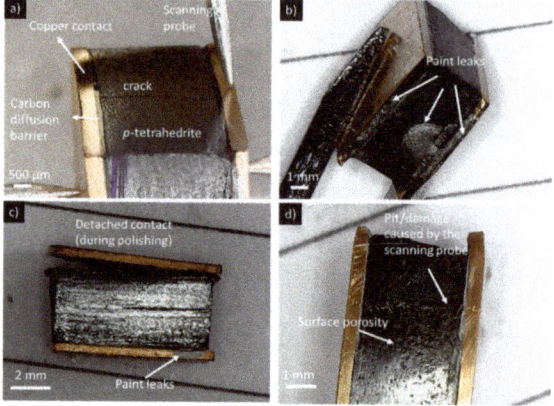

Figure 10. TE legs with (**a**) cracks (sample HP3), (**b**) paint or resin leaking (sample M1), (**c**) contacts detached (sample CP2), and (**d**) damage caused by the scanning probe (sample M2).

4. Discussion

The formation of Cu_2S, Cu_3SbS_3, and Sb, by reacting tetrahedrite with copper (sample A, Figure 3a) is in agreement with the reported Cu-Sb-S ternary phase diagram, where such phases are expected to appear at medium temperatures in the copper-rich region [44]. These assumptions were confirmed by the XRD and SEM-EDS analysis, presented in the Supplementary Materials File, Figures S2 and S4, and Table S3. In contrast, no extra phases in the graphite/tetrahedrite interface were observed (Figure 3c,d), which is also confirmed by the powder XRD analysis of the tetrahedrite green pellet presented in the Supplementary Materials File (Figure S3). This is in accordance with the inexistence of binary compounds between carbon and copper or antimony and with the development of carbon–sulfur phases only at higher temperatures [45]. Therefore, the inexistence of additional phases points to carbon as a good candidate to act as a tetrahedrite diffusion barrier.

In the case of the samples prepared with Ni paint and resin, the absence of other phases (Figure 4) is in agreement with the Ni-Cu binary phase diagram [46], where the formation of alloys between Ni and Cu just takes place at higher temperatures ($> 1100\ °C$) while at lower temperatures, a solid solution between Cu and Ni can exist. The absence of reactions between the Ni and C (Ni just presents a small solubility into graphite [47]) can also explain the inexistence of other compounds at their interface.

Contrary to the Ni resin and paint, for the case of the legs prepared with the Zn-Al 5 wt% solder, the formation of additional phases (between the solder and the Cu plates) is high, as many Cu-Zn intermetallic phases are stable in the medium-temperature range (200–400 °C) [48,49]. However, no reaction between the solder and the Cu plate was observed by SEM-EDS, with a continuous interlayer with good adhesion to copper being noticed. Nevertheless, the possibility of formation of CuZn intermetallic compounds at the copper/graphite interface cannot be discarded, especially if the legs are submitted to consecutive thermal cycles.

When using the Ag paint and resin, no additional phases are observed between the Cu contacts and the graphite diffusion barriers (Figure 6). The absence of additional phases, apart from the paint or resin, can be explained by the large immiscibility gap between Ag and Cu, with no solid solution or compounds according to the Cu-Ag binary phase diagram [50]. Moreover, the low miscibility between Ag and Cu and the good electrical conductivity provided by silver can explain why some of the lowest contact resistance values are observed when using Ag as a jointing material. Like in the optical microscopy analysis, no secondary phases are detected in the SEM-EDS observations of the HP legs using Ag paint (Figure S7 and Table S6, Supplementary Data File).

A summary of the contact resistances as a function of the jointing material and fixation method is displayed in Figure 9. The joint material that gives origin to the highest contact resistances (independently of the techniques used) is the Ni water-based resin, possibly due to the composition of the resin, as the other nickel-based jointings (Ni paint) show much lower values. Moreover, no evident differences were seen in the Ni particles when nickel-based jointings were used.

During the polishing of the samples prepared with the Ni and Ag resins, it was observed that the contacts of these legs where slightly more mechanically resistant than the others, possibly indicating a higher bonding strength. However, these joints gave origin to the highest contact resistances observed at this work. These observations are in agreement with the work of O.J. Mengali and M.R. Seiler [33,34], where it was reported that the bonding strength is not a critical parameter that affects the contact resistance in BiTe/Cu legs.

The high values and wide range of electrical resistances measured in the HP samples, when compared with the Manual, CP and root HP samples, are mainly related to the cracks formed by the compressive forces during the preparation process, which does not allow any reliable conclusions to be drawn. This effect is also seen in CP samples, but to a much lesser extent, and their resistance values were considered when establishing trends. Therefore, we can conclude that the fixation technique that normally resulted in the highest resistance values was the manual one. The low pressure applied is probably the main cause for these high resistances. Simultaneously, as there is no heat applied in this methodology, the bonding quality is probably lower due to the worse drying of the paints or resins, in comparison with methods where high temperatures and pressures are applied. In contrast, the lowest contact resistances and higher reproducibility were observed for the "root HP" legs, which, albeit prepared by HP, as they have started from the tetrahedrite powders, have no cracks and show good adhesion between the different materials. Low contact resistance values were also observed for the TE legs prepared with Ag water-based conductive paint and pressures of 37 MPa (sample HP2), which indicates that this material, which does not react with graphite or copper, is a good possibility for jointing materials using the HP fixation method if cracks could be avoided. Some of the technical difficulties described so far in the electrical contacts preparations are exemplified in Figure 10.

5. Computer Simulations

To obtain a better understanding of how the contact resistance of the TE legs can affect TEGs performance, computer simulations using the *COMSOL* software were performed. When a temperature gradient is established, the voltage of the device can be calculated through the expression: $V = \alpha(T_h - T_c)$, where α is the Seebeck coefficient difference (between the p and n legs), and $T_h - T_c$ is the temperature difference between the hot and cold sides, respectively. The generated electrical current depends on the voltage and on the resistance of the TEG plus the external load resistance. Since the p and n legs are connected in series, the TEG total resistance, R, is the sum of the element's resistance, or more specifically, it is the resistance of the copper electrodes plus the resistance of the contacts (top and bottom), plus the TE legs resistance. This way, the current output can only be varied by using an external load resistance, with the current being written as $I = V/(R + R_{load})$. As the power output is the current times voltage, the higher is the TEG internal resistance, the smaller is the current output and, consequently, the power.

Figure 11a presents *COMSOL* simulations of a TE pair containing a tetrahedrite and a magnesium silicide leg with contact resistances assumed to be of the same order as the ones found in commercial devices (~3 mΩ.mm^2). In Figure 11b, the simulations for different contact resistances in the tetrahedrite leg are presented. It can be seen that the power output of a TEG can be severely affected just by increasing the contact resistance in one of the elements. Using the above equations, the voltage of the device is not affected (if the thermal gradient is unchanged), but the current output is reduced as the contact resistances increases. Since the current and the voltage affect the power output, the performance of the device is reduced as the electrical contacts become worse.

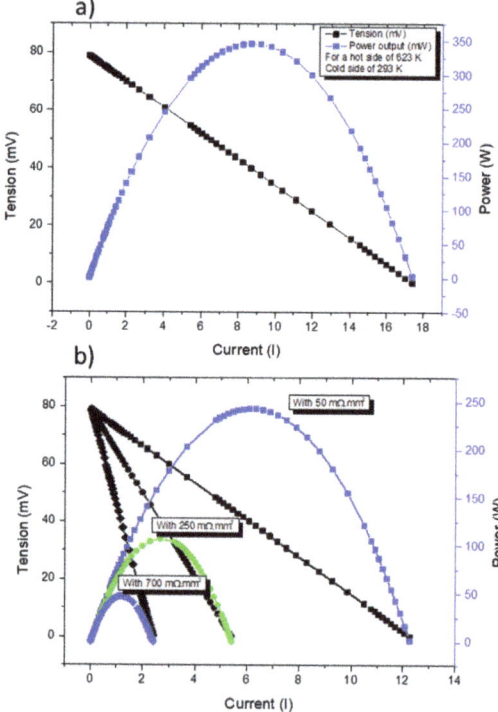

Figure 11. Current–Tension (IV) (black) and Current–Power (IP) (blue and green) curves for the thermocouple made with a tetrahedrite (p leg) and a MgSi material (n leg)—(**a**). IV and IP curves for the same thermocouple for different contact resistances on the p leg (50, 250 and 700 mΩ.mm^2)—(**b**).

In the present work, contact resistances as low as ~43–59 mΩ.mm^2 were obtained in tetrahedrite legs/copper junctions. These values are close to the range reported for commercial TE devices (typically from 0.075 to 3.7 mΩ.mm^2) [51–53] and to other TE materials currently being studied, such as the skutterudites (~10–12 mΩ.mm^2) [54], the CoSb$_3$ compounds (~ 15 mΩ.mm^2) [55], and the Mg$_2$(Si,Sn) legs (~1–20 mΩ.mm^2) [56,57].

6. Conclusions

In this work, the use of graphite as a diffusion barrier for the Cu$_{11}$Mn$_1$Sb$_4$S$_{13}$ tetrahedrite was investigated. No additional phases were detected after hot pressing graphite and Cu$_{11}$Mn$_1$Sb$_4$S$_{13}$ together, pointing to graphite as a good diffusion barrier for tetrahedrite.

Additionally, Cu$_{11}$Mn$_1$Sb$_4$S$_{13}$-based legs, to be used in thermoelectric devices, were assembled applying different techniques and multiple jointing materials. The contact resistance measurements allowed the identification of Ag water-based paint as the most suitable jointing material for the preparation of Cu$_{11}$Mn$_1$Sb$_4$S$_{13}$-based devices. However, it was the hot-pressing method without the use of paints or solders that proved to be the most reliable jointing methodology (this technique produced the lowest contact resistances and the highest reproducibility).

The effect of different pressures in the contacts preparation was also studied. Higher pressures tend to give origin to legs with lower contact resistances, but increasing the pressure proves to be beneficial just to up to some extent, as it also increases the probability of TE legs breaking or developing cracks.

Computer simulations clearly indicated that the increase in contact resistances can result in a strong reduction in the power output, as expected.

In conclusion, the results obtained in this work show that to build an efficient and competitive TE device, based on tetrahedrite legs, it is just not enough to have materials with good TE properties. A critical aspect is how the electrical contacts between the legs are made, as the direct contact between tetrahedrite and copper does not work due to the chemical reaction between the two materials. The low cost, high availability, and properties of tetrahedrites have the potential to be a game changer for thermoelectric industries and markets only if good and reliable electrical contacts are produced and effective diffusion barriers are found.

Supplementary Materials: The following supporting information can be downloaded at: https://www.mdpi.com/article/10.3390/ma15196698/s1, Figure S1: SEM micrographs of a sample. Different analyzed zones (a–d) acquired at 200x magnification in a secondary electron mode; Figure S2: XRD diffractogram of Sample A surface, corresponding to copper disks directly hot pressed to tetrahedrite powder; Figure S3: Powder XRD diffractogram of a green tetrahedrite pellet (hot pressed at 848 K and 56 MPa) bottom, and powder XRD diffractogram of a graphite layer after hot pressing with the tetrahedrite material; Figure S4: SEM-EDS analysis of the cross section of sample A: pellet edge (**a**); center of the pellet cross-section (**b**). Numbers correspond to the zones analyzed by EDS. BSE mode 1500× and 1200× magnification; Figure S5: SEM-EDS analysis of sample HP6. Numbers correspond to the zones analyzed by EDS: copper contact zone 1, Ni resin zone 2 and 3, graphite layer zone 4, and tetrahedrite leg zones 5 and 6. BSE mode 460× magnification; Figure S6: SEM-EDS analysis of sample HP8. Numbers correspond to the zones analyzed by EDS: copper contact zone 1, Zn-Al 5 wt% zones 2 and 3, graphite layer zone 4, and tetrahedrite leg zones 5. BSE mode 160× magnification; Figure S7: SEM-EDS analysis of sample HP4. Numbers correspond to the zones analyzed by EDS: copper contact zone 1, Ag paint zone 2, graphite layer zones 3 and 4, and tetrahedrite leg zone 5. BSE mode 440× magnification; Table S1: Porosity analysis of the zones presented in Figure S1 performed with *ImageJ* software; Table S2: Summary of the prepared samples indicating presence of visible cracks; Table S3: EDS analysis of sample A; Table S4: EDS analysis of sample HP6; Table S5: EDS analysis of sample HP8C; Table S6: EDS analysis of Sample HP4.

Author Contributions: Conceptualization, A.P.G. and E.B.L.; methodology, R.C.; software, E.B.L.; validation, R.C., E.B.L. and A.P.G.; formal analysis, R.C., E.B.L. and A.P.G.; investigation, R.C., Y.D.A. and F.C.; resources, E.B.L. and A.P.G.; data curation, R.C., Y.D.A. and F.C.; writing—original draft preparation, R.C., E.B.L. and A.P.G.; writing—review and editing, R.C., E.B.L. and A.P.G.; visualization, R.C., E.B.L. and A.P.G.; supervision, E.B.L. and A.P.G.; project administration, E.B.L. and A.P.G.; funding acquisition, R.C., E.B.L. and A.P.G. All authors have read and agreed to the published version of the manuscript.

Funding: This research was supported by Fundação para a Ciência e a Tecnologia (FCT), Portugal, through the contracts UID/Multi/04349/2020 and UI/BD/150713/2020.

Institutional Review Board Statement: Not applicable.

Informed Consent Statement: Not applicable.

Data Availability Statement: Not applicable.

Conflicts of Interest: The authors declare no conflict of interest.

References

1. Jänsch, D.; Köhne, M.; Altstedde, M.K.; Numus, J.N. Thermoelectrics: Power from waste heat Thermoelectric generators supply. *BINE-Themeninfo I* **2016**, *1*, 1–25.
2. Gonçalves, A.P.; Lopes, E.B.; Montemor, M.F.; Monnier, J. Bertrand Lenoir Oxidation studies of $Cu_{12}Sb_{3.9}Bi_{0.1}S_{10}Se_3$ tetrahedrite. *Electron. Mater.* **2018**, *47*, 2880–2889. [CrossRef]
3. Zhang, Q.H.; Huang, X.Y.; Bai, S.Q.; Shi, X.; Uher, C.; Chen, L.D. Thermoelectric devices for power generation: Recent progress and future challenges. *Adv. Eng. Mater.* **2016**, *18*, 194–213. [CrossRef]
4. Jaziri, N.; Boughamoura, A.; Müller, J.; Mezghani, B.; Tounsi, F.; Ismail, M. A comprehensive review of Thermoelectric Generators: Technologies and common applications. *Energy Rep.* **2020**, *6*, 264–287. [CrossRef]
5. Charilaou, K.; Kyratsi, T.; Louca, L.S. Design of an air-cooled thermoelectric generator system through modelling and simulations, for use in cement industries. *Mater. Today Proc.* **2020**, *44*, 3516–3524. [CrossRef]
6. Kuroki, T.; Kabeya, K.; Makino, K.; Kajihara, T.; Kaibe, H.; Hachiuma, H.; Matsuno, H.; Fujibayashi, A. Thermoelectric generation using waste heat in steel works. *J. Electron. Mater.* **2014**, *43*, 2405–2410. [CrossRef]
7. Yazawa, K.; Shakouri, A.; Hendricks, T.J. Thermoelectric heat recovery from glass melt processes. *Energy* **2017**, *118*, 1035–1043. [CrossRef]
8. Haras, M.; Skotnicki, T. Thermoelectricity for IoT—A review. *Nano Energy* **2018**, *54*, 461–476. [CrossRef]
9. Catalan, L.; Garacochea, A.; Casi, A.; Araiz, M.; Aranguren, P.; Astrain, D. Experimental evidence of the viability of thermoelectric generators to power volcanic monitoring stations. *Sensors* **2020**, *20*, 4839. [CrossRef]
10. Chen, A. *Thermal Energy Harvesting with Thermoelectrics for Self-Powered Sensors: With Applications to Implantable Medical Devices, Body Sensor Networks and Aging in Place*; UC Berkeley: Berkeley, CA, USA, 2020.
11. Lorenz, R.D.; Clarke, E.S. Influence of the Multi-Mission Radioisotope Thermoelectric Generator (MMRTG) on the local atmospheric environment. *Planet. Space Sci.* **2020**, *193*, 105075–105083. [CrossRef]
12. von Lukowicz, M.; Abbe, E.; Schmiel, T.; Tajmar, M. Thermoelectric generators on satellites—An approach for waste heat recovery in space. *Energies* **2016**, *9*, 541. [CrossRef]
13. Quoilin, S.; Van Den Broek, M.; Declaye, S.; Dewallef, P.; Lemort, V. Techno-economic survey of organic rankine cycle (ORC) systems. *Renew. Sustain. Energy Rev.* **2013**, *22*, 168–186. [CrossRef]
14. Gonçalves, A.P.; Lopes, E.B.; Monnier, J.; Alleno, E.; Godart, C.; Montemor, M.D.F.; Vaney, J.; Lenoir, B. Tetrahedrites for low cost and sustainable thermoelectrics. *Solid State Phenom.* **2017**, *257*, 135–138. [CrossRef]
15. Chen, Y.; Hou, X.; Ma, C.; Dou, Y.; Wu, W. Review of development status of Bi2Te3-based semiconductor thermoelectric power generation. *Adv. Mater. Sci. Eng.* **2018**, *2018*, 1210562. [CrossRef]
16. Cheng, X.; Farahi, N.; Kleinke, H. Mg_2Si-Based materials for the thermoelectric energy conversion. *Jom* **2016**, *68*, 2680–2687. [CrossRef]
17. Chetty, R.; Bali, A.; Mallik, R.C. Tetrahedrites as thermoelectric materials: An overview. *J. Mater. Chem. C* **2015**, *3*, 12364–12378. [CrossRef]
18. Lu, X.; Morelli, D.T.; Xia, Y.; Ozolins, V. Increasing the Thermoelectric Figure of Merit of Tetrahedrites by Co-doping with Nickel and Zinc. *Chemestry Mater.* **2015**, *27*, 408–413. [CrossRef]
19. Kumar, D.S.P.; Chetty, R.; Rogl, P.; Rogl, G.; Bauer, E.; Malar, P.; Chandra, R. Intermetallics Thermoelectric properties of Cd doped tetrahedrite: $Cu_{12-x}Cd_xSb_4S_{13}$. *Intermetallics* **2016**, *78*, 21–29. [CrossRef]
20. Wang, J.; Li, X.; Bao, Y. Thermoelectric properties of Mn dopbed Cu12-xMnxSb4S13 tetrahedrites. *Mater. Sci. Forum* **2016**, *847*, 161–165. [CrossRef]
21. Kwak, S.G.; Lee, G.E.; Kim, I.H. Effects of Se doping on thermoelectric properties of tetrahedrite $Cu_{12}Sb_4S_{13-z}Se_z$. *Electron. Mater. Lett.* **2021**, *17*, 164–171. [CrossRef]

22. Coelho, R.; Symeou, E.; Kyratsi, T.; Gonçalves, A. Tetrahedrite sintering conditions: The $Cu_{11}Mn_1Sb_4S_{13}$ case. *J. Electron. Mater.* **2020**, *49*, 5077–5083. [CrossRef]
23. Gu, M.; Bai, S.; Wu, J.; Liao, J.; Xia, X.; Liu, R.; Chen, L. A high-throughput strategy to screen interfacial diffusion barrier materials for thermoelectric modules. *J. Mater. Res.* **2019**, *34*, 1179–1187. [CrossRef]
24. Cheng, F.J.; Ma, Z.L.; Wang, Y.; Zhang, G.X.; Long, W.M. Microstructure and aging resistance of the joints between SAC305 solder and thermoelectric materials with different diffusion barriers. *Kov. Mater.* **2014**, *52*, 157–162. [CrossRef]
25. Li, J.; Zhao, S.; Chen, J.; Han, C.; Hu, L. Energy, Environmental, and Catalysis Applications Al-Si Alloy as a Diffusion Barrier for GeTe-Based Thermoelectric Legs with High Interfacial Reliability and Mechanical Strength Laboratory of Special Functional Materials; Shenzhen Engineering Laborator. *ACS Publ.* **2020**, *12*, 18562–18569. [CrossRef]
26. Lan, Y.C.; Wang, D.Z.; Chen, G.; Ren, Z.F. Diffusion of nickel and tin in p-type $(Bi,Sb)_2Te_3$ and n-type $Bi_2(Te,Se)_3$ thermoelectric materials. *Appl. Phys. Lett.* **2008**, *92*, 2–5. [CrossRef]
27. Coelho, R.; Lopes, E.B.; Gonçalves, A.P. Protective coatings for $Cu_{11}Mn_1Sb_4S_{13}$ and $Cu_{10.5}Ni_{1.5}Sb_4S_{13}$ tetrahedrites. *J. Electron. Mater.* **2021**, *50*, 467–477. [CrossRef]
28. Coelho, R.; Lopes, E.B.; Goncalves, A.P. Carbon and Gold as Effective Diffusion Barriers for Cu11Mn1Sb4S13 Tetrahedrites. In Proceedings of the 7th International Conference of Young Scientists on Energy and Natural Sciences Issues—CYSENI 2021, Kaunas, Lithuania, 24–28 May 2021; pp. 705–722.
29. Fu, Y.; Hou, M.; Liang, D.; Yan, X.; Fu, Y.; Shao, Z.; Hou, Z.; Ming, P.; Yi, B. The electrical resistance of flexible graphite as flowfield plate in proton exchange membrane fuel cells. *Carbon N. Y.* **2008**, *46*, 19–23. [CrossRef]
30. Ben-Ayoun, D.; Sadia, Y.; Gelbstein, Y. Compatibility between co-metallized PbTe thermoelectric legs and an Ag-Cu-In brazing alloy. *Materials* **2018**, *11*, 99. [CrossRef]
31. Kaszyca, K.; Schmidt, M.; Chmielewski, M.; Pietrzak, K.; Zybala, R. Joining of thermoelectric material with metallic electrode using spark plasma Sintering (SPS) technique. *Mater. Today Proc.* **2018**, *5*, 10277–10282. [CrossRef]
32. Kraemer, D.; Sui, J.; McEnaney, K.; Zhao, H.; Jie, Q.; Ren, Z.F.; Chen, G. High thermoelectric conversion efficiency of MgAgSb-based material with hot-pressed contacts. *Energy Environ. Sci.* **2015**, *8*, 1299–1308. [CrossRef]
33. Ren, Z.; Lan, Y.; Zhang, Q. (Eds.) *Advanced Thermoelectrics, Materials, Devices, Contacts and Systems*; CRC Press, Taylor & Francis Group, LLC.: Boca Raton, FL, USA, 2017. [CrossRef]
34. Mengali, O.J.; Seiler, M.R. Contact resistance studies on thermoelectric materials. *Adv. Energy Convers.* **1962**, *2*, 59–68. [CrossRef]
35. Kim, Y.; Yoon, G.; Park, S.H. Direct Contact Resistance Evaluation of Thermoelectric Legs. *Exp. Mech.* **2016**, *56*, 861–869. [CrossRef]
36. Thimont, Y.; Lognoné, Q.; Goupil, C.; Gascoin, F.; Guilmeau, E. Design of apparatus for Ni/Mg_2Si and $Ni/MnSi_{1.75}$ contact resistance determination for thermoelectric legs. *J. Electron. Mater.* **2014**, *43*, 2023–2028. [CrossRef]
37. Chetty, R.; Kikuchi, Y.; Bouyrie, Y.; Jood, P.; Yamamoto, A.; Suekuni, K.; Ohta, M. Power generation from the $Cu_{26}Nb_2Ge_6S_{32}$-based single thermoelectric element with Au diffusion barrier. *J. Mater. Chem. C* **2019**, *7*, 5184–5192. [CrossRef]
38. Brito, F.P.; Vieira, R.; Martins, J.; Goncalves, L.M.; Goncalves, A.P.; Coelho, R.; Lopes, E.B.; Symeou, E.; Kyratsi, T. Analysis of thermoelectric generator incorporating n-magnesium silicide and p-tetrahedrite materials. *Energy Convers. Manag.* **2021**, *236*, 114003. [CrossRef]
39. Markovic, S.; Bryan, J.L.; Rezaee, R.; Turakhanov, A. Application of XGBoost model for in-situ water saturation determination in Canadian oil-sands by LF-NMR and density data. *Sci. Rep.* **2022**, *12*, 13984. [CrossRef] [PubMed]
40. Hyndman, R.J.; Koehler, A.B. Another look at measures of forecast accuracy. *Int. J. Forecast.* **2006**, *22*, 679–688. [CrossRef]
41. Farahi, N. *Magnesium Silicide Based Thermoelectric Nanocomposites*; © Nader Farahi 2016; University of Waterloo: Waterloo, ON, Canada, 2016. [CrossRef]
42. Howard, L.; Tafone, D.; Grbovic, D.; Pollman, A. COMSOL Multiphysics® Simulation of TEGs for Waste Thermal Energy Harvesting. In Proceedings of the COMSOL Conference 2018, Boston, MA, USA, 3–5 October 2018; pp. 1–6.
43. Prasad, A.; Thiagarajan, R.C.N. Multiphysics Modeling and Multilevel Optimization of Thermoelectric Generator for Waste Heat Recovery. In Proceedings of the COMSOL Conference 2018, Bangalore, India, 9–10 August 2018; pp. 1–7.
44. Skinner, B.J.; Frederick, D.L.; Emil, M. Studies of the Sulfosalts of Copper III. Phases and Phase Relations in the System Cu-Sb-S. *Econ. Geol.* **1972**, *67*, 924–938. [CrossRef]
45. Arbeitsgestaltung, G.; Pg, L.; Warneeke, V.F.; Kl, Z.; Berufsgenos, O.; Krankheitsbfld, D.; Entdecker, D. Die gewerbliche Schwefelkohlenstoffvergiftung. *Arch. Gewerbepath.* **1941**, *11*, 198–248.
46. Zhang, Y.; Zuo, T.T.; Tang, Z.; Gao, M.C.; Dahmen, K.A.; Liaw, P.K.; Lu, Z.P. Microstructures and properties of high-entropy alloys. *Prog. Mater. Sci.* **2014**, *61*, 1–93. [CrossRef]
47. Fabuyide, A.A.; Apata, A.O.; Muobeleni, T.N.; Witcomb, M.J.; Jain, P.K.; Rading, G.O.; Borode, J.O.; Cornish, L.A. Experimental Liquidus Surface Projection and Isothermal Section at 1000 °C of the V-Ni-C System. *J. Phase Equilibria Diffus.* **2021**, *42*, 42–62. [CrossRef]
48. Kaprara, E.; Seridou, P.; Tsiamili, V.; Mitrakas, M.; Vourlias, G.; Tsiaoussis, I.; Kaimakamis, G.; Pavlidou, E.; Andritsos, N.; Simeonidis, K. Cu-Zn powders as potential Cr(VI) adsorbents for drinking water. *J. Hazard. Mater.* **2013**, *262*, 606–613. [CrossRef] [PubMed]
49. Kejzlar, P.; Machuta, J.; Nová, I. Comparison of the Structure of CuZn40MnAl Alloy Casted into Sand and Metal Moulds. *Manuf. Technol.* **2017**, *17*, 15. [CrossRef]
50. Elliott, R.P.; Shunk, F.A.; Giessen, W.C. The Ag-Cu (Silver-Copper) system. *Bull. Alloy Phase Diagrams* **1980**, *1*, 23. [CrossRef]

51. He, R.; Schierning, G.; Nielsch, K. Thermoelectric Devices: A Review of Devices, Architectures, and Contact Optimization. *Adv. Mater. Technol.* **2018**, *3*, 1700256. [CrossRef]
52. Salvador, J.R.; Cho, J.Y.; Ye, Z.; Moczygemba, J.E.; Thompson, A.J.; Sharp, J.W.; König, J.D.; Maloney, R.; Thompson, T.; Sakamoto, J.; et al. Thermal to electrical energy conversion of skutterudite-based thermoelectric modules. *J. Electron. Mater.* **2013**, *42*, 1389–1399. [CrossRef]
53. Liu, W.; Wang, H.; Wang, L.; Wang, X.; Joshi, G.; Chen, G.; Ren, Z. Understanding of the contact of nanostructured thermoelectric n-type $Bi_2Te_{2.7}Se_{0.3}$ legs for power generation applications. *J. Mater. Chem. A* **2013**, *1*, 13093–13100. [CrossRef]
54. Way, M.; Luo, D.; Tuley, R.; Goodall, R. A new high entropy alloy brazing filler metal design for joining skutterudite thermoelectrics to copper. *J. Alloys Compd.* **2021**, *858*, 157750. [CrossRef]
55. Wojciechowski, K.T.; Zybala, R.; Mania, R. High temperature $CoSb_3$-Cu junctions. *Microelectron. Reliab.* **2011**, *51*, 1198–1202. [CrossRef]
56. Camut, J.; Ayachi, S.; Castillo-Hernández, G.; Park, S.; Ryu, B.; Park, S.; Frank, A.; Stiewe, C.; Müller, E.; de Boor, J. Overcoming asymmetric contact resistances in al-contacted $Mg_2(Si,sn)$ thermoelectric legs. *Materials* **2021**, *14*, 6774. [CrossRef]
57. Jayachandran, B.; Prasanth, B.; Gopalan, R.; Dasgupta, T.; Sivaprahasam, D. Thermally stable, low resistance $Mg_2Si_{0.4}Sn_{0.6}$/Cu thermoelectric contacts using SS 304 interlayer by one step sintering. *Mater. Res. Bull.* **2021**, *136*, 111147. [CrossRef]

Excellent Thermoelectric Performance of 2D CuMN₂ (M = Sb, Bi; N = S, Se) at Room Temperature

Wenyu Fang [1,2,†], Yue Chen [1,†], Kuan Kuang [1] and Mingkai Li [1,*]

[1] Ministry-of-Education Key Laboratory of Green Preparation and Application for Functional Materials, Hubei Key Lab of Ferro & Piezoelectric Materials and Devices, Hubei Key Laboratory of Polymer Materials, and School of Materials Science & Engineering, Hubei University, Wuhan 430062, China
[2] Public Health and Management School, Hubei University of Medicine, Shiyan 442000, China
* Correspondence: mkli@hubu.edu.cn
† These authors contributed equally to this work.

Abstract: 2D copper-based semiconductors generally possess low lattice thermal conductivity due to their strong anharmonic scattering and quantum confinement effect, making them promising candidate materials in the field of high-performance thermoelectric devices. In this work, we proposed four 2D copper-based materials, namely $CuSbS_2$, $CuSbSe_2$, $CuBiS_2$, and $CuBiSe_2$. Based on the framework of density functional theory and Boltzmann transport equation, we revealed that the monolayers possess high stability and narrow band gaps of 0.57~1.10 eV. Moreover, the high carrier mobilities (10^2~10^3 $cm^2 \cdot V^{-1} \cdot s^{-1}$) of these monolayers lead to high conductivities (10^6~10^7 $\Omega^{-1} \cdot m^{-1}$) and high-power factors (18.04~47.34 mW/mK²). Besides, as the strong phonon-phonon anharmonic scattering, the monolayers also show ultra-low lattice thermal conductivities of 0.23~3.30 W/mK at 300 K. As results show, all the monolayers for both p-type and n-type simultaneously show high thermoelectric figure of merit (ZT) of about 0.91~1.53 at room temperature.

Keywords: 2D material; conductivity; power factor; spin-orbit effects; figure of merit

1. Introduction

Thermoelectric generators can directly convert heat into electrical power, thus attracting wide research interest. Generally, the thermal-electric conversion capacity can be ruled by the dimensionless figure of merit, $ZT = \frac{\sigma S^2 T}{\kappa_e + \kappa_l}$ [1], here S and σ are the Seebeck coefficient and electrical conductivity, T presents the temperature, κ_e and κ_l are the electron and lattice thermal conductivity, respectively. Clearly, the ideal thermoelectric material needs to have both high-power factor ($PF = \sigma S^2$) and low lattice thermal conductivity. However, this target is not easy to achieve simultaneously as the parameters above are tightly coupled, mutually restricted, and difficult to decouple. They can be regarded as the functions of the vector tensor $\mathbf{K}_n$, energy eigenvalue ε_i, and carrier relaxation time $\tau_i(\mathbf{k})$ [2–4]. Besides, the Seebeck coefficient is also closely related to the density of states effective mass (m_d^*) and intrinsic carrier concentration (n), $S = \frac{8\pi^2 k_B^2 T m_d^*}{3e\hbar^2} \left(\frac{\pi}{3n}\right)^{2/3}$ [5]. Additionally, electrical conductivity σ, and electron thermal conductivity κ_e are also restricted by the Wiedemann-Franz-Lorenz's law, $\kappa_e = L\sigma T$ [6], where L is Lorenz number.

In fact, the thermoelectric performance of traditional thermoelectric materials has been effectively improved over the past few decades. Among them, two-dimensional layered (2D) materials, as unique mechanical, electronic, thermal, and optoelectronic properties, as well as quantum confinement effects, make them as promising thermoelectric materials in a variety of applications. For example, quasi-two-dimensional SnSe transistors were revealed to have high Seebeck coefficient, and a field effect mobility of about 250 cm^2/Vs at 1.3 K, thus it was found to be a high-quality semiconductor ideal for thermoelectric applications [7]. The 2D Mg_3Sb_2 monolayer was proved to have a favorable ZT value of 2.5 at 900 K, which is higher than that of its bulk structure. These theoretical results also revealed that nano-engineering can effectively improve the thermoelectric conversion efficiency [8]. Additionally, quasi-two-dimensional GeSbTe compounds were observed by Wei et al. [9]. They found that the monolayer with maximal ZT values of 0.46~0.60 at 750 K, indicating that 2D GeSbTe is a promising mid-temperature thermoelectric material.

Recently, research on the thermoelectric properties of copper-based semiconductors has attracted much attention. For example, Yu et al. [10] proposed a novel phrase 2D σ-Cu_2S, which has a low lattice thermal conductivity of 0.10 W/mK, and a high ZT value of 1.33 at 800 K. Cao et al. [11] investigated the electronic structure and thermoelectric performance of β-Cu_2Se under strain of −4~4%, they found that its ZT values can reach 1.65~1.71 at 800 K. Other materials, such as Fm-3m Cu_2S [12], multi-scale Cu_2Se [13], and Cu-Se co-doped Ag_2S [14], all exhibited intrinsic low thermal conductivity and high ZT value. In fact, back in 2013 and 2016, Ma and Deng et al. [15,16] investigated the diffusion behavior of Cu in CdTe by density functional theory, they found that Cu s-d orbital coupling only occurred at asymmetric point, but instantly disappeared at symmetric location. These interesting properties make Cu based compound to possess a strong anharmonic scattering, resulting in low intrinsic lattice thermal conductivity. In addition, many semiconductors containing metallic atoms such as Bi have also been reported to possess non-negligible spin-orbit coupling (SOC) effects [17,18], which have the potential to be used in thermoelectric devices. For example, Kim et al. [19] prepared the bulk $Bi_{0.4}Sb_{1.6}Te_3$ alloy via an atomic-layer deposition (ALD) technique, and found that it possesses high ZT values of 1.50 at 329 K. Wu et al. [20] investigated the thermoelectric properties of β-BiAs and β-BiSb monolayers by first-principles calculation and Boltzmann transport theory. They concluded that the monolayers simultaneously exhibit ultra-low lattice thermal conductivities (0.6~0.8 W/mK), and high ZT values (0.78~0.82) at 300 K.

$CuMN_2$ (M = Sb, Bi; N = S, Se) are layered materials with narrow band gaps within ~1.38 eV. [21] Among them, $CuSbS_2$ exhibited the low lattice thermal conductivity of about 1.5~2 W/mK [22]. Additionally, both bulk and monolayer $CuSbS_2$ and $CuSbSe_2$ were revealed to have an excellent thermoelectric power factor at 300 K, reaching about 0.2~1.0 mW/mK^2 at constant relaxation time approximation (CRTA) of 11 fs [23]. Therefore, it is of high interest to see whether or not the monolayer $CuSbS_2$, $CuSbSe_2$, $CuBiS_2$, and $CuBiSe_2$, can also deliver good thermoelectric performance. To this end, using first principles calculations, we investigated the electronic structures, mechanical, and transport properties, of these four 2D materials. The calculations revealed that all the monolayers possess narrow band-gaps (0.57~1.10 eV), high power factors (18.04~47.34 mW/mK^2), and also low lattice thermal conductivities (0.23~3.30 W/mK). As a result, all the monolayers for both p-type and n-type simultaneously show high ZT values of about 0.91~1.53 at room temperature.

2. Calculation Details

We carried out the calculations in the Vienna Ab initio Simulation Package (VASP) [24], in which the Generalized Gradient Approximation (GGA) [25,26] and Perdew–Burke–Ernzerhof (PBE) functional was adopted to exchange-correlation approximation. To shorten the computation time, we used the HipHive [27] code to extract the second and third order force-constants (IFCs). Besides, we used the Phonopy [28] code to calculate the Grüneisen parameters and phonon dispersion, and employed the Phono3py [29] to evaluate the phonon scattering rate and lattice thermal conductivity. We also used the Wannier90 [30] code to solve Boltzmann transport equation, and then characterized the thermoelectric properties as a function of the chemical potential.

The transport properties, such as electrical conductivity, electron thermal conductivity and ZT value, are directly related to carrier relaxation time, therefore, we used the deformation potential theory (DPT) to calculate carrier mobility, and further corrected it by the acoustic phonon-limited method (APM), which is more suitable for anisotropic materials. See Supplementary Materials for more calculation details.

3. Results and Discussion

3.1. Crystal Structures

The crystal structures of CuMN$_2$ (M = Sb, Bi; N = S, Se) for bulk and single-layer are shown in Figure 1. As can be seen, the bulk CuMN$_2$ are layered structures, similar to graphite and MoS$_2$. Therefore, we first investigated the cleavage energy of their single-layers, and its calculation method was based on the latest Rigorous Method proposed by Jung et al. [31]. $E_f = \frac{E_{iso} - E_{bulk}/n}{\mathcal{A}}$, where E_{iso} and E_{bulk} are energies of single-layer and bulk unit-cell, $\mathcal{A}$ and n are the in-plane area and the number of the slab in a bulk-unit. The calculation results are listed in Table S1. The corresponding cleavage energies are within 0.68~0.93 J/m^2, which are higher than those of graphene (0.33 J/m^2), black phosphorus (BP) (0.36 J/m^2), and MoS$_2$ (0.27 J/m^2) [32], but still lower than those of single-layer Ca$_2$N (1.09 J/m^2) [33], GeP$_3$ (1.14 J/m^2), and InP$_3$ (1.32 J/m^2) [34]. All of these results indicate the feasibility of obtaining single-layer CuMN$_2$ by mechanical exfoliation in experiments. The lattice constants and thicknesses of the monolayers after structural relaxation are listed in Table 1. Owing to each atomic radius satisfies: S (1.03 Å) < Se (1.16 Å) < Cu (1.28 Å) < Sb (1.61 Å) < Bi (1.82 Å), both the lattice constants (a/b) and thickness (h) follow the order of CuSbS$_2$ < CuBiS$_2$ < CuSbSe$_2$ < CuBiSe$_2$.

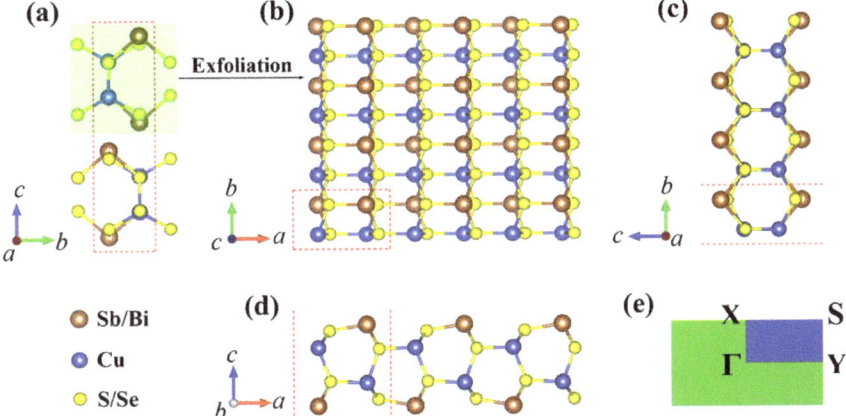

Figure 1. The structures of: (**a**) bulk; (**b**) top view; (**c,d**) side view; and (**e**) K-point path of 2D CuMN$_2$ (M = Sb, Bi; N = S, Se).

Table 1. The structural parameters, buckling height h, cleavage energies E_f, and band gaps E_g of 2D CuMN$_2$ (M = Sb, Bi; N = S, Se).

Materials	a (Å)	b (Å)	h (Å)	E_f (J/m^2)	E_g (eV)		
					PBE	PBE + SOC	HSE 06 + SOC
CuSbS$_2$	6.15	3.81	5.23	0.72	0.38	0.37	1.10
CuSbSe$_2$	6.48	4.03	5.43	0.68	0.20	0.19	0.68
CuBiS$_2$	6.21	3.95	5.24	0.93	0.35	0.26	0.83
CuBiSe$_2$	6.55	4.16	5.44	0.85	0.20	0.16	0.57

3.2. Elastic Properties and Stability

In general, for a new 2D material, we can identify its mechanical stability by its elastic constants C_{ij}. As listed in Table 2, all the monolayers satisfy the Born-Huang criterion, $C_{11}C_{22} - C_{12}^2 > 0$ and $C_{66} > 0$ [35], indicating that they all possess high mechanical stability. Additionally, since the structures of the monolayers are anisotropic, $C_{11} \neq C_{22}$. We further calculated the Young's moduli and Poisson's ratio of these materials [36], as shown in Figure 2. Here θ is the angle with respect to a-axis. As can be seen, their Young's moduli are relatively close, showing the maximum values of 58.20~66.52 N/m in the direction of 0° (180°) and the minimum values of 28.34~33.04 N/m in the direction of 90° (270°), which are obviously lower than those of graphene (350 ± 3.15 N/m) [37], h-BN ((270 N/m), [38] and MoS$_2$ (200 N/m) [39]. Such low Young's moduli are expected to exhibit low lattice thermal conductivity [40]. On the contrary, the Poisson's ratio minimizes in both 0° and 90° directions, and maximizes at 40° (140°) with values of 0.14~0.34, respectively. Fantastically, monolayer CuSbS$_2$ and CuBiS$_2$ are rare auxetic materials with negative Poisson's ratio (NPR) of −0.02 and −0.04 at 0° (180°). Such interesting NPR phenomenon is also observed in PN (−0.08) [41], and tetra-silicene (−0.06) [42], which have been revealed to hold high potential in medicine, defense, and the escalation of tensions [43].

Table 2. The Elastic constants C_{ij}, the maximums for Young's modulus Y and Poisson's ratio v, and Debye temperature Θ_D of 2D CuMN$_2$ (M = Sb, Bi; N = S, Se).

Materials	C_{11} (N/m)	C_{12} (N/m)	C_{22} (N/m)	C_{66} (N/m)	Y (N/m)	v	Θ_D (K)
CuSbS$_2$	66.53	−0.69	33.05	16.44	66.52	0.14	90.10
CuSbSe$_2$	66.58	7.92	29.28	13.81	64.44	0.29	77.30
CuBiS$_2$	58.26	−1.34	31.82	15.04	58.20	0.14	106.80
CuBiSe$_2$	60.83	8.49	33.70	12.94	58.69	0.34	75.40

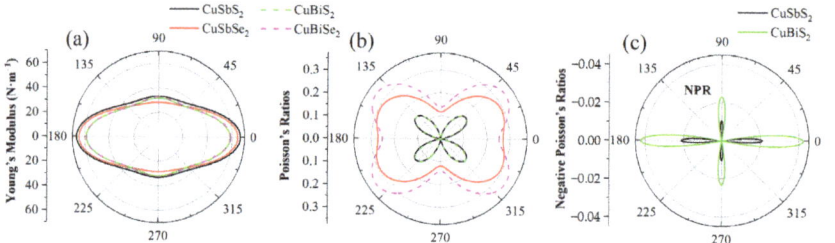

Figure 2. (a) The Young's moduli: (b) Poisson's ratio; and (c) NPR of 2D CuMN$_2$ (M = Sb, Bi; N = S, Se).

We also investigated the bonding strength of materials by calculating their electron localization function (ELF) and Bader charge analysis (see Figure S1 and Table S2 for more details). Clearly, the adjacent atoms exhibit a predominant ionic bond characteristic. However, the net charge transfer is relatively few, mostly within ~1.0 e, indicating that they have relatively weak bond strength and thus exhibit low Young's moduli. Besides, we analyzed the thermal stabilities of these four materials at different temperatures by using ab initio molecular dynamics (AIMD) simulations [44]. We revealed that they can remain high stability at 500 K, as their crystal structure does not bond breaking or undergo remodeling, as showed in Figure S2 in Supplementary Materials.

3.3. Electronic Structures

To accurately characterize the electronic structures of the monolayers, we first analyze the effect of spin-orbit coupling (SOC) on their electronic band structures. As shown in Figure S3, SOC has a non-ignorable impact on band structure, especially for CuBiS$_2$ and CuBiSe$_2$, so SOC was all considered in following calculations. As shown in Figure 3, all the monolayers are narrow band-gap semiconductors with band gaps of 0.57~1.10 eV, which are slightly smaller than or comparable to their bulk structures [21,23,45,46]. Since the valence band maximum (VBM) and conduction band minimum (CBM) are both located at Γ point, both CuSbS$_2$, CuSbSe$_2$, and CuBiS$_2$ belong to direct bandgap semiconductors. However, for CuBiSe$_2$, its VBM is transferred to between Y and Γ, so it is an indirect bandgap semiconductor. Moreover, as show in Figure S4, the VBMs are mainly composed of S-3p/Se-4p and Cu-3d electrons, while CBMs are mainly composed of Sb-5p/Bi-6p and S-3p/Se-4p electrons. In addition, the total density of states for both VBMs and CBMs showed relatively steep distribution, indicating that the monolayers have high density of states effective mass, and thus to possess both high p-type and n-type Seebeck coefficients, as shown in Figure 4.

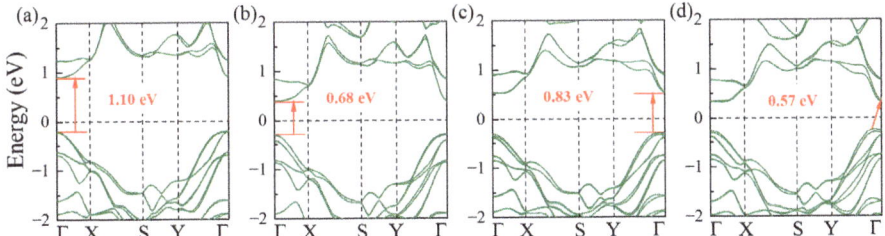

Figure 3. Band structures of monolayer: (**a**) $CuSbS_2$; (**b**) $CuSbSe_2$; (**c**) $CuBiS_2$; and (**d**) $CuBiSe_2$ at HSE 06 + SOC functional.

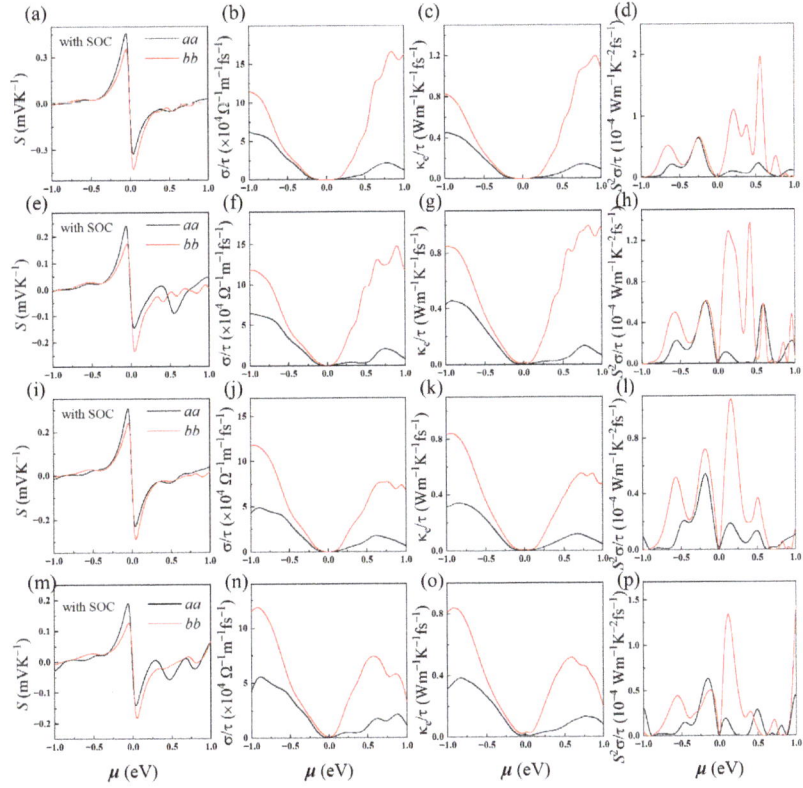

Figure 4. Electron transport properties of 2D $CuMN_2$ (M = Sb, Bi; N = S, Se): (**a,e,i,m**) Seebeck coefficients; (**b,f,j,n**) electrical conductivities; (**c,g,k,o**) electron thermal conductivities; and (**d,h,l,p**) power factors along a—(black line) and b—directions (red line) with PBE + SOC functional.

Next, we explored the carrier mobility of electrons and holes along the a- and b-axis of the monolayers (see Figures S5 and S6, and Table S3 for more details). For monolayer CuSbS$_2$, CuSbSe$_2$, and CuBiS$_2$, their effective masses (m_e) of electrons are higher than those of holes (m_h) along a-axis, corresponding to their flatter band near CBMs than VBMs along the Γ–X direction. However, the opposite is true for CuBiSe$_2$, the m_e (0.93 m_0) is indeed lower than m_h (2.83 m_0), which can be interpreted as the steeper dispersion curve corresponding to its VBM between Γ and X. Besides, for all monolayers, their elastic modulus (C^{2D}) along the a-axis is larger than that along b-axis, which is consistent with their elastic constants. Further, the deformation potential constant (E_l) fluctuates widely from 0.75 eV to 4.48 eV, and the lower E_l occur simultaneously for the hole along the a-axis. As a result, the hole mobilities in the a-axis are higher than those in other cases, and the highest is even up to 4562.47 cm$^2\cdot$V$^{-1}\cdot$s^{-1} for CuSbS$_2$. Since DPT method tends to overestimate the mobility of semiconductor, especially when the E_l is relatively small [35], we adopted the APT method to correct the results, as listed in Table 3. Clearly, after corrected by APT, the carrier mobility increases when the E_l is relatively large, and decreases otherwise. As can be seen, the mobilities of both electrons (μ_e) and holes (μ_h) are basically in the range of 10^2~10^3 cm$^2\cdot$V$^{-1}\cdot$s^{-1}, in which CuSbS$_2$ and CuSbSe$_2$ exhibit the highest μ_h and μ_e of 1661.49 and 937.12 cm$^2\cdot$V$^{-1}\cdot$s^{-1}, which are far higher than that of MoS$_2$ (μ_h ~200 cm$^2\cdot$V$^{-1}\cdot$s^{-1}) [47], but lower than those of silicene (μ_e ~10^5 cm$^2\cdot$V$^{-1}\cdot$s^{-1}) [48], and phosphorene (μ_h ~10^4 cm$^2\cdot$V$^{-1}\cdot$s^{-1}) [49].

Table 3. The carrier effective mass (m^*/m_0), deformation potential constant (E_l/eV), plane stiffness (C^{2D}/N·m^{-1}), hole (μ_h/cm$^2\cdot$V$^{-1}\cdot$s^{-1}) and electron mobility (μ_e/cm$^2\cdot$V$^{-1}\cdot$s^{-1}), and relaxation time (τ/fs) of the monolayers under PBE + SOC functional at 300 K.

Materials	Direction	Type	m^*	C^{2D}	E_l	DPT	APT	
						μ	μ	τ
CuSbS$_2$	a-axis	electron	1.83	63.50	1.38	584.78	206.92	215.29
		hole	0.77		0.75	4562.47	1661.49	732.56
	b-axis	electron	0.24	34.15	3.27	424.59	895.32	122.89
		hole	0.60		3.05	192.61	684.04	232.33
CuSbSe$_2$	a-axis	electron	1.97	51.83	2.08	176.33	166.61	186.53
		hole	0.93		1.22	1085.34	248.49	132.20
	b-axis	electron	0.27	26.30	2.28	532.97	937.12	146.58
		hole	0.57		4.48	66.84	196.34	63.82
CuBiS$_2$	a-axis	electron	3.11	53.75	2.40	54.62	58.96	104.40
		hole	0.78		1.20	1511.80	820.27	366.20
	b-axis	electron	0.44	33.55	1.83	415.29	412.27	103.07
		hole	0.57		2.08	430.36	745.06	242.74
CuBiSe$_2$	a-axis	electron	0.99	44.13	2.67	138.35	196.09	110.83
		hole	2.83		0.89	495.73	176.14	283.82
	b-axis	electron	0.93	27.99	1.94	178.15	213.08	112.36
		hole	0.25		2.63	404.18	1071.33	153.80

3.4. Electrical Transport Properties

Furthermore, we explored the electron transport properties of the monolayers by solving Boltzmann transport equation, as show in Figure 4 (see Figure S7 for more details about maximally localized Wannier functions (MLWFs)). Obviously, the maximums of Seebeck coefficient (S) satisfy CuSbS$_2$ > CuBiS$_2$ > CuSbSe$_2$ > CuBiSe$_2$, which consist with their band-gaps ordering, as small gap implies high carrier concentration. Coincidentally, the p-type Seebeck coefficients are higher than those of n-type in a-axis, but opposite in the b-axis, which may be caused by the anisotropy of the density of states effective mass. Besides, constrained by the Wiedemann-Franz-Lorenz's law [3], the electronic thermal conductivity and electrical conductivity have similar curves. In general, electron transport properties are directly related to the carrier relaxation time (τ). Therefore, we further calculated it by using the formula $\tau = \frac{m^*\mu}{e}$ [50], here m^*, μ, and e are carrier effective mass, mobility, and electron charge, as listed in Table 3. After taking the relaxation time, the monolayers exhibit high electrical conductivity of up to $10^6 \sim 10^7$ $\Omega^{-1} \cdot m^{-1}$, also high PF of 18.04~47.34 mW·K^{-2}·m^{-1} at 300 K (see Table 4), which are higher than or comparable to those of their bulk structures (0.8~1.0 mW/mK^2) [23], Pd$_2$Se$_3$ (1.21~1.61 mW·K^{-2}·m^{-1}) [6], PdSe$_2$ (5~25 mW·K^{-2}·m^{-1}) [51], and Tl$_2$O (10~33 mW·K^{-2}·m^{-1}) [52].

Table 4. The maximums of Seebeck coefficient S (mV·K), electron thermal conductivity κ_e (W·m^{-1}·K^{-1}), electrical conductivity σ ($\times 10^6$ Ω^{-1}·m^{-1}), lattice thermal conductivity κ_l (W·m^{-1}·K^{-1}), PF (mW·K^{-2}·m^{-1}) and ZT values in the chemical potential of −1 eV~1 eV at 300 K.

Monolayers	Direction	Carriertype	S	σ	κ_e	κ_l	PF	ZT
CuSbS$_2$	aa	p-type	0.46	45.96	329.64	2.96	47.34	1.17
		n-type	0.33	4.66	31.01		5.07	0.16
	bb	p-type	0.36	26.56	191.40	2.38	15.50	0.65
		n-type	0.43	20.47	147.55		24.43	0.74
CuSbSe$_2$	aa	p-type	0.24	8.52	60.52	0.49	8.05	1.09
		n-type	0.15	3.80	24.82		10.84	0.91
	bb	p-type	0.17	7.49	53.62	0.12	3.89	0.84
		n-type	0.23	21.64	145.33		20.42	1.53
CuBiS$_2$	aa	p-type	0.31	17.87	125.18	3.30	19.79	0.76
		n-type	0.23	1.84	12.27		1.96	0.14
	bb	p-type	0.24	28.63	204.23	1.32	17.45	0.91
		n-type	0.29	7.96	57.16		11.15	0.94
CuBiSe$_2$	aa	p-type	0.19	15.79	108.79	0.23	17.93	1.03
		n-type	0.14	2.40	14.66		5.00	0.55
	bb	p-type	0.13	18.19	128.59	0.21	7.74	0.50
		n-type	0.18	8.33	57.81		18.04	1.03

3.5. Phonon Transport Properties

The phonon dispersions are shown in Figure 5, where the red, green, blue, and pink curves denote the out-of-plane acoustic (ZA), longitudinal acoustic (LA), transverse acoustic (TA), and optical phonons, respectively. As can be seen, these phonon dispersions have no virtual frequencies, indicating that these four monolayers have high kinetic stability. As the Sb(Bi) atoms are heavier than the others, they show lower phonon frequencies, while the lighter S(Se) atoms possess higher frequencies. Meanwhile, there is some coupling between in-plane (XY) and out-of-plane phonons (ZZ) for all atoms, which can be attributed to the fact that each atom is dispersed in multiple layers (see Figure 1), which breaks the plane symmetry of their structure and allows more phonons to participate in scattering [53]. Additionally, the lowest optical mode boundary frequencies at Γ point of the monolayers are within 0.42~0.89 THz, which are close to those of SnSe (~0.99 THz) [54], KAgS (~1.20 THz) [50], and PbSe (~0.63 THz) [55], indicating that their optical modes softening is relatively severe, as in these materials with intrinsic low thermal conductivity. Further, the low-frequency optical modes at Γ points are caused by the antiparallel motions of the outer Sb/Bi and S/Se atoms, which can effectively increase the phonon dissipation and further reduce the phonon lifetime [55]. Comparatively, for monolayer CuSbSe$_2$ and CuBiSe$_2$, their phonon frequencies are relatively lower, and more coupling occurs in the low frequency range, resulting in their scattering free path is shorter, and thus have lower phonon lifetimes.

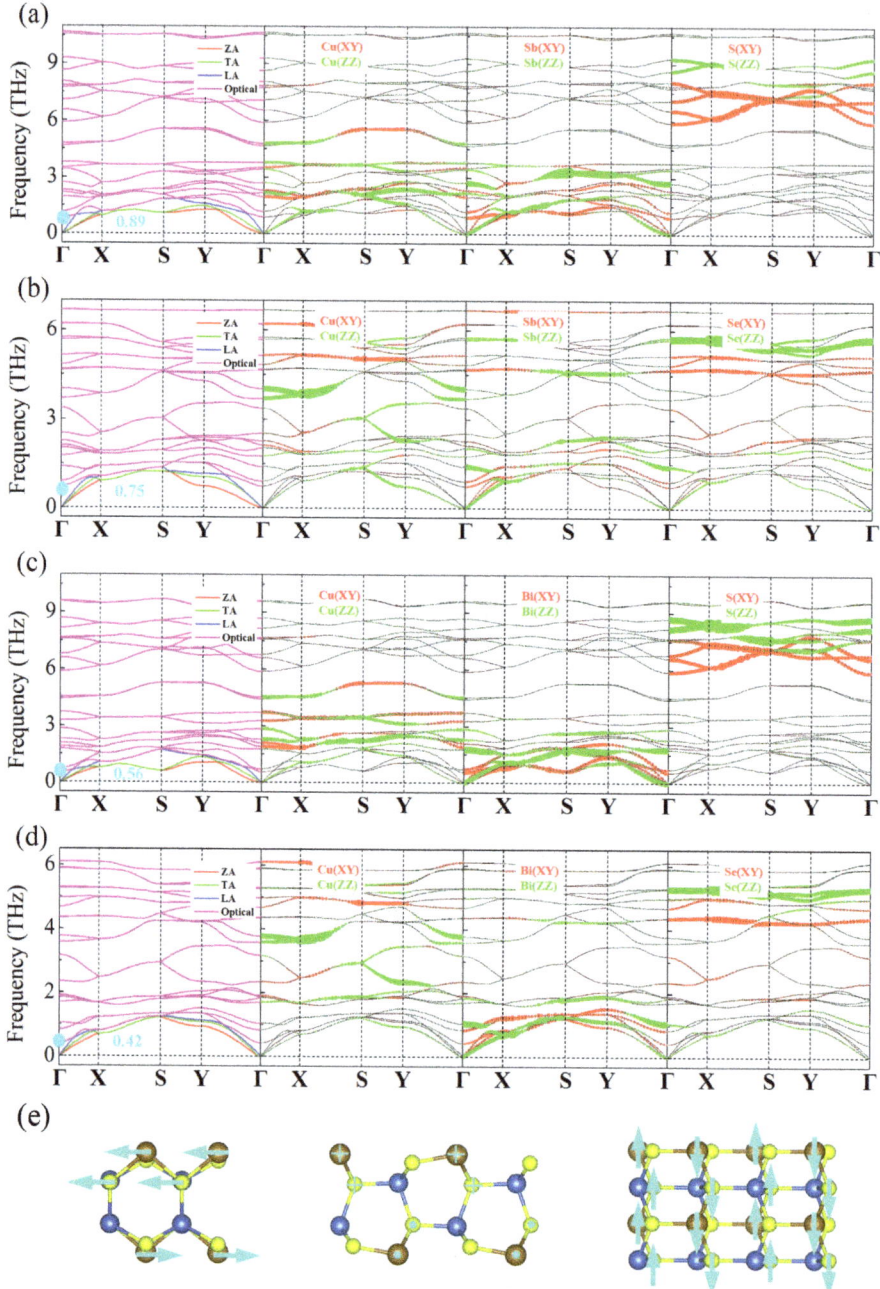

Figure 5. The orbital-resolved phonon dispersion of: (**a**) CuSbS$_2$; (**b**) CuSbSe$_2$; (**c**) CuBiS$_2$; (**d**) CuBiSe$_2$; where the XY and ZZ denote the in-plane and out-of-plane phonons; and (**e**) the schematic diagram of low frequency optical phonon at Γ point for monolayer CuSbS$_2$, where the arrows represent the direction of the vibration.

The lattice thermal conductivity, $k_l = \sum_\lambda c_{ph,\lambda} v_{\alpha,\lambda}^2 \tau_\lambda$, can be expressed as the volumetric specific heat $c_{ph,\lambda}$, group velocity $v_{\alpha,\lambda}$, and phonon lifetime τ_λ, respectively. The group velocity $v_{\alpha,\lambda} = \partial \omega(q)/\partial q$, can also be calculated by the first derivative of frequency $\omega(q)$ with respect to the wave vector q [38]. As seen in Figure 6a–d, the LA modes for all the monolayers exhibit maximum group velocities of 2.06~3.59 km/s, smaller than those of Arsenene and Antimonene (~4.5 km/s) [56], BP (~8.6 km/s) [57], and MoS$_2$ (~6.5 km/s) [58]. Although the optical modes also exhibit large group velocities in high frequency region, their phonon lifetime is very small, almost zero, as shown in Figure 6e–h, so the k_l of these monolayers are mainly contributed by acoustic modes. In addition, for monolayer CuSbSe$_2$ and CuBiSe$_2$, their phonon lifetimes are significantly shorter than those of the others, which is mainly due to their strong coupling at low frequency phonons, as analyzed above.

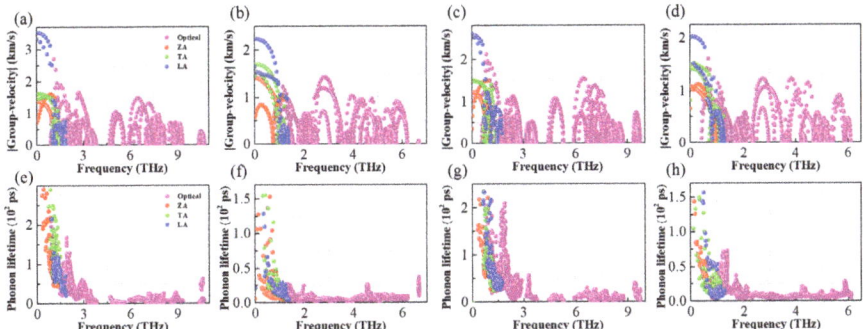

Figure 6. Group velocities (**a–d**) and phonon lifetimes (**e–h**) of 2D CuMN$_2$ (M = Sb, Bi; N = S, Se), here the red, blue, green, and pink dot correspond to ZA, LA, TA, and Optical phonons, respectively.

Generally, we can use the Grüneisen parameters γ to describe the anharmonic interactions of a material, which is effective mean to analyze the physical nature of lattice thermal conductivity. It can be obtained from the relationship of phonon frequency $\omega(q)$ and volume V as $\gamma = \frac{V}{\omega(q)} \frac{\partial \omega(q)}{\partial V}$ [59]. For a large $|\gamma|$ indicates the strong phonon-phonon anharmonic scattering, resulting in a low intrinsic k_l. As shown in Figure 7, all the monolayers exhibited the high $|\gamma|$ in the low frequency range, which are similar to that of KAgX (X = S, Se) [50]. Obviously, CuSbSe$_2$ and CuBiSe$_2$ exhibited larger values than those of CuSbS$_2$ and CuBiS$_2$, and thus have inherently stronger anharmonic interactions, as well as lower k_l. Moreover, the negative γ indicate that these materials may have negative thermal expansion (NTE) properties [4].

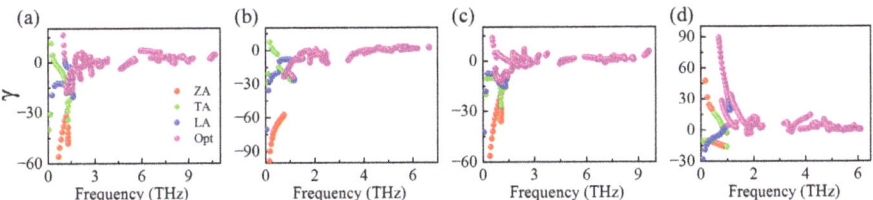

Figure 7. The Grüneisen parameters γ for monolayer: (**a**) CuSbS$_2$; (**b**) CuSbSe$_2$; (**c**) CuBiS$_2$; and (**d**) CuBiSe$_2$.

Although the volumetric specific heat $c_{ph,\lambda}$ is also directly related to the k_l, the difference is very small, especially as the temperature increases, as seen in Figure 8a. As a result, all the materials exhibit low k_l within ~3.30 W·m^{-1}·K^{-1}, with CuSbSe$_2$ and CuBiSe$_2$ having lower values due to stronger phonon anharmonic interactions and low phonon lifetimes. Additionally, we can notice that the k_l is higher in a-axis, which can be attributed to the stronger bonding, as well as higher Young's moduli in this direction, and thus better heat transport. As shown in Figure 8b and Table 4, the monolayers show the low lattice thermal conductivities of 0.23~3.30 W·m^{-1}·K^{-1} at 300 K, which are comparable to or lower than those of bilayer SnSe (0.9 W·m^{-1}·K^{-1}) [60], Tl$_2$O (0.9~1.2 W·m^{-1}·K^{-1}) [52], Tetradymites (1.2~2.1 W·m^{-1}·K^{-1}) [61], and Antimonene (5 W·m^{-1}·K^{-1}) [56].

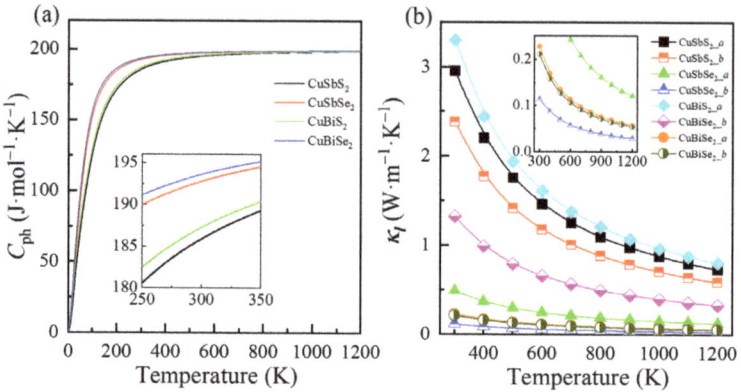

Figure 8. (a) The volumetric specific heat, and (b) lattice thermal conductivity of 2D CuMN$_2$ (M = Sb, Bi; N = S, Se).

3.6. Thermoelectric Figure of Merit

Finally, we fitted the thermoelectric figure of merit (ZT) of these four monolayers at 300 K, as shown in Figure 9 (see Figures S8 and S9 for more details about the thermoelectric properties at without SOC functional). Obviously, the monolayers exhibit higher p-type ZT values in a-axis, while higher n-type ZT in b-axis, which is consistent with the results of higher hole mobility in the a-axis, while higher electron mobility in b-axis. As listed in Table 4, the monolayers simultaneously exhibit high ZT values for p-type of 0.91~1.17, and n-type of 0.74~1.53 at 300 K, which are higher than or comparable to those of many 2D thermoelectric materials, such as Pd$_2$Se$_3$ (0.9) [6], Tellurene (0.6) [62], InSe (0.5) [63], and SnSe (0.5) [64].

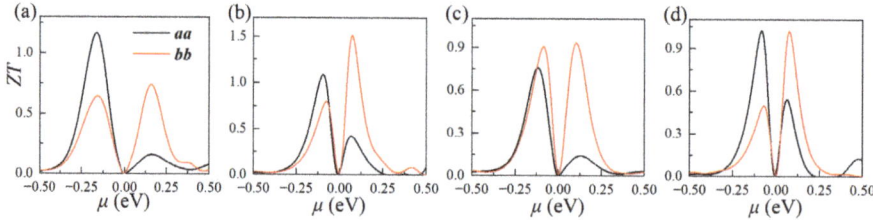

Figure 9. ZT values of the monolayer: (a) CuSbS$_2$; (b) CuSbSe$_2$; (c) CuBiS$_2$; and (d) CuBiSe$_2$ at 300 K.

4. Conclusions

In this work, we investigated the stability, mechanical, electrical, and phonon transport properties of 2D CuMN$_2$ (M = Sb, Bi; N = S, Se). We found that monolayers possess the acceptable cleavage energies of 0.68~0.93 J/m^2, and narrow band-gaps of 0.57~1.10 eV, respectively. Based on the acoustic phonon-limited method, we revealed that the electron and hole mobility are basically in the range of 10^2~10^3 cm^2·V^{-1}·s^{-1}. Besides, they also have high electrical conductivity of 10^6~10^7 $\Omega^{-1} \cdot m^{-1}$, high PF of 18.04~47.34 mW·K^{-2}·m^{-1} at 300 K. Furthermore, due to the stronger phonon anharmonic interactions and low phonon lifetimes, their lattice thermal conductivities are as low as 0.23~3.30 Wm^{-1} K^{-1}. As a result, all the monolayers simultaneously exhibit high ZT values for p-type of 0.91~1.17, and n-type of 0.74~1.53 at 300 K, indicating that they have potential applications in nanoelectronic and thermoelectric devices.

Supplementary Materials: The following supporting information can be downloaded at: https://www.mdpi.com/article/10.3390/ma15196700/s1, Table S1: The cleavage energies, Figure S1: the Electron Localization Functions, Figure S2: the ab initio molecules dynamics simulation, Table S2: Bader charge analysis results, Figure S3: the band structures at PBE functional without and with SOC functional, Figure S4: the partial density of states, Figure S5, Table S3, Figure S6: the details on carrier mobility calculations, Figure S7: the MLWFs calculated by Wannnier90 code, Figures S8 and S9: the thermoelectric properties of monolayer $CuMN_2$ (M = Sb, Bi; N = S, Se) without SOC functional.

Author Contributions: W.F. and Y.C.: conceptualization, formal analysis, validation, visualization, writing—review and editing. K.K.: formal analysis, validation, visualization, data curation, methodology. M.L.: supervision, funding acquisition, software, project administration, writing—review and editing. All authors have read and agreed to the published version of the manuscript.

Funding: We gratefully acknowledge the financial support from the National Key R&D Program of China (Grant No. 2019YFB1503500), the National Natural Science Foundation of China (Grant Nos. 61874040, 11774082 and 11975093), the Natural Science Foundation of Hubei Province (Grant Nos. 2019CFA006 and 2021EHB005), and the Program for Science and Technology Innovation Team in Colleges of Hubei Province (Grant No. T201901).

Institutional Review Board Statement: Not applicable.

Informed Consent Statement: Not applicable.

Data Availability Statement: The data presented in this study are available on request from the corresponding authors. The data are not publicly available due to ongoing research in the project.

Conflicts of Interest: The authors declare no conflict of interest.

References

1. Wang, S.F.; Zhang, Z.G.; Wang, B.T.; Zhang, J.R.; Wang, F.W. Intrinsic Ultralow Lattice Thermal Conductivity in the Full-Heusler Compound Ba_2AgSb. *Phys. Rev. Appl.* **2022**, *17*, 034023. [CrossRef]
2. Gao, Z.; Liu, G.; Ren, J. High Thermoelectric Performance in Two-Dimensional Tellurium: An Ab Initio Study. *ACS Appl. Mater. Interfaces* **2018**, *10*, 40702–40709. [CrossRef] [PubMed]
3. Fang, W.; Wei, H.; Xiao, X.; Chen, Y.; Kuang, K.; Li, M.; He, Y. XTlO (X = K, Rb, Cs): Novel 2D semiconductors with high electron mobilities, ultra-low lattice thermal conductivities and high thermoelectric figures of merit at room temperature. *Appl. Surf. Sci.* **2022**, *599*, 153924. [CrossRef]
4. Fang, W.; Wei, H.; Xiao, X.; Chen, Y.; Li, M.; He, Y. Monolayer SnX (X = O, S, Se): Two-Dimensional Materials with Low Lattice Thermal Conductivities and High Thermoelectric Figures of Merit. *ACS Appl. Energy Mater.* **2022**, *5*, 7802–7812. [CrossRef]
5. Wu, Y.; Xu, K.; Ma, C.; Chen, Y.; Lu, Z.; Zhang, H.; Fang, Z.; Zhang, R. Ultrahigh carrier mobilities and high thermoelectric performance at room temperature optimized by strain-engineering to two-dimensional aw-antimonene. *Nano Energy* **2019**, *63*, 103870. [CrossRef]
6. Naghavi, S.S.; He, J.; Xia, Y.; Wolverton, C. Pd_2Se_3 Monolayer: A Promising Two-Dimensional Thermoelectric Material with Ultralow Lattice Thermal Conductivity and High Power Factor. *Chem. Mater.* **2018**, *30*, 5639–5647. [CrossRef]
7. Tayari, V.; Senkovskiy, B.V.; Rybkovskiy, D.; Ehlen, N.; Fedorov, A.; Chen, C.Y.; Avila, J.; Asensio, M.; Perucchi, A.; di Pietro, P.; et al. Quasi-two-dimensional thermoelectricity in SnSe. *Phys. Rev. B* **2018**, *97*, 136. [CrossRef]
8. Huang, S.; Wang, Z.; Xiong, R.; Yu, H.; Shi, J. Significant enhancement in thermoelectric performance of Mg_3Sb_2 from bulk to two-dimensional mono layer. *Nano Energy* **2019**, *62*, 212–219. [CrossRef]
9. Wei, T.-R.; Hu, P.; Chen, H.; Zhao, K.; Qiu, P.; Shi, X.; Chen, L. Quasi-two-dimensional GeSbTe compounds as promising thermoelectric materials with anisotropic transport properties. *Appl. Phys. Lett.* **2019**, *114*, 053903. [CrossRef]
10. Yu, J.; Li, T.; Nie, G.; Zhang, B.P.; Sun, Q. Ultralow lattice thermal conductivity induced high thermoelectric performance in the delta-Cu_2S monolayer. *Nanoscale* **2019**, *11*, 10306–10313. [CrossRef]
11. Cao, W.; Wang, Z.; Miao, L.; Shi, J.; Xiong, R. Thermoelectric Properties of Strained beta-Cu_2Se. *ACS Appl. Mater. Interfaces* **2021**, *13*, 34367–34373. [CrossRef] [PubMed]

12. Yao, Y.; Zhang, B.P.; Pei, J.; Liu, Y.C.; Li, J.F. Thermoelectric performance enhancement of Cu_2S by Se doping leading to a simultaneous power factor increase and thermal conductivity reduction. *J. Mater. Chem. C* **2017**, *5*, 7845–7852. [CrossRef]
13. Bo, L.; Li, F.J.; Hou, Y.B.; Zuo, M.; Zhao, D.G. Enhanced Thermoelectric Performance of Cu_2Se via Nanostructure and Compositional Gradient. *Nanomaterials* **2022**, *12*, 640. [CrossRef] [PubMed]
14. Li, L.W.; Peng, C.X.; Chen, J.; Ma, Z.; Chen, Y.Q.; Li, S.Y.; Wang, J.L.; Wang, C. Study the effect of alloying on the phase transition behavior and thermoelectric properties of Ag_2S. *J. Alloys Compd.* **2021**, *886*, 161241. [CrossRef]
15. Ma, J.; Wei, S.H. Origin of Novel Diffusions of Cu and Ag in Semiconductors: The Case of CdTe. *Phys. Rev. Lett.* **2013**, *110*, 235901. [CrossRef]
16. Deng, H.X.; Luo, J.W.; Li, S.S.; Wei, S.H. Origin of the Distinct Diffusion Behaviors of Cu and Ag in Covalent and Ionic Semiconductors. *Phys. Rev. Lett.* **2016**, *117*, 165901. [CrossRef]
17. Freitas, R.R.Q.; Mota, F.D.; Rivelino, R.; de Castilho, C.M.C.; Kakanakova-Georgieva, A.; Gueorguiev, G.K. Spin-orbit-induced gap modification in buckled honeycomb XBi and XBi_3 (X = B, Al, Ga, and In) sheets. *J. Phys. Condens. Matter* **2015**, *27*, 485306. [CrossRef]
18. dos Santos, R.B.; Rivelino, R.; Mota, F.D.; Kakanakova-Georgieva, A.; Gueorguiev, G.K. Feasibility of novel $(H_3C)(n)X(SiH_3)(3-n)$ compounds (X = B, Al, Ga, In): Structure, stability, reactivity, and Raman characterization from ab initio calculations. *Dalton Trans.* **2015**, *44*, 3356–3366. [CrossRef]
19. Kim, K.C.; Lim, S.S.; Lee, S.H.; Hong, J.; Cho, D.Y.; Mohamed, A.Y.; Koo, C.M.; Baek, S.H.; Kim, J.S.; Kim, S.K. Precision Interface Engineering of an Atomic Layer in Bulk $Bi2Te3$ Alloys for High Thermoelectric Performance. *ACS Nano* **2019**, *13*, 7146–7154. [CrossRef]
20. Wu, C.Y.; Sun, L.; Han, J.C.; Gong, H.R. Band structure, phonon spectrum, and thermoelectric properties of beta-BiAs and beta-BiSb monolayers. *J. Mater. Chem. C* **2020**, *8*, 581–590. [CrossRef]
21. Gassoumi, A.; Alfaify, S.; Ben Nasr, T.; Bouarissa, N. The investigation of crystal structure, elastic and optoelectronic properties of $CuSbS_2$ and $CuBiS_2$ compounds for photovoltaic applications. *J. Alloys Compd.* **2017**, *725*, 181–189. [CrossRef]
22. Parker, D.; Singh, D.J. Transport properties of hole-doped $CuBiS_2$. *Phys. Rev. B* **2011**, *83*, 233206. [CrossRef]
23. Alsaleh, N.M.; Singh, N.; Schwingenschlögl, U. Role of interlayer coupling for the power factor of $CuSbS_2$ and $CuSbSe_2$. *Phys. Rev. B* **2016**, *94*, 125440. [CrossRef]
24. Hafner, J. Ab-initio simulations of materials using VASP: Density-functional theory and beyond. *J. Comput. Chem.* **2008**, *29*, 2044–2078. [CrossRef]
25. Blochl, P.E. Projector augmented-wave method. *Phys. Rev. B* **1994**, *50*, 17953–17979. [CrossRef]
26. Zhang, T.; Li, M.K.; Chen, J.; Wang, Y.; Miao, L.S.; Lu, Y.M.; He, Y.B. Multi-component ZnO alloys: Bandgap engineering, hetero-structures, and optoelectronic devices. *Mater. Sci. Eng. R* **2022**, *147*, 100661. [CrossRef]
27. Eriksson, F.; Fransson, E.; Erhart, P. The Hiphive Package for the Extraction of High-Order Force Constants by Machine Learning. *Adv. Theory Simul.* **2019**, *2*, 1800184. [CrossRef]
28. Togo, A.; Tanaka, I. First principles phonon calculations in materials science. *Scripta Mater.* **2015**, *108*, 1–5. [CrossRef]
29. Togo, A.; Chaput, L.; Tanaka, I. Distributions of phonon lifetimes in Brillouin zones. *Phys. Rev. B* **2015**, *91*, 094306. [CrossRef]
30. Pizzi, G.; Vitale, V.; Arita, R.; Blugel, S.; Freimuth, F.; Geranton, G.; Gibertini, M.; Gresch, D.; Johnson, C.; Koretsune, T.; et al. Wannier90 as a community code: New features and applications. *J. Phys. Condens. Matter* **2020**, *32*, 165902. [CrossRef]
31. Jung, J.H.; Park, C.H.; Ihm, J. A Rigorous Method of Calculating Exfoliation Energies from First Principles. *Nano Lett.* **2018**, *18*, 2759–2765. [CrossRef] [PubMed]
32. Song, Y.Q.; Yuan, J.H.; Li, L.H.; Xu, M.; Wang, J.F.; Xue, K.H.; Miao, X.S. KTlO: A metal shrouded 2D semiconductor with high carrier mobility and tunable magnetism. *Nanoscale* **2019**, *11*, 1131–1139. [CrossRef] [PubMed]
33. Zhao, S.; Li, Z.; Yang, J. Obtaining two-dimensional electron gas in free space without resorting to electron doping: An electride based design. *J. Am. Chem. Soc.* **2014**, *136*, 13313–13318. [CrossRef] [PubMed]
34. Zhang, P.; Yuan, J.-H.; Fang, W.-Y.; Li, G.; Wang, J. Two-dimensional V-shaped PdI_2: Auxetic semiconductor with ultralow lattice thermal conductivity and ultrafast alkali ion mobility. *Appl. Surf. Sci.* **2022**, *601*, 154176. [CrossRef]
35. Fang, W.Y.; Li, P.A.; Yuan, J.H.; Xue, K.H.; Wang, J.F. Nb_2SiTe_4 and Nb_2GeTe_4: Unexplored 2D Ternary Layered Tellurides with High Stability, Narrow Band Gap and High Electron Mobility. *J. Electron. Mater.* **2020**, *49*, 959–968. [CrossRef]
36. Xiong, W.Q.; Huang, K.X.; Yuan, S.J. The mechanical, electronic and optical properties of two-dimensional transition metal chalcogenides MX_2 and M_2X_3 (M = Ni, Pd; X = S, Se, Te) with hexagonal and orthorhombic structures. *J. Mater. Chem. C* **2019**, *7*, 13518–13525. [CrossRef]
37. Lajevardipour, A.; Neek-Amal, M.; Peeters, F.M. Thermomechanical properties of graphene: Valence force field model approach. *J. Phys. Condens. Matter* **2012**, *24*, 175303. [CrossRef]
38. Fang, W.Y.; Xiao, X.L.; Wei, H.R.; Chen, Y.; Li, M.K.; He, Y.B. The elastic, electron, phonon, and vibrational properties of monolayer XO_2 (X = Cr, Mo, W) from first principles calculations. *Mater. Today Commun.* **2022**, *30*, 103183. [CrossRef]
39. Fang, W.Y.; Yue, C.; Pan, Y.; Wei, H.R.; Xiao, X.L.; Li, M.K.; Rajeev, A.; He, Y.B. Elastic constants, electronic structures and thermal conductivity of monolayer XO_2 (X = Ni, Pd, Pt). *Acta Phys. Sin.* **2021**, *70*, 246301. [CrossRef]
40. Xie, Q.Y.; Liu, P.F.; Ma, J.J.; Kuang, F.G.; Zhang, K.W.; Wang, B.T. Monolayer SnI_2: An Excellent p-Type Thermoelectric Material with Ultralow Lattice Thermal Conductivity. *Materials* **2022**, *15*, 3147. [CrossRef]

41. Xiao, W.Z.; Xiao, G.; Rong, Q.Y.; Wang, L.L. Theoretical discovery of novel two-dimensional V-A-N binary compounds with auxiticity. *Phys. Chem. Chem. Phys.* **2018**, *20*, 22027–22037. [CrossRef] [PubMed]
42. Qiao, M.; Wang, Y.; Li, Y.F.; Chen, Z.F. Tetra-silicene: A Semiconducting Allotrope of Silicene with Negative Poisson's Ratios. *J. Phys. Chem. C* **2017**, *121*, 9627–9633. [CrossRef]
43. Peng, R.; Ma, Y.; He, Z.; Huang, B.; Kou, L.; Dai, Y. Single-Layer Ag_2S: A Two-Dimensional Bidirectional Auxetic Semiconductor. *Nano Lett.* **2019**, *19*, 1227–1233. [CrossRef]
44. Yuan, J.H.; Yu, N.N.; Xue, K.H.; Miao, X.S. Stability, electronic and thermodynamic properties of aluminene from first-principles calculations. *Appl. Surf. Sci.* **2017**, *409*, 85–90. [CrossRef]
45. Wada, T.; Maeda, T. Optical properties and electronic structures of $CuSbS_2$, $CuSbSe_2$, and $CuSb(S_{1-x}Se_x)_2$ solid solution. *Phys. Status Solidi C* **2017**, *14*, 1600196. [CrossRef]
46. Prudhvi Raju, N.; Thangavel, R. Theoretical investigation of spin–orbit coupling on structural, electronic and optical properties for $CuAB_2$ (A = Sb, Bi; B = S, Se) compounds using Tran–Blaha-modified Becke–Johnson method: A first-principles approach. *J. Alloys Compd.* **2020**, *830*, 154621. [CrossRef]
47. Zeng, L.; Xin, Z.; Chen, S.W.; Du, G.; Kang, J.F.; Liu, X.Y. Phonon-Limited Electron Mobility in Single-Layer MoS_2. *Chin. Phys. Lett.* **2014**, *31*, 027301. [CrossRef]
48. Shao, Z.G.; Ye, X.S.; Yang, L.; Wang, C.L. First-principles calculation of intrinsic carrier mobility of silicene. *J. Appl. Phys.* **2013**, *114*, 093712. [CrossRef]
49. Pu, C.Y.; Yu, J.H.; Yu, R.M.; Tang, X.; Zhou, D.W. Hydrogenated PtP_2 monolayer: Theoretical predictions on the structure and charge carrier mobility. *J. Mater. Chem. C* **2019**, *7*, 12231–12239. [CrossRef]
50. Zhu, X.L.; Yang, H.Y.; Zhou, W.X.; Wang, B.T.; Xu, N.; Xie, G.F. KAgX (X = S, Se): High-Performance Layered Thermoelectric Materials for Medium-Temperature Applications. *ACS Appl. Mater. Interfaces* **2020**, *12*, 36102–36109. [CrossRef]
51. Qin, D.; Yan, P.; Ding, G.; Ge, X.; Song, H.; Gao, G. Monolayer $PdSe_2$: A promising two-dimensional thermoelectric material. *Sci. Rep.* **2018**, *8*, 2764. [CrossRef] [PubMed]
52. Huang, H.H.; Xing, G.; Fan, X.; Singh, D.J.; Zheng, W.T. Layered Tl_2O: A model thermoelectric material. *J. Mater. Chem. C* **2019**, *7*, 5094–5103. [CrossRef]
53. Huang, L.F.; Gong, P.L.; Zeng, Z. Correlation between structure, phonon spectra, thermal expansion, and thermomechanics of single-layer MoS_2. *Phys. Rev. B* **2014**, *90*, 045409. [CrossRef]
54. Li, C.W.; Hong, J.; May, A.F.; Bansal, D.; Chi, S.; Hong, T.; Ehlers, G.; Delaire, O. Orbitally driven giant phonon anharmonicity in SnSe. *Nat. Phys.* **2015**, *11*, 1063. [CrossRef]
55. Liu, P.F.; Bo, T.; Xu, J.P.; Yin, W.; Zhang, J.R.; Wang, F.W.; Eriksson, O.; Wang, B.T. First-principles calculations of the ultralow thermal conductivity in two-dimensional group-IV selenides. *Phys. Rev. B* **2018**, *98*, 235426. [CrossRef]
56. Sharma, S.; Kumar, S.; Schwingenschlögl, U. Arsenene and Antimonene: Two-Dimensional Materials with High Thermoelectric Figures of Merit. *Phys. Rev. Appl.* **2017**, *8*, 044013. [CrossRef]
57. Zhu, L.Y.; Zhang, G.; Li, B.W. Coexistence of size-dependent and size-independent thermal conductivities in phosphorene. *Phys. Rev. B* **2014**, *90*, 214302. [CrossRef]
58. Hong, Y.; Zhang, J.C.; Zeng, X.C. Thermal Conductivity of Monolayer $MoSe_2$ and MoS_2. *J. Phys. Chem. C* **2016**, *120*, 26067–26075. [CrossRef]
59. Xie, Q.Y.; Ma, J.J.; Liu, Q.Y.; Liu, P.F.; Zhang, P.; Zhang, K.W.; Wang, B.T. Low thermal conductivity and high performance anisotropic thermoelectric properties of XSe (X = Cu, Ag, Au) monolayers. *Phys. Chem. Chem. Phys.* **2022**, *24*, 7303–7310. [CrossRef]
60. Hung, N.T.; Nugraha, A.R.T.; Saito, R. Designing high-performance thermoelectrics in two-dimensional tetradymites. *Nano Energy* **2019**, *58*, 743–749. [CrossRef]
61. Nag, S.; Saini, A.; Singh, R.; Kumar, R. Ultralow lattice thermal conductivity and anisotropic thermoelectric performance of AA stacked SnSe bilayer. *Appl. Surf. Sci.* **2020**, *512*, 145640. [CrossRef]
62. Sharma, S.; Singh, N.; Schwingenschlögl, U. Two-Dimensional Tellurene as Excellent Thermoelectric Material. *ACS Appl. Energy Mater.* **2018**, *1*, 1950–1954. [CrossRef]
63. Hung, N.T.; Nugraha, A.R.T.; Saito, R. Two-dimensional InSe as a potential thermoelectric material. *Appl. Phys. Lett.* **2017**, *111*, 092107. [CrossRef]
64. Wang, F.Q.; Zhang, S.H.; Yu, J.B.; Wang, Q. Thermoelectric properties of single-layered SnSe sheet. *Nanoscale* **2015**, *7*, 15962–15970. [CrossRef] [PubMed]

Article

Microstructure Evolution in Plastic Deformed Bismuth Telluride for the Enhancement of Thermoelectric Properties

Haishan Shen [1], In-Yea Kim [2], Jea-Hong Lim [2], Hong-Baek Cho [1] and Yong-Ho Choa [1,*]

[1] Department of Materials Science and Chemical Engineering, Hanyang University, 55 Hanyangdaehak-ro, Sangnok-gu, Ansan 15588, Korea; seadheart@hanyang.a.kr (H.S.); hongbaek@hanyang.ac.kr (H.-B.C.)
[2] Department of Materials Science and Engineering, Gachon University, 1342 Seongnamdaero, Sujeong-gu, Seongnam-si 13120, Korea; kiy7484@gachon.ac.kr (I.-Y.K.); limjh@gachon.ac.kr (J.-H.L.)
* Correspondence: choa15@hanyang.ac.kr; Tel.: +82-31-400-5650

Abstract: Thermoelectric generators are solid-state energy-converting devices that are promising alternative energy sources. However, during the fabrication of these devices, many waste scraps that are not eco-friendly and with high material cost are produced. In this work, a simple powder processing technology is applied to prepare n-type Bi_2Te_3 pellets by cold pressing (high pressure at room temperature) and annealing the treatment with a canning package to recycle waste scraps. High-pressure cold pressing causes the plastic deformation of densely packed pellets. Then, the thermoelectric properties of pellets are improved through high-temperature annealing (500 °C) without phase separation. This enhancement occurs because tellurium cannot escape from the canning package. In addition, high-temperature annealing induces rapid grain growth and rearrangement, resulting in a porous structure. Electrical conductivity is increased by abnormal grain growth, whereas thermal conductivity is decreased by the porous structure with phonon scattering. Owing to the low thermal conductivity and satisfactory electrical conductivity, the highest ZT value (i.e., 1.0) is obtained by the samples annealed at 500 °C. Hence, the proposed method is suitable for a cost-effective and environmentally friendly way.

Keywords: n-type Bi_2Te_3; powder processing; cold pressing; canning package; recycled waste scraps

Citation: Shen, H.; Kim, I.-Y.; Lim, J.-H.; Cho, H.-B.; Choa, Y.-H. Microstructure Evolution in Plastic Deformed Bismuth Telluride for the Enhancement of Thermoelectric Properties. *Materials* 2022, 15, 4204. https://doi.org/10.3390/ma15124204

Academic Editors: Bao-Tian Wang and Peng-Fei Liu

Received: 19 May 2022
Accepted: 9 June 2022
Published: 14 June 2022

Publisher's Note: MDPI stays neutral with regard to jurisdictional claims in published maps and institutional affiliations.

Copyright: © 2022 by the authors. Licensee MDPI, Basel, Switzerland. This article is an open access article distributed under the terms and conditions of the Creative Commons Attribution (CC BY) license (https://creativecommons.org/licenses/by/4.0/).

1. Introduction

Thermoelectric generators (TEGs) can directly convert thermal energy to electricity. They are excellent materials for renewable energy applications to reduce the environmental impact of CO_2 emissions and achieve net-zero emissions by 2050 under the Paris Agreement [1,2]. These materials recover waste heat from sources, such as manufacturing plants, combustion engines, and even the human body, and convert it to electrical energy. In addition, they are noiseless, have a long service life, and do not require large-scale systems. Accordingly, TEGs have a wide range of possible applications, including Internet-of-Things sensors, wearable devices, internal combustion engine vehicles, and manufacturing plants [3–5]. The performance of TEGs is typically evaluated using the thermoelectric figure of merit (ZT), defined as

$$ZT = \frac{\alpha^2 \sigma}{\kappa} T, \quad (1)$$

where α is the Seebeck coefficient (μV/K); σ is the electrical conductivity (S/m); κ is the thermal conductivity (W/(m·K)); and T is the absolute temperature (K). A high ZT value indicates that thermoelectric materials have a high electric conductivity and Seebeck coefficient; however, their thermal conductivity is low. Many researchers endeavor to improve the ZT value by reducing the thermal conductivity of materials (e.g., phonon-glass electron crystals [6,7] and nanostructured material [8–11]), conducting electronic band engineering [12,13], examining the energy-filtering effect and defect [14–16], and implementing grain boundary engineering [17–19]. Bismuth–tellurium-based materials are excellent TEGs for near-room temperature (RT) applications and are widely used commercially. Typically, the fabrication of TEG devices involves a series of processes, such as the synthesis of bulk ingots via zone melting, dicing, plating cleaning, and soldering [20–23]. During device fabrication, expensive material resources are wasted because many scraps are unavoidably produced [24–26]. The recycling of waste scraps by grinding them into powders and then synthesizing them into high-performance samples is an excellent industrial achievement with reduced costs, energy consumption, and environmental friendliness. Conventionally, the powder sintering process involves hot pressing (HP) [27–29] and a hot-pressing texture (HPT) [27,30] to achieve fully dense ceramics. However, the cost and contamination limit the general applicability of the process to a few ceramic systems [31]. Recently, to compact powders, spark plasma sintering (SPS) has been applied to many Bi–Te-based materials [19,28,32,33] because it enables rapid heating and cooling. However, the foregoing techniques consume considerable energy and are sometimes applied in a vacuum system. In contrast, cold sintering implemented at RT and high pressure (hundreds of megapascals to gigapascals) produces high-density pellets without the necessity of energy-consuming technologies, such as HP, HPT, and SPS [34–36]. In this work, we recycled commercial n-type Bi_2Te_3 scraps and synthesized the compounds by cold pressing (high pressure at room temperature), followed by annealing treatment using a coin cell canning package. Moreover, optimizing the annealing treatment using a canning package improved the thermoelectric performance. Because the electrical conductivity was excellent and the thermal conductivity was low, the highest ZT value (i.e., 1.0) was obtained at a high annealing temperature (500 °C) without phase deformation. The foregoing technique can be applied to mass production with no energy consumption.

2. Experiments

2.1. Experimental Section

For this study, n-type thermoelectric waste scraps were obtained from a commercial Bi_2Te_3 ingot (Kryotherm Co., Ltd., Saint Petersburg, Russia). The scraps were ground into fine powder using a mortar. To exclude the effect of particle size on the thermoelectric properties of the material during the sintering process, the powdered scraps were sieved such that the particle size was between 45 and 53 µm. At RT, the sieved powder was loaded into a hydraulic press die (Ø10 mm) with high-pressure compression (1.5 GPa) for 5 min to produce a highly dense compacted pellet with approximately 1 mm of thickness. To avoid phase separation, a coin cell canning package between two graphite foils was utilized due to the thermal and chemical stabilities of graphite foil (Figure 1). The pellets packed by canning were annealed at different temperatures (300, 400, and 500 °C) for 1 h in a tube furnace with Ar gas flow to enhance the thermoelectric properties.

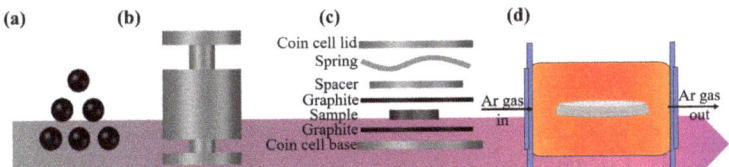

Figure 1. Schematic of experimental process: (**a**) powder grinding and sieving (45–53 µm); (**b**) cold pressing; (**c**) coin cell canning package; and (**d**) annealing process.

2.2. Characterization

The crystal structure and morphology of the pellets were analyzed by X-ray diffraction (XRD, Cu Kα, Rigaku Co., Tokyo, Japan) at 40 kV and 100 mA. The morphology of the pellets was observed by field-emission scanning electron microscopy (FE-SEM, Hitachi Ltd.,

Tokyo, Japan). High-resolution images and microstructure morphologies were obtained using transmission electron microscopy (TEM, JEM-2100F, JEOL, Tokyo, Japan) at 200 kV. To further analyze the microstructure, electron backscattered diffraction (EBSD, Velocity Super, Ametek, Berwyn, PA, USA) was implemented using a field emission gun scanning electron microscope (SU5000, Hitachi, Tokyo, Japan). All EBSD data were analyzed using the TSL-OIM software. Data points with a confidence index of less than 0.1 were removed from the EBSD data. Thermoelectric properties (electrical and Seebeck coefficient) of pellets were measured in an in-plane direction, which possesses a high performance [37]. Electrical conductivity (σ) was measured by the four-point probe method (CMT-SR 1000N, Advanced Instrument Technology, Suwon, Korea) and hall measurement (HMS-5000, Ecopia, Chandler Hall, AZ, USA) at room temperature. Before measuring the electrical conductivity, the pellet thickness was determined using a micrometer caliper (MDC-25MJ, Mitutoyo, Kawasaki, Japan). The Seebeck coefficient (α) was measured at RT using customized measurement techniques. The total thermal conductivity (κ) was calculated as $\kappa = DC_p\rho$, where D, C_p, and ρ are the thermal diffusivity coefficient, specific heat capacity, and density, respectively. The thermal diffusivity coefficient was measured using a laser flash apparatus (Netzch LFA 467, NETZCH, Selb, Germany), and the specific heat capacity was measured using a differential scanning calorimeter (DSC-60 plus, Shimadzu, Kyoto, Japan). The pellet densities were determined using an immersion technique (Archimedes principle) at RT with ethanol as a medium.

3. Results and Discussions

The XRD patterns of the ingot at RT and annealed Bi_2Te_3 pellets (with and without coin cell canning packages) are shown in Figure 2. For the pellets without canning packages, phase separation was observed as the annealing temperature increased. In particular, for the sample without the coin cell package, $Bi_{4-x}Te_{3+x}$ peaks appeared at 500 °C (annealing temperature). This explains the dissociation and sublimation of tellurium during annealing at high temperatures (Figure 2a) [38]. The peak split at (1 0 10) of the XRD pattern was detected at annealing temperatures of 400 and 500 °C, indicating that phase separation occurred at 400 °C and above (Figure 2c). Correspondingly, no phase separation or other peaks were observed in the pellet of the sample with the coin cell canning package, although the pellets were annealed at a high temperature value (500 °C) (Figure 2b,d). This indicates that the closely packed graphite foil surrounding the cold-pressed pellet blocked the escape of tellurium from the samples. Interestingly, the RT sample exhibits slight peak shifts from 37.92° to 37.96°, 38.0°, and 38.02° at 300, 400, and 500 °C, respectively. Moreover, the peak at 500 °C virtually matches the ingot peak (i.e., 38.04°). This may be the result of the residual compressive strain and stress in the test sample due to the applied high pressure and the release of residual energy via the annealing treatment. For the cold-pressed sample prepared at RT (i.e., no annealing) and pellets annealed at 300 °C, the main peaks are at (0 0 6); their intensity was considerably higher than that of the peak at (0 1 5) (the main peak of Bi_2Te_3 and ingot based on standard data (JCPDS no. 85-439)). The relative peak intensity ratios (I_{006}/I_{015}) are 1.237, 1.105, 0.839, and 0.836 at RT, 300, 400, and 500 °C (with the canning package), respectively. These intensity ratios are considerably higher than that of ingot (i.e., 0.775). These results suggest that the cold-pressed pellet subjected to a 1.5-GPa pressure value preferred the (0 0 1) orientation perpendicular to the press direction by high-pressure stress. After the annealing process, the (0 0 1) texture orientation decreased, and the intensity of the main peak of (0 1 5) increased with the annealing temperature. Such a texture reorientation is attributed to the recrystallization process, releasing residual energy from cold pressing [39]. Jun et al. observed the same recrystallization process at the annealing treatment when they applied strain energy to stoichiometric ingots ($Bi_{0.45}Sb_{1.55}Te_3$) by cold pressing (pressure at gigapascal scale) and then annealing at 300 °C in a vacuum [39].

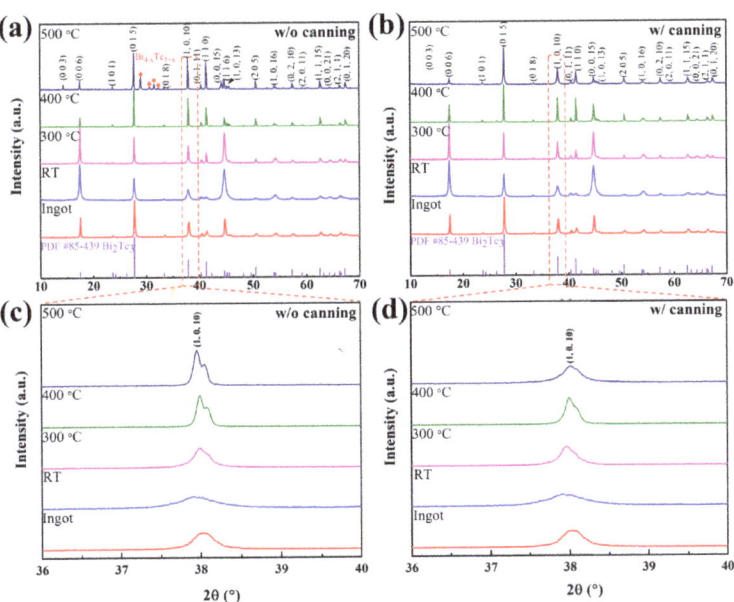

Figure 2. XRD patterns of ingot, Bi$_2$Te$_3$ pellet prepared at RT, and pellets annealed at various temperatures (300, 400, and 500 °C) (**a**) without and (**b**) with coin cell canning package; magnified view of peaks at (1 0 10) (**c**) without and (**d**) with canning package accompanied by peak shift.

The top view and FE-SEM images of the fractured surfaces of specimens prepared at RT, 300, 400, and 500 °C are shown in Figure 3. The top view shows numerous pores and microcracks on the surface of the RT pellet, indicating that interparticle bonds by van der Waals forces are produced via high pressure without any other heat energy. When heat energy was introduced into the cold-pressed pellet, the microcracks disappeared, and the number of large pores were decreased leading to the smoother surface morphology (Figure 3a–d). For the fractured surface, lamellar grains were densely stacked parallel to the pressure direction. As heat energy was applied (i.e., by annealing) to the cold-pressed pellets, grain growth and grain realignment occurred in the annealed samples; the grain growth was considerable with increasing annealing temperature. In addition, many pores among the grains are observed in the pellet annealed at 500 °C; this is consistent with the top view of surface images. The foregoing leads to the conclusion that the densely packed pellets (cold-pressed at high pressure) display rapid grain growth involving the disappearance of pores and microcracks during the annealing process. However, the high-temperature annealing contributes to rapid grain growth and grain realignment, generating many pores and low-density pellets, as shown by the samples annealed at 300 and 400 °C.

The density of each sample was also measured during annealing (with and without the canning process). All pressed pellets have high relative densities exceeding 90% with respect to the ingot density of 7.721 g/cm^3 (Figure S1). For the pellets prepared at RT, the relative density reached 97.9%, indicating that they were densely packed at a high pressure. Further, the relative density of pellets slightly increased at 300 °C and then decreased at 400 and 500 °C. This trend was observed regardless of the canning package and agreeing well with the FE-SEM images, in which some pores and microcracks shrunk during the abnormal grain growth at low-temperature annealing. The rapid grain growth and grain realignment during high-temperature annealing led to the generation of pores and low-density pellets.

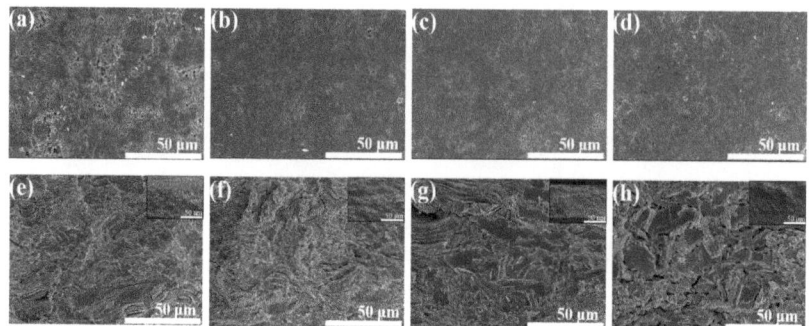

Figure 3. (**a–d**) Top view images; (**e–h**) fractured surfaces and full FE-SEM images of samples at RT, 300, 400, and 500 °C.

The effect of annealing on the electrical conductivity at various annealing temperatures is shown in Figure 4a. As the annealing temperature increased, the electrical conductivity gradually increased from $3.45 \pm 0.04 \times 10^4$ (at RT) to $12.22 \pm 1 \times 10^4$ S/m (at 500 °C) due to the grain growth induced by annealing. In Figure S2, the hall measurement about electrical conductivity also displayed the same trend with 4-probe measurement. The electron mobility at room temperature increased from 56 to 101.22 cm^2/V·s, implying the grain growth for the pellets. The carrier concentration decreased at 300 °C and increased again for high annealing treatment pellets. The Seebeck coefficient of the annealing-treated pellet was slightly smaller than that of the cold-pressed pellet (Figure 4b). The thermal conductivity first increased from 0.61 (at RT) to 1.63 W/m·K (at 400 °C) and then decreased to 1.17 W/m·K at 500 °C (Figure 4c). This may be due to the grain growth and presence of large grains, which induce less electron scattering, leading to high electrical and thermal conductivities. However, the porous structure in pellets may cause considerable phonon scattering, causing the thermal conductivity at 500 °C to be lower than that at 400 °C. Owing to the significantly lower thermal conductivity (0.6 W/m·K), at RT, the ZT value is slightly higher than those of the pellets annealed at 300 and 400 °C. Then, the pellet annealed at 500 °C, the ZT value substantially increased to approximately 1.0, leading to high electrical conductivity and low thermal conductivity. These results indicate that dense cold-pressed pellets compacted by high pressure have numerous microcracks that facilitate low electrical and thermal conductivities due to electron scattering. However, with grain growth and grain realignment during annealing, the microcracks disappeared, thus increasing the electrical and thermal conductivities. In addition, rapid grain growth and grain realignment created porous pellets that reduced the thermal conductivity by phonon scattering.

The crystal orientation mapping images from the EBSD data, with an image quality (IQ) map, a color-coded inverse pole figure (IPF) map, and grain size distribution, clearly verify the present state. The overall IQ map shown in Figure 5a distinctly exhibits the grain morphology of each sample. For the samples at RT and annealed at 300 °C, elongated and non-equiaxed grains were observed. The dark gray shades indicate the grains that are considerably deformed by high pressure and may be highly compacted by small grains. Because of the high pressure in the z-axis direction, the elongated grains preferred the [0 0 1] orientation, as shown in Figure 5b: the red, green, and blue colors represent the [0 0 1], [−1 −2 0], and [1 −1 0] directions, respectively. Moreover, the maximum intensity of the (0 0 l) plane was 13.11 for the samples prepared at RT; the intensity decreased to 8.29 for the samples annealed at 300 °C (Figure S3a,b). However, the grains of the samples annealed at 400 and 500 °C were recrystallized by grain growth and changed to random orientation (Figure 5b). The maximum intensity changed to 4.012 at 400 °C and then increased to 7.227 in the (0 1 5) plane at 500 °C (Figure S3c,d). The grains were recrystallized during the annealing process and changed to a random orientation. In particular, at high annealing temperatures, the grains rapidly grow and rearrange in the (0 1 5) plane, which is the main plane of Bi_2Te_3. Because of the rapid grain growth, a porous structure is observed in the IQ map of the samples annealed at 500 °C without applied pressure (Figure 5a) [40,41].

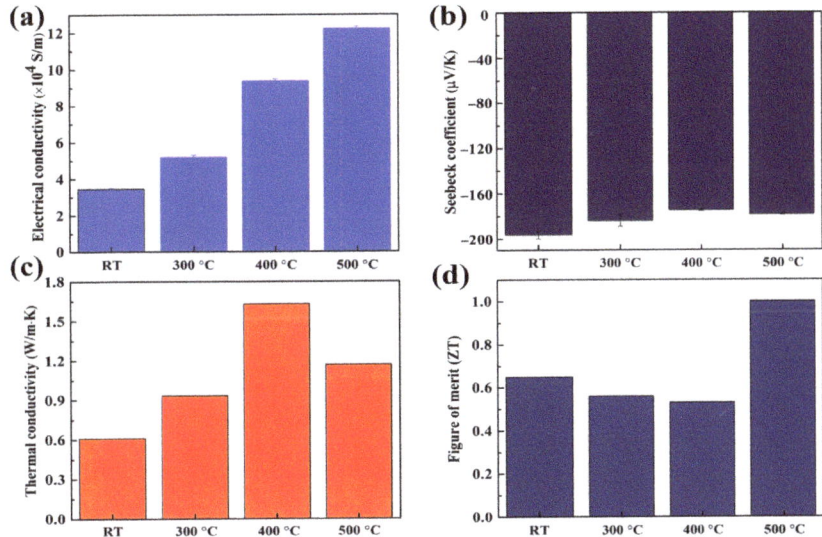

Figure 4. Measurement of thermoelectric properties at RT: (**a**) Electrical conductivity; (**b**) Seebeck coefficient; (**c**) thermal conductivity; and (**d**) ZT at RT, 300, 400, and 500 °C.

The grain size distribution in the samples that have been cold-pressed and annealed at different temperatures after cold pressing is shown in Figure 5c. Most of the grain sizes are less than 1 µm. In the sample cold-pressed at RT, the grain size exceeds 30 µm. These large grains may be elongated due to the distinct local plastic deformation caused by high pressure. The TEM analysis results shown in Figure S4a also indicate the existence of a lamellar structure with elongated subgrains perpendicular to the pressing direction. The highly distributed low-angle grain boundaries in the sample prepared at RT also exhibit local plastic deformation, as shown in Figure S5a [42,43]. These results are based on the unit cell of Bi_2Te_3 that is a trigonal crystal structure consisting of quintuple atomic stacked layers of Bi and Te atoms in the sequence $Te^{(1)}$–Bi–$Te^{(2)}$–Bi–$Te^{(1)}$ [44–46]. The five stacked atomic layers interact by van der Waals forces because of the facile plastic deformation by basal slip [47–49]. In addition, thermo-mechanically treated polycrystalline bulk thermoelectric materials introduce dislocations through plastic deformation. These dislocations improve the densification and control the crystallographic texture of the material, as demonstrated by the case of samples prepared at RT [47]. In contrast, in the less-compacted area packed with small random grains, large pores and microcracks were detected at the interfacial grains (Figure S4b). Overall, in the absence of heat energy, the particles were physically compressed by high pressure through plastic deformation; further, the formation of pores and microcracks resulted in a densely compacted sample.

After the introduction of heat energy, the plastically deformed grains began to coarsen. Consequently, the degree of distribution of low-angle grain boundaries slightly decreased at 300 °C. Then, at 400 and 500 °C, the samples gained higher angle grain boundaries (such as 60°) (Figure S5). At high annealing temperatures (400 and 500 °C), the main grain size distribution trend is bimodal, indicating abnormal grain growth (Figure 5c) [50,51]. The kinetics of abnormal growth is based on the recrystallization and dynamic recovery from the plastic deformation that occurs at elevated temperatures [52,53]. In some instances, grain boundaries were also observed at the interfaces between two large grains; this explains the abnormal grain growth upon isothermal annealing, as shown in Figure S4c. However, at high temperatures, rapid grain growth and nanosized pores were detected in the grains (Figure S4d). Note that this grain size increase was more rapid than the pore elimination during the abnormal grain growth at high annealing temperatures. This growth leads to the coalescence of large grains and production of pores along the large grains; this is consistent with the SEM and EBSD images.

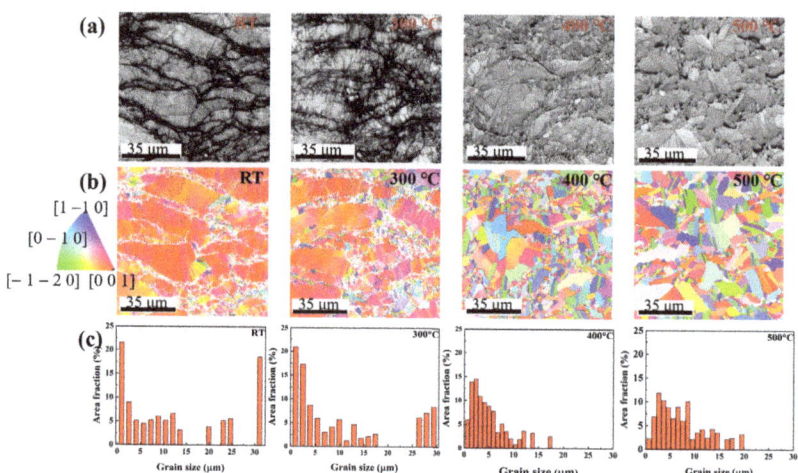

Figure 5. EBSD data with (**a**) IQ map, (**b**) color-coded IPF map, and (**c**) grain size distribution at RT, 300, 400, and 500 °C.

In summary, our results strongly indicate that the pristine powders were highly compacted by plastic deformation under high pressure; then, the particles were coarsened through annealing. At high annealing temperatures, rapid grain growth results in the generation of porous structures (Figure 6a). As shown in Figure 6b, the texture orientation from the (0 0 1) plane in the samples highly compacted at RT is changed into that of random structures, implying grain recrystallization. Finally, the recrystallized subgrains underwent abnormal grain growth at high annealing temperatures for crystallization oriented at (0 1 5). Most Bi_2Te_3 thermoelectric materials (mainly oriented at (0 1 5)) have high thermoelectric properties because they are anisotropic [54–56]. Hence, the electrical conductivity of samples annealed at 500 °C was considerably higher than those of the other samples although their porous structure affected electron scattering. As a result, the best thermoelectric properties were obtained from the samples treated at 500 °C due to their low thermal conductivity and satisfactory electrical conductivity. This results from the grain rearrangement and grain growth through the annealing treatment with canning package after high-pressure cold pressing.

The ingot parameters and various powder processing methods on n-type Bi_2Te_3 pellets reported in previous studies are compared in Table 1. Our work demonstrates that low thermal conductivity is achieved because of the porous structure resulting from grain rearrangement. Moreover, high electrical conductivity via abnormal grain growth is realized. The simple and rapid powder processing technique consumes less energy than other methods. Therefore, it is suitable for mass production and cost-effective fabrication.

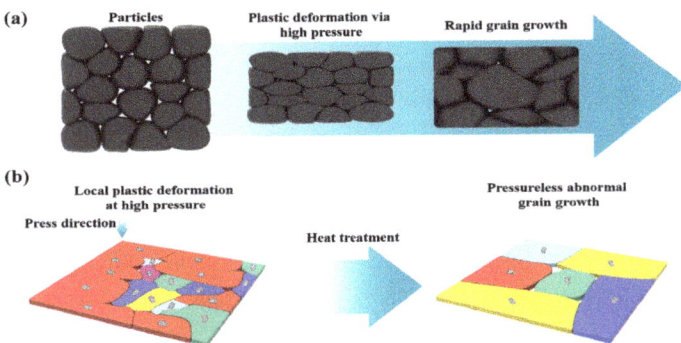

Figure 6. Schematic of (**a**) particle deformation and (**b**) texture change and grain growth from powder compaction with cold sintering at high pressure to rapid grain growth via high-temperature annealing.

Table 1. Comparison of thermoelectric parameters of n-type Bi_2Te_3 pellet with various powder processing methods. All the thermoelectric properties and ZT were measured and obtained at room temperature.

Sample with Process	σ (×10^5 S/m)	α (µV/K)	κ (W/m·K)	ZT	Reference
Bulk material (zone melting)	1.9	−180	1.8	1.03	From ingot
Microwave-activated hot-press sintering	0.56	−160	1.1	0.39	[29]
SPS and hot-forging	1.67	−140	1.15	0.85	[33]
Plasma-activated sintering	1.65	−125	1.38	0.56	[28]
High-pressure sintering	1.00	−147	0.8	0.81	[49]
Cold pressing and canning	1.22	−179	1.17	1.0	This work

4. Conclusions

This study demonstrates the recycling of n-type Bi_2Te_3 waste scraps via a simple powder processing method involving cold pressing at high pressure and annealing treatment with canning package. The samples pressed by high pressure at RT have a (0 0 l) texture orientation with elongated grains and high packing density caused by the local plastic deformation. Subsequently, different annealing temperatures were applied to enhance the thermoelectric properties through the canning process. The sample that underwent the canning package (compared with that without this process) exhibited no phase transformation despite the high annealing temperature. The highest ZT value, i.e., 1.0, was attained by the samples annealed at 500 °C. The rapid grain growth and rearrangement induced a porous structure, leading to low thermal conductivity through phonon scattering without a decrease in electrical conductivity. The technique proposed in this work is deemed to be a new and advantageous approach to achieve a cost-effective and environmentally friendly production.

Supplementary Materials: The following supporting information can be downloaded at: https://www.mdpi.com/article/10.3390/ma15124204/s1, Figure S1: Comparison of pellet densities at RT, 300, 400, and 500 °C; Figure S2: (a) comparison of 4-point and hall measurement about electrical conductivity and (b) carrier concentration and mobility from hall measurement; Figure S3: EBSD pole figure images of samples for (0 0 1) and (0 1 5) planes at (a) RT, (b) 300 °C, (c) 400 °C and (d) 500 °C. Electron diffraction intensity is represented by color marks in scale bar; Figure S4: TEM images of cold-sintered samples (a,b) at RT and (c,d) annealed at 500 °C with canning package; Figure S5: Misorientation angle distribution based on EBSD data at RT, 300, 400, and 500 °C.

Author Contributions: Conceptualization, H.S. and Y.-H.C.; methodology, H.S.; validation, Y.-H.C.; formal analysis, H.S., I.-Y.K. and J.-H.L.; investigation, H.S.; resources, H.S.; data curation, H.S.; writing—original draft preparation, H.S.; writing—review and editing, H.S. and H.-B.C. All authors have read and agreed to the published version of the manuscript.

Funding: This work was supported by a research fund from Hanyang University (HY-2020) and Korea Institute for Advancement of Technology (KIAT) grant funded by the Korea Government (MOTIE) (P0008425, The Competency Development Program for Industry Specialist).

Institutional Review Board Statement: Not applicable.

Informed Consent Statement: Not applicable.

Data Availability Statement: Not applicable.

Conflicts of Interest: The authors declare no conflict of interest.

References

1. Zhu, Y.X.; Newbrook, D.W.; Dai, P.; de Groot, C.H.K.; Huang, R.M. Artificial neural network enabled accurate geometrical design and optimisation of thermoelectric generator. *Appl. Energy* **2022**, *305*, 117800. [CrossRef]
2. IEA. *Net Zero by 2050—A Roadmap for the Global Energy Sector*; IEA: Paris, France, 2021.
3. Bahk, J.H.; Fang, H.Y.; Yazawa, K.; Shakouri, A. Flexible thermoelectric materials and device optimization for wearable energy harvesting. *J. Mater. Chem. C* **2015**, *3*, 10362–10374. [CrossRef]
4. Hsiao, Y.Y.; Chang, W.C.; Chen, S.L. A mathematic model of thermoelectric module with applications on waste heat recovery from automobile engine. *Energy* **2010**, *35*, 1447–1454. [CrossRef]
5. Casi, A.; Araiz, M.; Catalan, L.; Astrain, D. Thermoelectric heat recovery in a real industry: From laboratory optimization to reality. *Appl. Therm. Eng.* **2021**, *184*, 116275. [CrossRef]
6. Nam, W.H.; Kim, B.B.; Lim, Y.S.; Dae, K.S.; Seo, W.S.; Park, H.H.; Lee, J.Y. Phonon-glass electron-crystals in ZnO-multiwalled carbon nanotube nanocomposites. *Nanoscale* **2017**, *9*, 12941–12948. [CrossRef]
7. Daniels, L.M.; Savvin, S.N.; Pitcher, M.J.; Dyer, M.S.; Claridge, J.B.; Ling, S.; Slater, B.; Cora, F.; Alaria, J.; Rosseinsky, M.J. Phonon-glass electron-crystal behaviour by A site disorder in n-type thermoelectric oxides. *Energy Environ. Sci.* **2017**, *10*, 1917–1922. [CrossRef]
8. Son, J.S.; Choi, M.K.; Han, M.K.; Park, K.; Kim, J.Y.; Lim, S.J.; Oh, M.; Kuk, Y.; Park, C.; Kim, S.J.; et al. n-Type Nanostructured Thermoelectric Materials Prepared from Chemically Synthesized Ultrathin Bi_2Te_3 Nanoplates. *Nano Lett.* **2012**, *12*, 640–647. [CrossRef]
9. Hong, M.; Chasapis, T.C.; Chen, Z.G.; Yang, L.; Kanatzidis, M.G.; Snyder, G.J.; Zou, J. n-Type $Bi_2Te_{3-x}Se_x$ Nanoplates with Enhanced Thermoelectric Efficiency Driven by Wide-Frequency Phonon Scatterings and Synergistic Carrier Scatterings. *ACS Nano* **2016**, *10*, 4719–4727. [CrossRef]
10. Fang, H.Y.; Feng, T.L.; Yang, H.R.; Ruan, X.L.; Wu, Y. Synthesis and thermoelectric properties of compositional-modulated lead telluride-bismuth telluride nanowire heterostructures. *Abstr. Pap. Am. Chem. Soc.* **2013**, *246*, 2058–2063. [CrossRef]
11. Luo, C.; Yu, K.H.; Wu, X.; Sun, L.T. In Situ Interfacial Manipulation of Metastable States Between Nucleation and Decomposition of Single Bismuth Nanoparticle. *Phys. Status Solidi B-Basic Solid State Phys.* **2019**, *256*, 1800442. [CrossRef]
12. Pei, Y.Z.; Heinz, N.A.; LaLonde, A.; Snyder, G.J. Combination of large nanostructures and complex band structure for high performance thermoelectric lead telluride. *Energy Environ. Sci.* **2011**, *4*, 3640–3645. [CrossRef]
13. Morelli, D.T.; Jovovic, V.; Heremans, J.P. Intrinsically minimal thermal conductivity in cubic I-V-VI2 semiconductors. *Phys. Rev. Lett.* **2008**, *101*, 035901. [CrossRef] [PubMed]
14. Ghodke, S.; Yamamoto, A.; Hu, H.C.; Nishino, S.; Matsunaga, T.; Byeon, D.; Ikuta, H.; Takeuchi, T. Improved Thermoelectric Properties of Re-Substituted Higher Manganese Silicides by Inducing Phonon Scattering and an Energy-Filtering Effect at Grain Boundary Interfaces. *ACS Appl. Mater. Interfaces* **2019**, *11*, 31169–31175. [CrossRef]
15. Cai, B.W.; Zhuang, H.L.; Cao, Q.; Hu, H.H.; Dong, J.F.; Asfandiyar; Li, J.F. Practical High-Performance $(Bi,Sb)(2)Te_3$-Based Thermoelectric Nanocomposites Fabricated by Nanoparticle Mixing and Scrap Recycling. *ACS Appl. Mater. Interfaces* **2020**, *12*, 16426–16435. [CrossRef] [PubMed]
16. Liu, Y.F.; Sahoo, P.; Makongo, J.P.A.; Zhou, X.Y.; Kim, S.J.; Chi, H.; Uher, C.; Pan, X.Q.; Poudeu, P.F.P. Large Enhancements of Thermopower and Carrier Mobility in Quantum Dot Engineered Bulk Semiconductors. *J. Am. Chem. Soc.* **2013**, *135*, 7486–7495. [CrossRef] [PubMed]
17. Li, S.K.; Huang, Z.Y.; Wang, R.; Wang, C.Q.; Zhao, W.G.; Yang, N.; Liu, F.S.; Luo, J.; Xiao, Y.G.; Pan, F. Precision grain boundary engineering in commercial $Bi_2Te_{2.7}Se_{0.3}$ thermoelectric materials towards high performance. *J. Mater. Chem. A* **2021**, *9*, 11442–11449. [CrossRef]
18. Kim, K.C.; Lim, S.S.; Lee, S.H.; Hong, J.; Cho, D.Y.; Mohamed, A.Y.; Koo, C.M.; Baek, S.H.; Kim, J.S.; Kim, S.K. Precision Interface Engineering of an Atomic Layer in Bulk Bi_2Te_3 Alloys for High Thermoelectric Performance. *ACS Nano* **2019**, *13*, 7146–7154. [CrossRef]

19. Kim, S.I.; Lee, K.H.; Mun, H.A.; Kim, H.S.; Hwang, S.W.; Roh, J.W.; Yang, D.J.; Shin, W.H.; Li, X.S.; Lee, Y.H.; et al. Dense dislocation arrays embedded in grain boundaries for high-performance bulk thermoelectrics. *Science* **2015**, *348*, 109–114. [CrossRef]
20. Jo, S.; Choo, S.; Kim, F.; Heo, S.H.; Son, J.S. Ink Processing for Thermoelectric Materials and Power-Generating Devices. *Adv. Mater.* **2019**, *31*, 1804930. [CrossRef]
21. Champness, C.H.; Chiang, P.T.; Parekh, P. Thermoelectric Properties of Bi2te3-Sb2te3 Alloys. *Can. J. Phys.* **1965**, *43*, 653–669. [CrossRef]
22. Jiang, J.; Chen, L.D.; Bai, S.Q.; Yao, Q.; Wang, Q. Thermoelectric properties of p-type (Bi2Te3)(x)(Sb2Te3)(1-x) crystals prepared via zone melting. *J. Cryst. Growth* **2005**, *277*, 258–263. [CrossRef]
23. Greifzu, M.; Tkachov, R.; Stepien, L.; Lopez, E.; Bruckner, F.; Leyens, C. Laser Treatment as Sintering Process for Dispenser Printed Bismuth Telluride Based Paste. *Materials* **2019**, *12*, 3453. [CrossRef] [PubMed]
24. Xiang, Q.S.; Fan, X.; Han, X.W.; Zhang, C.C.; Hu, J.; Feng, B.; Jiang, C.P.; Li, G.Q.; Li, Y.W.; He, Z. Bi2Te3 based bulks with high figure of merit obtained from cut waste fragments of cooling crystal rods. *Mater. Chem. Phys.* **2017**, *201*, 57–62. [CrossRef]
25. Xiang, Q.S.; Fan, X.; Han, X.W.; Zhang, C.C.; Hu, J.; Feng, B.; Jiang, C.P.; Li, G.Q.; Li, Y.W.; He, Z. Preparation and optimization of thermoelectric properties of Bi2Te3 based alloys using the waste particles as raw materials from the cutting process of the zone melting crystal rods. *J. Phys. Chem. Solids* **2017**, *111*, 34–40. [CrossRef]
26. Fan, X.; Cai, X.Z.; Han, X.W.; Zhang, C.C.; Rong, Z.Z.; Yang, F.; Li, G.Q. Evolution of thermoelectric performance for (Bi,Sb)(2)Te-3 alloys from cutting waste powders to bulks with high figure of merit. *J. Solid State Chem.* **2016**, *233*, 186–193. [CrossRef]
27. Wang, H.X.; Xiong, C.L.; Luo, G.Q.; Hu, H.Y.; Yu, B.; Shao, H.Z.; Tan, X.J.; Xu, J.T.; Liu, G.Q.; Noudem, J.G.; et al. Texture Development and Grain Alignment of Hot-Pressed Tetradymite Bi0.48Sb1.52Te3 via Powder Molding. *Energy Technol.* **2019**, *7*, 1900814. [CrossRef]
28. Lee, J.K.; Son, J.H.; Park, S.D.; Park, S.; Oh, M.W. Control of oxygen content of n-type Bi2Te3 based compounds by sintering process and their thermoelectric properties. *Mater. Lett.* **2018**, *230*, 211–214. [CrossRef]
29. Fan, X.A.; Rong, Z.Z.; Yang, F.; Cai, X.Z.; Han, X.W.; Li, G.Q. Effect of process parameters of microwave activated hot pressing on the microstructure and thermoelectric properties of Bi2Te3-based alloys. *J. Alloys Compd.* **2015**, *630*, 282–287. [CrossRef]
30. Wang, H.X.; Luo, G.Q.; Tan, C.; Xiong, C.L.; Guo, Z.; Yin, Y.N.; Yu, B.; Xiao, Y.K.; Hu, H.Y.; Liu, G.Q.; et al. Phonon Engineering for Thermoelectric Enhancement of p-Type Bismuth Telluride by a Hot-Pressing Texture Method. *ACS Appl. Mater. Interfaces* **2020**, *12*, 31612–31618. [CrossRef]
31. Messing, G.L.; Stevenson, A.J. Toward pore-free ceramics. *Science* **2008**, *322*, 383–384. [CrossRef]
32. Delaizir, G.; Bernard-Granger, G.; Monnier, J.; Grodzki, R.; Kim-Hak, O.; Szkutnik, P.D.; Soulier, M.; Saunier, S.; Goeuriot, D.; Rouleau, O.; et al. A comparative study of Spark Plasma Sintering (SPS), Hot Isostatic Pressing (HIP) and microwaves sintering techniques on p-type Bi2Te3 thermoelectric properties. *Mater. Res. Bull.* **2012**, *47*, 1954–1960. [CrossRef]
33. Zhao, L.D.; Zhang, B.P.; Li, J.F.; Zhang, H.L.; Liu, W.S. Enhanced thermoelectric and mechanical properties in textured n-type Bi2Te3 prepared by spark plasma sintering. *Solid State Sci.* **2008**, *10*, 651–658. [CrossRef]
34. Maria, J.P.; Kang, X.Y.; Floyd, R.D.; Dickey, E.C.; Guo, H.Z.; Guo, J.; Baker, A.; Funihashi, S.; Randall, C.A. Cold sintering: Current status and prospects. *J. Mater. Res.* **2017**, *32*, 3205–3218. [CrossRef]
35. Yu, T.; Cheng, J.; Li, L.; Sun, B.S.; Bao, X.J.; Zhang, H.T. Current understanding and applications of the cold sintering process. *Front. Chem. Sci. Eng.* **2019**, *13*, 654–664. [CrossRef]
36. Grasso, S.; Biesuz, M.; Zoli, L.; Taveri, G.; Duff, A.I.; Ke, D.Y.; Jiang, A.; Reece, M.J. A review of cold sintering processes. *Adv. Appl. Ceram.* **2020**, *119*, 115–143. [CrossRef]
37. Wang, Y.; Liu, W.D.; Gao, H.; Wang, L.J.; Li, M.; Shi, X.L.; Hong, M.; Wang, H.; Zou, J.; Chen, Z.G. High Porosity in Nanostructured n-Type Bi2Te3 Obtaining Ultralow Lattice Thermal Conductivity. *ACS Appl. Mater. Interfaces* **2019**, *11*, 31237–31244. [CrossRef]
38. Ohsugi, I.J.; Tokunaga, D.; Kato, M.; Yoneda, Y.; Isoda, Y. Dissociation and sublimation of tellurium from the thermoelectric tellurides. *Mater. Res. Innov.* **2015**, *19*, 301–303. [CrossRef]
39. Jung, S.J.; Kim, J.H.; Kim, D.I.; Kim, S.K.; Park, H.H.; Kim, J.S.; Hyun, D.B.; Baek, S.H. Strain-assisted, low-temperature synthesis of high-performance thermoelectric materials. *Phys. Chem. Chem. Phys.* **2014**, *16*, 3529–3533. [CrossRef]
40. Dudina, D.V.; Bokhonov, B.B.; Olevsky, E.A. Fabrication of Porous Materials by Spark Plasma Sintering: A Review. *Materials* **2019**, *12*, 541. [CrossRef]
41. Olevsky, E.A.; Dudina, D.V. Microwave Sintering. In *Field-Assisted Sintering*; Springer: Cham, Switzerland, 2018.
42. Kumar, D.; Idapalapati, S.; Wang, W.; Narasimalu, S. Effect of Surface Mechanical Treatments on the Microstructure-Property-Performance of Engineering Alloys. *Materials* **2019**, *12*, 2503. [CrossRef]
43. Bobylev, S.V.; Gutkin, M.Y.; Ovid'ko, I.A. Transformations of grain boundaries in deformed nanocrystalline materials. *Acta Mater.* **2004**, *52*, 3793–3805. [CrossRef]
44. Li, Z.; Xiao, C.; Xie, Y. Layered thermoelectric materials: Structure, bonding, and performance mechanisms. *Appl. Phys. Rev.* **2022**, *9*, 011303. [CrossRef]
45. Beneking, H. Hogarth Ca—Materials Used in Semiconductor Devices. *Arch. Elektr. Ubertragung* **1966**, *20*, 426.
46. Kagarakis, C.A. Chemical Bonding in Bismuth Telluride. *J. Mater. Sci.* **1978**, *13*, 1594–1596. [CrossRef]
47. Srinivasan, R.; McReynolds, K.; Gothard, N.W.; Spowart, J.E. Texture development during deformation processing of the n-type bismuth telluride alloy Bi2Se0.3Te2.7. *Mater. Sci. Eng.-Struct. Mater. Prop. Microstruct. Process.* **2013**, *588*, 376–387. [CrossRef]

48. Medlin, D.L.; Yang, N.; Spataru, C.D.; Hale, L.M.; Mishin, Y. Unraveling the dislocation core structure at a van der Waals gap in bismuth telluride. *Nat. Commun.* **2019**, *10*, 1820. [CrossRef]
49. Yu, F.R.; Xu, B.; Zhang, J.J.; Yu, D.L.; He, J.L.; Liu, Z.Y.; Tian, Y.J. Structural and thermoelectric characterizations of high pressure sintered nanocrystalline Bi2Te3 bulks. *Mater. Res. Bull.* **2012**, *47*, 1432–1437. [CrossRef]
50. Najafkhani, F.; Kheiri, S.; Pourbahari, B.; Mirzadeh, H. Recent advances in the kinetics of normal/abnormal grain growth: A review. *Arch. Civ. Mech. Eng.* **2021**, *21*, 1–20. [CrossRef]
51. Zhao, P.C.; Chen, B.; Zheng, Z.G.; Guan, B.; Zhang, X.C.; Tu, S.T. Microstructure and Texture Evolution in a Post-dynamic Recrystallized Titanium During Annealing, Monotonic and Cyclic Loading. *Metall. Mater. Trans.-Phys. Metall. Mater. Sci.* **2021**, *52*, 394–412. [CrossRef]
52. Khodabakhshi, F.; Mohammadi, M.; Gerlich, A.P. Stability of ultra-fine and nano-grains after severe plastic deformation: A critical review. *J. Mater. Sci.* **2021**, *56*, 15513–15537. [CrossRef]
53. Sakai, T.; Belyakov, A.; Kaibyshev, R.; Miura, H.; Jonas, J.J. Dynamic and post-dynamic recrystallization under hot, cold and severe plastic deformation conditions. *Prog. Mater. Sci.* **2014**, *60*, 130–207. [CrossRef]
54. Bao, D.Y.; Chen, J.; Yu, Y.; Liu, W.D.; Huang, L.S.; Han, G.; Tang, J.; Zhou, D.L.; Yang, L.; Chen, Z.G. Texture-dependent thermoelectric properties of nano-structured Bi2Te3. *Chem. Eng. J.* **2020**, *388*, 124295. [CrossRef]
55. Hu, L.P.; Liu, X.H.; Xie, H.H.; Shen, J.J.; Zhu, T.J.; Zhao, X.B. Improving thermoelectric properties of n-type bismuth-telluride-based alloys by deformation-induced lattice defects and texture enhancement. *Acta Mater.* **2012**, *60*, 4431–4437. [CrossRef]
56. Hu, L.P.; Wu, H.J.; Zhu, T.J.; Fu, C.G.; He, J.Q.; Ying, P.J.; Zhao, X.B. Tuning Multiscale Microstructures to Enhance Thermoelectric Performance of n-Type Bismuth-Telluride-Based Solid Solutions. *Adv. Energy Mater.* **2015**, *5*, 1500411. [CrossRef]

Article

Strain-Enhanced Thermoelectric Performance in GeS₂ Monolayer

Xinying Ruan [1,†], Rui Xiong [1,†], Zhou Cui [1], Cuilian Wen [1], Jiang-Jiang Ma [2,3], Bao-Tian Wang [2,3,4,*] and Baisheng Sa [1,*]

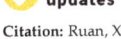

Citation: Ruan, X.; Xiong, R.; Cui, Z.; Wen, C.; Ma, J.-J.; Wang, B.-T.; Sa, B. Strain-Enhanced Thermoelectric Performance in GeS₂ Monolayer. *Materials* 2022, 15, 4016. https://doi.org/10.3390/ma15114016

Academic Editor: Alexandros Lappas

Received: 16 May 2022
Accepted: 3 June 2022
Published: 6 June 2022

Publisher's Note: MDPI stays neutral with regard to jurisdictional claims in published maps and institutional affiliations.

Copyright: © 2022 by the authors. Licensee MDPI, Basel, Switzerland. This article is an open access article distributed under the terms and conditions of the Creative Commons Attribution (CC BY) license (https://creativecommons.org/licenses/by/4.0/).

[1] Key Laboratory of Eco-Materials Advanced Technology, College of Materials Science and Engineering, Fuzhou University, Fuzhou 350100, China; ruanxinyingjg@163.com (X.R.); rxiong421@163.com (R.X.); cuizhoufzu@163.com (Z.C.); clwen@fzu.edu.cn (C.W.)
[2] Institute of High Energy Physics, Chinese Academy of Sciences (CAS), Beijing 100049, China; majj88@ihep.ac.cn
[3] Spallation Neutron Source Science Center (SNSSC), Dongguan 523803, China
[4] Collaborative Innovation Center of Extreme Optics, Shanxi University, Taiyuan 030006, China
* Correspondence: wangbt@ihep.ac.cn (B.-T.W.); bssa@fzu.edu.cn (B.S.)
† These authors contributed equally to this work.

Abstract: Strain engineering has attracted extensive attention as a valid method to tune the physical and chemical properties of two-dimensional (2D) materials. Here, based on first-principles calculations and by solving the semi-classical Boltzmann transport equation, we reveal that the tensile strain can efficiently enhance the thermoelectric properties of the GeS₂ monolayer. It is highlighted that the GeS₂ monolayer has a suitable band gap of 1.50 eV to overcome the bipolar conduction effects in materials and can even maintain high stability under a 6% tensile strain. Interestingly, the band degeneracy in the GeS₂ monolayer can be effectually regulated through strain, thus improving the power factor. Moreover, the lattice thermal conductivity can be reduced from 3.89 to 0.48 W/mK at room temperature under 6% strain. More importantly, the optimal ZT value for the GeS₂ monolayer under 6% strain can reach 0.74 at room temperature and 0.92 at 700 K, which is twice its strain-free form. Our findings provide an exciting insight into regulating the thermoelectric performance of the GeS₂ monolayer by strain engineering.

Keywords: GeS₂ monolayer; strain engineering; first-principles calculations; thermoelectric materials; thermal conductivity

1. Introduction

Thermoelectric technology is one of the most fantastic energy-conversion technologies that can convert heat energy and electrical energy into each other directly [1–3]. Thermoelectric materials have recently gained extensive attention as a critical factor for thermoelectric technology. The figure of merit ZT can be directly used to visualize the thermoelectric conversion efficiency of thermoelectric materials and can be calculated by [4–7]:

$$ZT = \frac{S^2 \sigma T}{\kappa} \quad (1)$$

where S stands for the Seebeck coefficient, σ is electrical conductivity, and T represents temperature. κ is the thermal conductivity, consisting of both electronic and lattice parts. Herein, the thermoelectric power factor (PF) can be defined as PF = $S^2\sigma$. Apparently, a higher PF and lower κ can contribute to an immense ZT value.

The development of 2D materials provides an excellent platform for discovering novel high-performance thermoelectric materials [8–13]. Previous studies have reported graphene [14,15], phosphorene (BP) [16–18], IVA–VIA compounds [19–21], and transition metal dichalcogenides (TMDs) [22–24], and all show excellent thermoelectric performance. In particular, IVA–VIA compounds exhibit high ZT values due to their ultralow lattice thermal conductivities [19,20]. Recently, the 1T-GeS$_2$ monolayer has been reported as a potential thermoelectric material due to its relatively high electronic fitness function (EFF) value from high-through computational screening [21]. Moreover, the high-power factor of the GeS$_2$ monolayer further reveals its great potential application in the field of thermoelectrics [25]. However, the ZT value of the 1T-GeS$_2$ monolayer is only 0.23 when the thermal transport property is considered [25], which significantly hinders its further application. Therefore, it is of great significance to improve its thermoelectric performance by adjusting the thermal transport properties of GeS$_2$ monolayers. It is worth mentioning that the electronic structures of 2D materials are easily affected by applied strains [26–28]. Strain engineering has been theoretically and experimentally proposed as a valid way to enhance the thermoelectric properties of 2D thermoelectric materials [29,30]. Experimentally, the thermal conductivity of the Bi$_2$Te$_3$ monolayer can be reduced by 50% by applying a tensile strain of 6% [31]. Theoretically, tensile strain can significantly enhance Seebeck coefficients while reducing thermal conductivity, and this has been observed in the PtSe$_2$ monolayer [32]. Therefore, it is very interesting to investigate the strain effect on the electronic and thermoelectric properties of the GeS$_2$ monolayer.

In the present work, based on first-principles calculations and by solving the semi-classical Boltzmann transport equation, we systematically studied the tensile strain effects on the thermoelectric properties of the GeS$_2$ monolayer, including electronic structures, electronic transport properties, and phonon transport properties. It was found that the valence band near the Fermi level of the GeS$_2$ monolayer will degenerate under tensile strain, which leads to an improvement in the power factor. Meanwhile, the phonon group velocities and phonon relaxation times decrease with an increasing tensile strain, resulting in a reduction in the lattice thermal conductivity, thereby enhancing the thermoelectric performance. Our results provided a new tactic for improving the thermoelectric properties of the GeS$_2$ monolayer.

2. Methods

Our simulation works were based on first-principles calculations with the projector augmented-wave (PAW) [33] method, which is executed by the VASP [34] code, and the corresponding results were dealt with the ALKEMIE platform [35]. The generalized gradient approximation [36] with the Perdew–Burke–Ernzerhof functional (GGA-PBE) [37] was used to deal with the interaction between electronics and ions. The structure of the GeS$_2$ monolayer was completely optimized until the energy and force convergence criteria were less than 10^{-6} eV and -0.01 eV, respectively. The cutoff energy was set to 600 eV, and a k-point mesh of $15 \times 15 \times 1$ was adopted [38]. A vacuum thickness of 20 Å perpendicular to the in-plane direction of the GeS$_2$ monolayer was built. The Heyd–Scuseria–Ernzerhof (HSE06) [39] hybrid functional with a range-separation parameter of 0.2 and mixing parameter of 0.25 was also adopted to obtain more accurate band structures and electronic transport properties of the GeS$_2$ monolayer. The ab initio molecular dynamics (AIMD) simulations with the Nosé–Hoover thermostat (NVT) ensemble and a time step of 2ps were performed to investigate the thermal stability of the GeS$_2$ monolayer [40,41].

A denser k-point mesh of $35 \times 35 \times 1$ was used for static calculations to obtain more accurate electronic structures to solve semi-classical Boltzmann transport equations, which is realized in the BoltzTraP code [42]. The phonon spectrum and second-order anharmonic force constants were calculated by the Phonopy package [43] with a $6 \times 6 \times 1$ supercell, while a $4 \times 4 \times 1$ supercell was used to calculate third-order interatomic force constants. The sixth nearest neighbors were selected to obtain the third-order interatomic force constants to ensure the accuracy of lattice thermal conductivity and save the calculation time. Combing with second-order anharmonic force constants and third-order interatomic force constants as input files, the lattice thermal conductivity of the GeS$_2$ monolayer can be obtained through the ShengBTE code [44].

3. Results and Discussion

3.1. Structural Stability and Band Structure

Similar to the 1T-MoS$_2$ monolayer [45], each unit cell of the GeS$_2$ monolayer consists of one Ge atom and two S atoms with the Ge sublayer sandwiched between two S sublayers. The side and top views of the GeS$_2$ monolayer are plotted in Figure 1a,b, respectively. The relaxed lattice parameters are $a = b = 3.44$ Å, which agree with previous theoretical predictions [21,25]. Figure 1c describes the atom orbitals project band structure of the GeS$_2$ monolayer. It is clear that the GeS$_2$ monolayer demonstrates indirect band gap semiconductor features with a band gap of 1.50 eV. It is noted that the relatively large band gap can effectively prevent the bipolar conduction behavior in the materials and thus prevents the thermoelectric performance from being destroyed. Moreover, the VBM is mainly contributed by the S-p orbital, while the CBM is occupied by both Ge-s and S-p orbitals. Our results are in accordance with the previous theoretical predicated [25,46], indicating that our calculation parameters are reasonable.

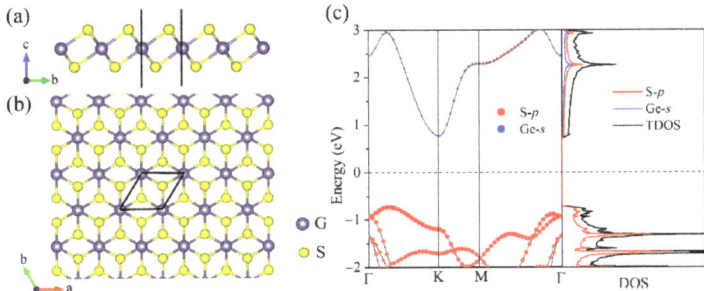

Figure 1. The structure of GeS$_2$ monolayer's (**a**) side and (**b**) top views. (**c**) The atom orbitals' project band structure and DOS for GeS$_2$ monolayer.

To understand the stability of the GeS$_2$ monolayer, we then conducted phonon spectrum calculations and AIMD simulations to explore the lattice and thermal dynamic stabilities, respectively. Figure 2a describes the phonon spectrum for the GeS$_2$ monolayer. Obviously, there are nine dispersion curves with three acoustic branches and six optical branches since a GeS$_2$ unit cell contains three atoms. Moreover, no imaginary frequency can be found in phonon dispersion curves, indicating that the GeS$_2$ monolayer possesses a good lattice dynamic stability. It is noted that the ZA mode for the GeS$_2$ monolayer near the Γ point is quadratically converged, which can be usually observed in 2D materials systems [47]. Furthermore, from the PhDOS of the GeS$_2$ monolayer, we know that the low- and high-frequency regions are mainly contributed by Ge and S atoms, respectively. Moreover, the phonon spectrum of the GeS$_2$ monolayer under 2% compressive strain was also calculated, as shown in Figure S1. A negative frequency was observed in the phonon spectrum, indicating the instability of the GeS$_2$ monolayer under compressive strain. Hence, in our study, we mainly concentrated on the tensile strain effects on the thermoelectric properties of the GeS$_2$ monolayer. Figure 2b illustrates the energy evolution and structure snapshot of the GeS$_2$ monolayer for 10 ps at 300 K. It is clear that the changes in total energy are minimal, and atoms are slightly vibrating around their equilibrium positions, suggesting that the GeS$_2$ monolayer exhibits excellent thermal dynamic stability as well.

Figure 3 illustrates the band structures of the GeS$_2$ monolayer at different biaxial tensile strains. Herein, the tensile strains can be calculated by $\varepsilon = (a - a_0)/a_0 \times 100\%$, where a_0 stands for the lattice constant when unstrained, while a represents the lattice constant under strain. Obviously, within our investigated strain range (0~6%), the band gap of the GeS$_2$ monolayer increases gradually with tensile strain since CBM moves toward the higher energy level. Additionally, with the increases in tensile strain, the valence bands between K and Γ points move toward the Fermi level, which can enhance the degeneracy of the valence band and thus improve the Seebeck coefficient. Moreover, the band structure of the GeS_2 monolayer under 8% tensile strain was also calculated, as shown in Figure S2. However, the valence band maximum shifts to the position between Γ and K under 8% tensile strain. This phenomenon will decrease band degeneracy in the GeS_2 monolayer, which is not conducive to the thermoelectric application. Hence, in our study, we mainly concentrate on the 2–6% tensile strain effects on the thermoelectric properties of the GeS_2 monolayer. These consequences indicate that the tensile strain can effectively regulate the electronic structures of the GeS_2 monolayer. Therefore, an improvement in thermoelectric performance in the GeS_2 monolayer is anticipated [48,49].

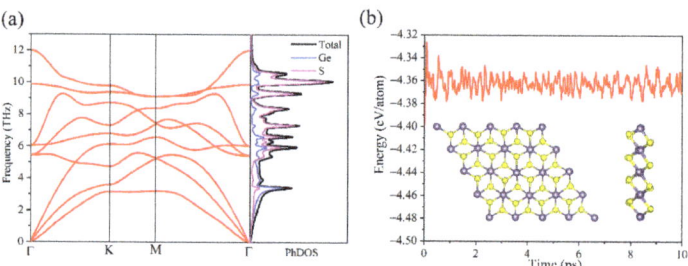

Figure 2. The (**a**) phonon spectrum and PhDOS of GeS_2 monolayer and (**b**) total energies evolution and structure snapshots after 10 ps AIMD simulations at 300 K for GeS_2 monolayer.

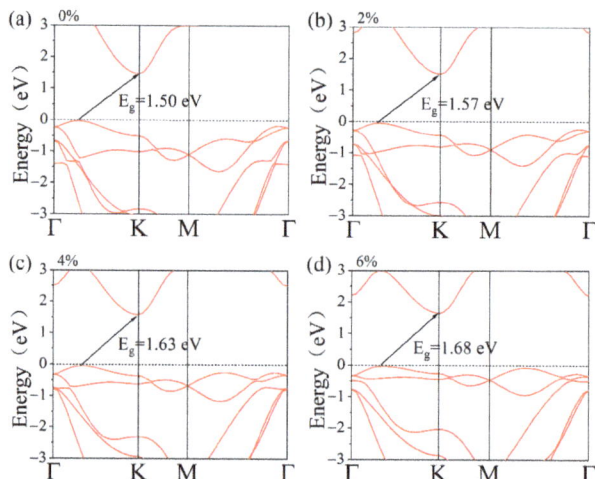

Figure 3. Band structures under different biaxial tensile strains of (**a**) 0%, (**b**) 2%, (**c**) 4% and (**d**) 6% for GeS_2 monolayer.

3.2. Electronic Transport Properties

We next investigate the effect of biaxial tensile strains on the electronic transport properties of the GeS_2 monolayer, including the Seebeck coefficient (S), electric conductivity (σ), electronic thermal conductivity (κ_e), and the power factor (PF). Figure 4 shows the contour maps of the Seebeck coefficient with respect to chemical potential under different biaxial tensile strains. Clearly, the S increases with an increasing tensile strain and decreases with an increasing temperature. The maximum S increases from 2386 μVK^{-1} (2318 μVK^{-1}) to 2697 μVK^{-1} (2605 μVK^{-1}) under p-type (n-type) doping, as the tensile strain augments from 0 to 6%. This phenomenon is mainly contributed by enlarging the band gap and band degeneracy in the GeS_2 monolayer with the increase in tensile strain.

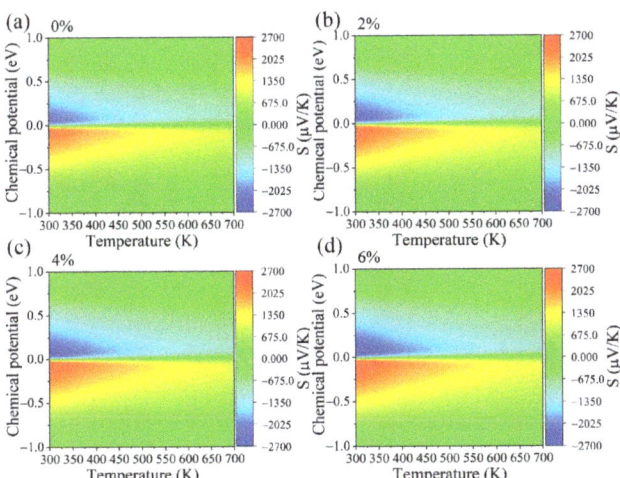

Figure 4. The contour maps of the Seebeck coefficient S with respect to chemical potential under different biaxial tensile strains of (**a**) 0%, (**b**) 2%, (**c**) 4% and (**d**) 6% for GeS$_2$ monolayer.

On the other hand, Figure 5a–d shows the electrical conductivity divided by the relaxation time (σ/τ) of the GeS$_2$ monolayer under different tensile strains. Contrary to the Seebeck coefficients, electrical conductivity is insensitive to the temperature and decreases with an increasing tensile strain. A similar tendency as σ/τ can be observed in electronic thermal conductivity (Figure 6a–d) since it can be calculated by [50]: $\kappa_e = L\sigma T$, where L represents the Lorenz number. Our results above show that the S and σ/τ exhibit opposite trends under tensile strain. Hence, we also calculated the power factor (PF) under different tensile strains, and the corresponding results are shown in Figure 7a–d. Apparently, the optimal value of the PF under p-type doping is much higher than n-type doping for all cases. More importantly, the PF gradually increases as the tensile strain is applied, which is due to the fact that the applied tensile strain has a more significant effect on the S than the σ/τ. The power factor as a function of carrier concentrations is also plotted in Figure S3.

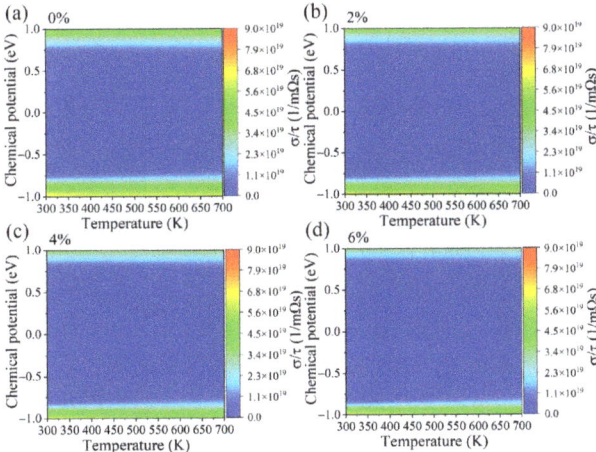

Figure 5. The contour map of the electrical conductivity divided by relaxation time (σ/τ) with respect to chemical potential under different biaxial tensile strains of (**a**) 0%, (**b**) 2%, (**c**) 4% and (**d**) 6% for GeS$_2$ monolayer.

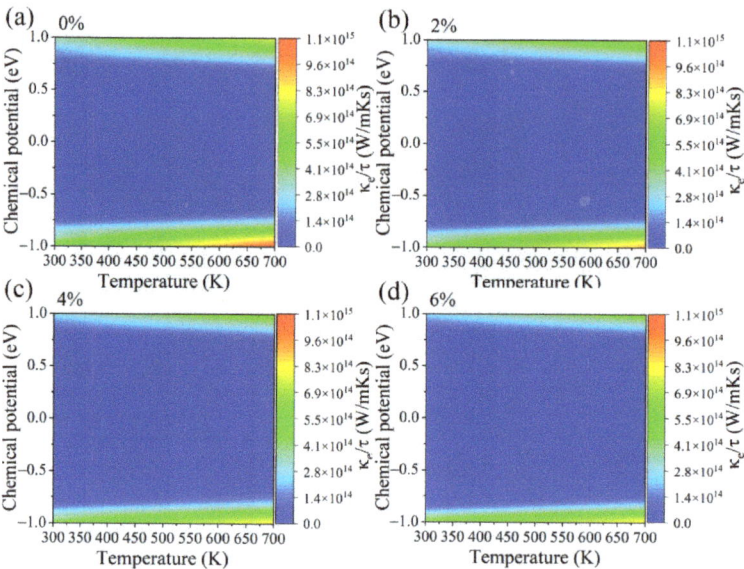

Figure 6. The contour map of the electronic thermal conductivity divided by relaxation time (κ_e/τ) with respect to chemical potential under different biaxial tensile strains of (**a**) 0%, (**b**) 2%, (**c**) 4% and (**d**) 6% for GeS$_2$ monolayer.

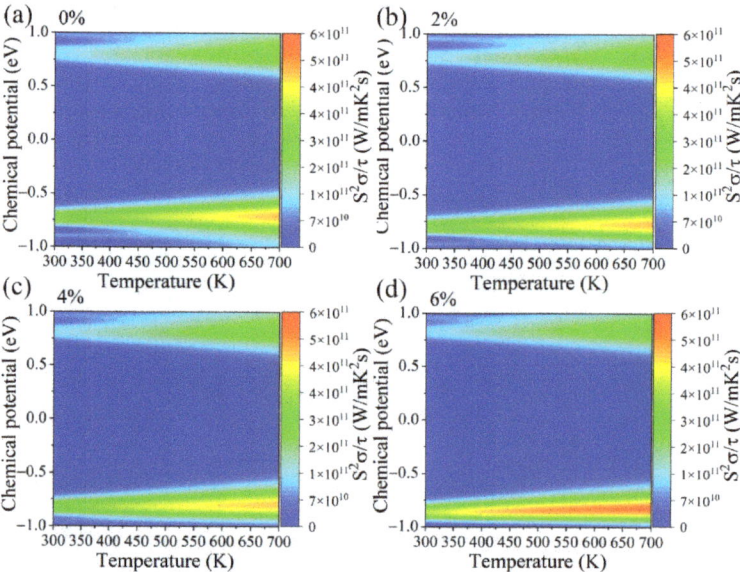

Figure 7. The contour map of the power factor divided by relaxation time ($S^2\sigma/\tau$) with respect to chemical potential under different biaxial tensile strains of (**a**) 0%, (**b**) 2%, (**c**) 4% and (**d**) 6% for GeS$_2$ monolayer.

3.3. Phonon Dispersion Curves and Transport Properties

Phonon thermal transport property is another critical factor for thermoelectric materials. Hence, the effect of tensile strain on the phonon transport properties of the GeS$_2$ monolayer was investigated in the following. The phonon dispersion curves under different strains are illustrated in Figure 8. Clearly, no negative frequency was observed in any of the cases, suggesting that the GeS$_2$ monolayer's lattice is dynamically stable under these tensile strains. Furthermore, the frequencies of both optical and acoustic phonon modes gradually decrease with the increase in the tensile strain, leading to reducing phonon group velocities and thus a lower lattice thermal conductivity. This phenomenon is beneficial for the application of GeS$_2$ monolayer in the fields of thermoelectrics.

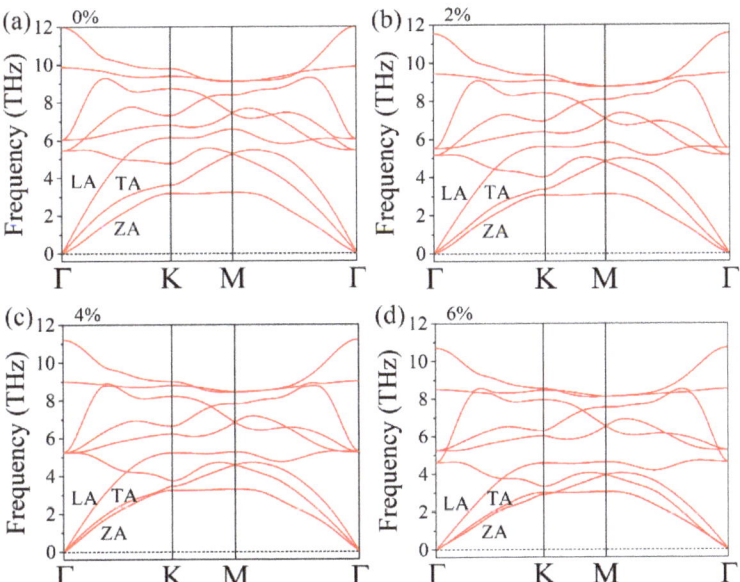

Figure 8. The phonon spectrum of GeS$_2$ monolayer under different biaxial tensile strains of (**a**) 0%, (**b**) 2%, (**c**) 4% and (**d**) 6%.

To evaluate the convergence of the lattice's thermal conductivity, we calculated the lattice thermal conductivity as a function of the nearest neighbor atomic, which is plotted in Figure S4. It is noted that the lattice thermal conductivity can reach good convergence criteria when the nearest neighbor atom is up to six. Figure 9a describes the lattice thermal conductivity (κ_l) of the GeS$_2$ monolayer with respect to temperature under different tensile strains. It is interesting to note that κ_l decreases with both increasing temperature and tensile strain. For example, the κ_l of the unstrained GeS$_2$ monolayer reduces from 3.89 to 1.13 W/mK when the temperature increases from 300 K to 1000 K. More importantly, the κ_l will reduce to 0.48 W/mK when 6% strain is applied at 300 K. Such a small κ_l is comparable to some recently reported novel 2D thermoelectric materials, such as a SnTe monolayer (0.67 W m^{-1} K^{-1}) [51], Sb$_2$Te$_2$Se monolayer (0.46 W m^{-1} K^{-1}) [52], and HfSe$_2$ monolayer (0.7 W m^{-1} K^{-1}) [53]. To unravel the strain-induced reduced lattice thermal conductivity behavior in the GeS$_2$ monolayer, we also calculated the phonon group velocities (v_λ) and phonon relaxation times (τ_λ) since κ_l can be obtained by [54]:

$$\kappa_l = \frac{\sum_\lambda C_\lambda v_\lambda^2 \tau_\lambda}{V} \qquad (2)$$

where V represents the volume, which can be defined as $V = Sh$, where S is the cross-sectional area and h is the layer thickness of the GeS$_2$ monolayer. The layer thickness is obtained by the distance between the top and bottom surface atoms plus the Van der Waals radii of the surface atoms. C_λ is capacity heat. At room temperature, the capacity heat follows the Dulong–Petit limit; thus, κ_l is mainly contributed by v_λ and τ_λ. Figure 9b,c show

v_λ and τ_λ of the GeS$_2$ monolayer under different tensile strains, respectively. Both v_λ and τ_λ decrease with an increasing tensile strain. This phenomenon leads to a decrease in the κ_l with an increasing tensile strain, which agrees with our previous results. Moreover, the calculated average value of v_λ is reduced from 1.14 to 1.08 Km/s, while the average value of τ_λ decreases from 0.94 to 0.25 ps when the strain rises from 0 to 6%. Such small v_λ and τ_λ further guarantee the low κ_l of the GeS$_2$ monolayer. Furthermore, we also calculated the Grüneisen parameters of the GeS$_2$ monolayer, as shown in Figure 9d. Interestingly, when the strain rises to 6%, the average value of Grüneisen parameters is enhanced from 1.11 to 3.15, indicating that anharmonic phonon interaction of the GeS$_2$ monolayer is strengthened under tensile strain.

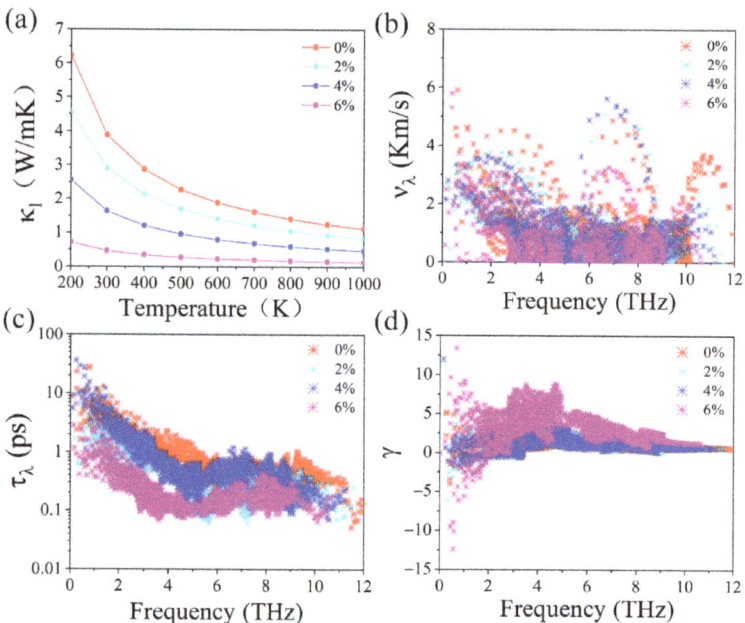

Figure 9. (**a**) The calculated lattice thermal conductivity κ_l with respect to temperature under different tensile strains for GeS$_2$ monolayer. The (**b**) phonon group velocity, (**c**) phonon relaxation time, and (**d**) Grüneisen constants for GeS$_2$ monolayer at different tensile strains.

3.4. Thermoelectric Performance

Due to the relaxation time approximation in Boltzmann transport theory, we calculated the electron relaxation time before evaluating the quality factor ZT of the GeS$_2$ monolayer. The carrier relaxation time can be defined as:

$$\tau = \frac{\mu m*}{e} \tag{3}$$

where the μ is carrier mobility, which can be estimated through deformation potential theory [55,56]:

$$\mu = \frac{2e\hbar^3 C_{2D}}{3k_B T |m*|^2 E_i^2} \tag{4}$$

where e, $\hbar$, k_B, T, and $m*$ stand for the electron charge, reduced Planck constant, Boltzmann constant, temperature, and electron (hole) effective mass, respectively. The effective mass can be defined by: $m* = \hbar^2/(\partial^2 E/\partial k^2)$, where $\hbar$ is the reduced Planck constant and E is the energy of the electron (hole) at wavevector k in the band. Therefore, the electron effective mass can be obtained from the second-order derivatives of the energy band near the conduction band minimum, while the hole's effective mass is obtained from the energy band near the valence band maximum, and the corresponding fitting parameters are shown in Table S1. C_{2D} and E_i are the elastic modulus and deformation potential constant for 2D systems, respectively. Here, $C_{2D} = 2(\partial^2(E - E_0)/\partial \varepsilon^2)/S$, where S is the cross-sectional area. Herein, the orthorhombic lattice of the GeS$_2$ monolayer was built for the carrier mobility calculation, as plotted in Figure 10a. The band structure, total energy, and E_{edge} vs. strain for GeS$_2$ monolayer in the orthorhombic unit cell are illustrated in Figure 10b–d, respectively. The corresponding parameters calculated and mentioned above are summarized in Table 1.

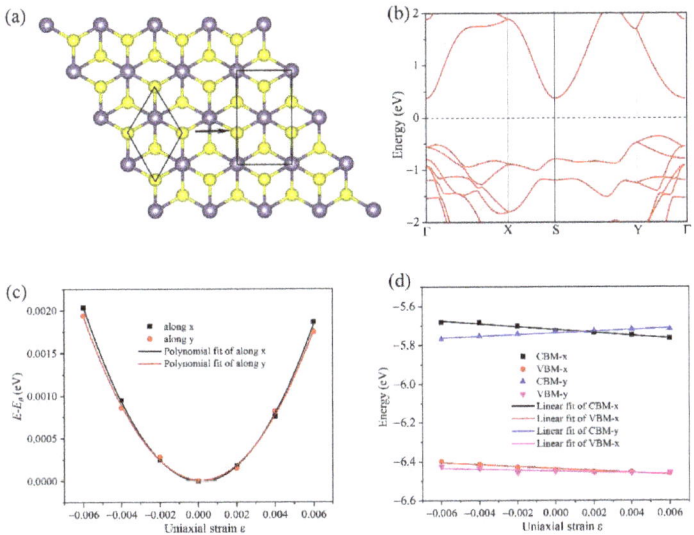

Figure 10. (a) The orthorhombic lattice of GeS$_2$ monolayer. The calculated (b) electronic band structure, (c) total energy shift, and (d) band alignment for orthorhombic lattice GeS$_2$ monolayer with respect to the uniaxial strain ε by PBE functional.

Table 1. Calculated deformation potentials (E_1), effective mass (m^*), elastic modulid (C_{2D}), carrier mobility (μ), and electronic relaxation time (τ) of GeS$_2$ monolayer under different directions.

Direction	Carrier Type	E_1 (eV)	C_{2D} (N m^{-1})	m^*/m_0	μ (cm^2 V^{-1} s^{-1})	τ (ps)
x	e	7.310	52.9	0.21	321.52	0.04
	h	5.065	52.9	0.88	37.41	0.02
y	e	4.359	49.9	0.68	79.80	0.03
	h	2.302	49.9	1.19	93.43	0.07

Finally, based on the thermoelectric parameters we obtained, the figure of merit ZT of the GeS$_2$ monolayer under different tensile strains is plotted in Figure 11. Additionally, the figure of merit ZT as a function of carrier concentrations is also shown in Figure S5. Clearly, the tensile strain greatly enhances the ZT value of the GeS$_2$ monolayer. The optimal ZT value at 300 K is 0.74 under a 6% strain, which is twice the strain-free GeS$_2$ monolayer (ZT = 0.37). This phenomenon is mainly because the tensile strain enhances the PF while reducing both κ_l and κ_e. More importantly, the ZT value will be increased from 0.74 to 0.92 with temperature increases from 300 to 700K. This value is comparable with the SiP$_2$ monolayer (0.9 at 700 K) [57], TiS$_2$ monolayer (0.95 at 300 K and an 8% tensile strain) [58], and WSSe monolayer (1.08 at 1500K and a 6% compressive strain) [48].

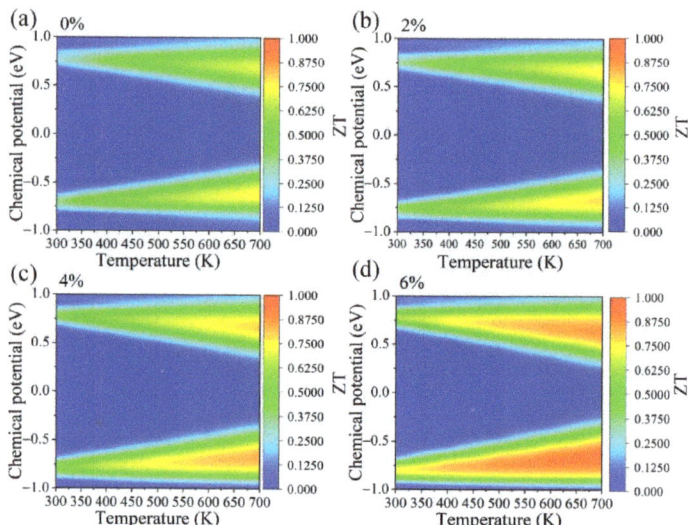

Figure 11. The contour map of the figure of merit ZT with respect to chemical potential under different biaxial tensile strains of (**a**) 0%, (**b**) 2%, (**c**) 4% and (**d**) 6% for GeS$_2$ monolayer.

4. Conclusions

In summary, by employing DFT calculations combined with semi-classical Boltzmann transport theory, the influence of tensile strain on the thermoelectric properties of the GeS$_2$ monolayer was theoretically studied. Our findings manifest that the GeS$_2$ monolayer exhibits indirect band gap semiconductor characteristics, and the band gap gradually increases with tensile strain. Moreover, the electronic and thermal transport properties of the GeS$_2$ monolayer can be efficiently tuned by tensile strain. The tensile strain can significantly enhance the power factor while decreasing thermal conductivity, leading to the enhancement of the ZT value of the GeS$_2$ monolayer. The lattice thermal conductivity of the GeS$_2$ monolayer at 300 K is only 0.48 W/mK under 6% tensile strain. This phenomenon is mainly attributed to the ultralow phonon group velocities and phonon relaxation times of GeS$_2$ monolayer under 6% strain. More importantly, the optimal ZT value of the 6% strained GeS$_2$ monolayer at room temperature is about twice more significant than the case without strain. Our results give a new insight into the strain-modulated thermoelectric performance of the GeS$_2$ monolayer.

Supplementary Materials: The following supporting information can be downloaded at: https://www.mdpi.com/article/10.3390/ma15114016/s1. Figure S1: The phonon spectrum of GeS$_2$ under 2% compressive strain; Figure S2: The band structure of GeS$_2$ monolayer under 8% tensile strain; Figure S3: The power factor of GeS$_2$ monolayer as a function of carrier concentrations under different tensile strains; Figures S4: The lattice thermal conductivity of GeS$_2$ monolayer as a function of the nearest neighbor atomic; Figure S5: The figure of merit ZT of GeS$_2$ monolayer as a function of carrier concentrations under different tensile strains; Table S1. The calculated parameters for effective mass of the GeS$_2$ monolayer, the number of band for quadratic function fitting (N$_b$), k-cutoff, band extrema points (B$_p$), fitting points (F$_p$).

Author Contributions: Conceptualization, R.X. and B.S.; methodology, R.X. and B.-T.W.; software, B.-T.W. and B.S.; validation, R.X., Z.C. and C.W., J.-J.M.; formal analysis, X.R. and R.X.; investigation, X.R. and R.X.; resources, B.-T.W. and B.S.; data curation, X.R. and R.X.; writing—original draft preparation, X.R. and R.X.; writing—review and editing, C.W., B.-T.W. and B.S.; visualization, R.X., Z.C., J.-J.M.; supervision, B.S.; project administration, B.S.; funding acquisition, B.-T.W. and B.S. All authors have read and agreed to the published version of the manuscript.

Funding: This work was supported by the National Natural Science Foundation of China (No. 21973012 and No. 12074381), the Natural Science Foundation of Fujian Province (grant No. 2021J06011, 2020J01351, 2020J01474, and 2021J01590), the Guangdong Basic and Applied Basic Research Foundation (grant No. 2021A1515110587), and the "Qishan Scholar" Scientific Research Project of Fuzhou University.

Institutional Review Board Statement: Not applicable.

Informed Consent Statement: Not applicable.

Data Availability Statement: The data presented in this study are available on request from the corresponding authors.

Conflicts of Interest: The authors declare no conflict of interest.

References

1. Bell, L.E. Cooling, heating, generating power, and recovering waste heat with thermoelectric systems. *Science* **2008**, *321*, 1457–1461. [CrossRef] [PubMed]
2. He, J.; Tritt Terry, M. Advances in thermoelectric materials research: Looking back and moving forward. *Science* **2017**, *357*, eaak9997. [CrossRef]
3. Mbaye, M.T.; Pradhan, S.K.; Bahoura, M. Data-driven thermoelectric modeling: Current challenges and prospects. *J. Appl. Phys.* **2021**, *130*, 190902. [CrossRef]
4. Yang, J.; Xi, L.; Qiu, W.; Wu, L.; Shi, X.; Chen, L.; Yang, J.; Zhang, W.; Uher, C.; Singh, D.J. On the tuning of electrical and thermal transport in thermoelectrics: An integrated theory—Experiment perspective. *Npj Comput. Mater.* **2016**, *2*, 15015. [CrossRef]
5. Xiong, R.; Sa, B.; Miao, N.; Li, Y.-L.; Zhou, J.; Pan, Y.; Wen, C.; Wu, B.; Sun, Z. Structural stability and thermoelectric property optimization of Ca_2Si. *RSC Adv.* **2017**, *7*, 8936–8943. [CrossRef]
6. Zhu, X.-L.; Yang, H.; Zhou, W.-X.; Wang, B.; Xu, N.; Xie, G. KAgX (X = S, Se): High-performance layered thermoelectric materials for medium-temperature applications. *ACS Appl. Mater. Inter.* **2020**, *12*, 36102–36109. [CrossRef] [PubMed]
7. Gutiérrez Moreno, J.J.; Cao, J.; Fronzi, M.; Assadi, M.H.N. A review of recent progress in thermoelectric materials through computational methods. *Mater. Renew. Sustain. Energy* **2020**, *9*, 16. [CrossRef]
8. Li, D.; Gong, Y.; Chen, Y.; Lin, J.; Khan, Q.; Zhang, Y.; Li, Y.; Zhang, H.; Xie, H. Recent progress of two-dimensional thermoelectric materials. *Nanomicro Lett.* **2020**, *12*, 36. [CrossRef] [PubMed]
9. Gan, Y.; Wang, G.; Zhou, J.; Sun, Z. Prediction of thermoelectric performance for layered IV-V-VI semiconductors by high-throughput ab initio calculations and machine learning. *Npj Comput. Mater.* **2021**, *7*, 176. [CrossRef]
10. Li, Z.; Miao, N.; Zhou, J.; Sun, Z.; Liu, Z.; Xu, H. High thermoelectric performance of few-quintuple Sb_2Te_3 nanofilms. *Nano Energy* **2018**, *43*, 285–290. [CrossRef]
11. Zhu, X.L.; Liu, P.F.; Zhang, J.; Zhang, P.; Zhou, W.X.; Xie, G.; Wang, B.T. Monolayer SnP_3: An excellent p-type thermoelectric material. *Nanoscale* **2019**, *11*, 19923–19932. [CrossRef]
12. Xie, Q.-Y.; Liu, P.-F.; Ma, J.-J.; Kuang, F.-G.; Zhang, K.-W.; Wang, B.-T. Monolayer SnI_2: An excellent p-type thermoelectric material with ultralow lattice thermal conductivity. *Materials* **2022**, *15*, 3147. [CrossRef] [PubMed]
13. Kaur, K.; Khandy, S.A.; Dhiman, S.; Sharopov, U.B.; Singh, J. Computational prediction of thermoelectric properties of 2D materials. *Electron. Struct.* **2022**, *4*, 023001. [CrossRef]
14. Sevinçli, H.; Cuniberti, G. Enhanced thermoelectric figure of merit in edge-disordered zigzag graphene nanoribbons. *Phys. Rev. B* **2010**, *81*, 113401. [CrossRef]
15. Xu, X.; Gabor, N.M.; Alden, J.S.; van der Zande, A.M.; McEuen, P.L. Photo-thermoelectric effect at a graphene interface junction. *Nano Lett.* **2010**, *10*, 562–566. [CrossRef]
16. Zhang, J.; Liu, H.J.; Cheng, L.; Wei, J.; Liang, J.H.; Fan, D.D.; Jiang, P.H.; Sun, L.; Shi, J. High thermoelectric performance can be achieved in black phosphorus. *J. Mater. Chem. C* **2016**, *4*, 991–998. [CrossRef]
17. Saito, Y.; Iizuka, T.; Koretsune, T.; Arita, R.; Shimizu, S.; Iwasa, Y. Gate-Tuned Thermoelectric Power in Black Phosphorus. *Nano Lett.* **2016**, *16*, 4819–4824. [CrossRef] [PubMed]
18. Fei, R.; Faghaninia, A.; Soklaski, R.; Yan, J.-A.; Lo, C.; Yang, L. Enhanced thermoelectric efficiency via orthogonal electrical and thermal conductances in phosphorene. *Nano Lett.* **2014**, *14*, 6393–6399. [CrossRef]
19. Chen, Z.-G.; Shi, X.; Zhao, L.-D.; Zou, J. High-performance SnSe thermoelectric materials: Progress and future challenge. *Prog. Mater. Sci.* **2018**, *97*, 283–346. [CrossRef]
20. Xu, P.; Fu, T.; Xin, J.; Liu, Y.; Ying, P.; Zhao, X.; Pan, H.; Zhu, T. Anisotropic thermoelectric properties of layered compound $SnSe_2$. *Sci. Bull.* **2017**, *62*, 1663–1668. [CrossRef]
21. Sarikurt, S.; Kocabaş, T.; Sevik, C. High-throughput computational screening of 2D materials for thermoelectrics. *J. Mater. Chem. A* **2020**, *8*, 19674–19683. [CrossRef]
22. Zhang, G.; Zhang, Y.-W. Thermoelectric properties of two-dimensional transition metal dichalcogenides. *J. Mater. Chem. C* **2017**, *5*, 7684–7698. [CrossRef]

23. Hong, J.; Lee, C.; Park, J.-S.; Shim, J.H. Control of valley degeneracy in MoS$_2$ by layer thickness and electric field and its effect on thermoelectric properties. *Phys. Rev. B* **2016**, *93*, 035445. [CrossRef]
24. Wickramaratne, D.; Zahid, F.; Lake, R.K. Electronic and thermoelectric properties of few-layer transition metal dichalcogenides. *J. Chem. Phys.* **2014**, *140*, 124710. [CrossRef]
25. Wang, X.; Feng, W.; Shen, C.; Sun, Z.; Qi, H.; Yang, M.; Liu, Y.; Wu, Y.; Wu, X. The verification of thermoelectric performance obtained by high-throughput calculations: The case of GeS$_2$ monolayer from first-principles calculations. *Front. Mater.* **2021**, *8*, 709757. [CrossRef]
26. Sa, B.; Li, Y.-L.; Qi, J.; Ahuja, R.; Sun, Z. Strain engineering for phosphorene: The potential application as a photocatalyst. *J. Phys. Chem. C* **2014**, *118*, 26560–26568. [CrossRef]
27. Yang, S.; Chen, Y.; Jiang, C. Strain engineering of two-dimensional materials: Methods, properties, and applications. *InfoMat* **2021**, *3*, 397–420. [CrossRef]
28. Du, J.; Yu, H.; Liu, B.; Hong, M.; Liao, Q.; Zhang, Z.; Zhang, Y. Strain engineering in 2D material-based flexible optoelectronics. *Small Methods* **2021**, *5*, 2000919. [CrossRef]
29. Panneerselvam, I.R.; Kim, M.H.; Baldo, C.; Wang, Y.; Sahasranaman, M. Strain engineering of polar optical phonon scattering mechanism—An effective way to optimize the power-factor and lattice thermal conductivity of ScN. *Phys. Chem. Chem. Phys.* **2021**, *23*, 23288–23302. [CrossRef] [PubMed]
30. Lou, X.; Li, S.; Chen, X.; Zhang, Q.; Deng, H.; Zhang, J.; Li, D.; Zhang, X.; Zhang, Y.; Zeng, H.; et al. Lattice strain leads to high thermoelectric performance in polycrystalline SnSe. *ACS Nano* **2021**, *15*, 8204–8215. [CrossRef]
31. Yu, C.; Zhang, G.; Zhang, Y.-W.; Peng, L.-M. Strain engineering on the thermal conductivity and heat flux of thermoelectric Bi$_2$Te$_3$ nanofilm. *Nano Energy* **2015**, *17*, 104–110. [CrossRef]
32. Guo, S.-D. Biaxial strain tuned thermoelectric properties in monolayer PtSe$_2$. *J. Mater. Chem. C* **2016**, *4*, 9366–9374. [CrossRef]
33. Blochl, P.E. Projector augmented-wave method. *Phys. Rev. B* **1994**, *50*, 17953–17979. [CrossRef]
34. Hafner, J. Ab-initio simulations of materials using VASP: Density-functional theory and beyond. *J. Comput. Chem.* **2008**, *29*, 2044–2078. [CrossRef] [PubMed]
35. Wang, G.; Peng, L.; Li, K.; Zhu, L.; Zhou, J.; Miao, N.; Sun, Z. ALKEMIE: An intelligent computational platform for accelerating materials discovery and design. *Comput. Mater. Sci.* **2021**, *186*, 110064. [CrossRef]
36. Perdew, J.P.; Burke, K.; Ernzerhof, M. Generalized Gradient Approximation Made Simple. *Phys. Rev. Lett.* **1996**, *77*, 3865–3868. [CrossRef] [PubMed]
37. Perdew, J.; Burke, K.; Ernzerhof, M. Perdew, burke, and ernzerhof reply. *Phys. Rev. Lett.* **1998**, *80*, 891. [CrossRef]
38. Monkhorst, H.J.; Pack, J.D. Special points for Brillouin-zone integrations. *Phys. Rev. B* **1976**, *13*, 5188–5192. [CrossRef]
39. Heyd, J.; Scuseria, G.E.; Ernzerhof, M. Hybrid functionals based on a screened Coulomb potential. *J. Chem. Phys.* **2003**, *118*, 8207–8215. [CrossRef]
40. Hoover, W.G. Canonical dynamics: Equilibrium phase-space distributions. *Phys. Rev. A* **1985**, *31*, 1695–1697. [CrossRef]
41. Nosé, S. A unified formulation of the constant temperature molecular dynamics methods. *J. Chem. Phys.* **1984**, *81*, 511–519. [CrossRef]
42. Madsen, G.K.H.; Singh, D.J. BoltzTraP. A code for calculating band-structure dependent quantities. *Comput. Phys. Commun.* **2006**, *175*, 67–71. [CrossRef]
43. Togo, A.; Oba, F.; Tanaka, I. First-principles calculations of the ferroelastic transition between rutile-type and CaCl$_2$-type SiO$_2$ at high pressures. *Phys. Rev. B* **2008**, *78*, 134106. [CrossRef]
44. Li, W.; Carrete, J.; Katcho, N.A.; Mingo, N. ShengBTE: A solver of the Boltzmann transport equation for phonons. *Comput. Phys. Commun.* **2014**, *185*, 1747–1758. [CrossRef]
45. Ding, Y.; Hu, L.; Dai, J.; Tang, X.; Wei, R.; Sheng, Z.; Liang, C.; Shao, D.; Song, W.; Liu, Q.; et al. Highly ambient-sTable 1T-MoS$_2$ and 1T-WS$_2$ by hydrothermal synthesis under high magnetic fields. *ACS Nano* **2019**, *13*, 1694–1702. [CrossRef]
46. Nandi, P.; Rawat, A.; Ahammed, R.; Jena, N.; De Sarkar, A. Group-IV(A) Janus dichalcogenide monolayers and their interfaces straddle gigantic shear and in-plane piezoelectricity. *Nanoscale* **2021**, *13*, 5460–5478. [CrossRef] [PubMed]
47. Taheri, A.; Pisana, S.; Singh, C.V. Importance of quadratic dispersion in acoustic flexural phonons for thermal transport of two-dimensional materials. *Phys. Rev. B* **2021**, *103*, 235426. [CrossRef]
48. Chaurasiya, R.; Tyagi, S.; Singh, N.; Auluck, S.; Dixit, A. Enhancing thermoelectric properties of Janus WSSe monolayer by inducing strain mediated valley degeneracy. *J. Alloy. Compd.* **2021**, *855*, 157304. [CrossRef]
49. Li, Y.; Ma, K.; Fan, X.; Liu, F.; Li, J.; Xie, H. Enhancing thermoelectric properties of monolayer GeSe via strain-engineering: A first principles study. *Appl. Surf. Sci.* **2020**, *521*, 146256. [CrossRef]
50. Jonson, M.; Mahan, G.D. Mott's formula for the thermopower and the Wiedemann-Franz law. *Phys. Rev. B* **1980**, *21*, 4223–4229. [CrossRef]
51. Wei, Q.-L.; Zhu, X.-L.; Liu, P.-F.; Wu, Y.-Y.; Ma, J.-J.; Liu, Y.-B.; Li, Y.-H.; Wang, B.-T. Quadruple-layer group-IV tellurides: Low thermal conductivity and high performance two-dimensional thermoelectric materials. *Phys. Chem. Chem. Phys.* **2021**, *23*, 6388–6396. [CrossRef] [PubMed]
52. Xu, B.; Xia, Q.; Zhang, J.; Ma, S.; Wang, Y.; Xu, Q.; Li, J.; Wang, Y. High figure of merit of monolayer Sb$_2$Te$_2$Se of ultra low lattice thermal conductivity. *Comput. Mater. Sci.* **2020**, *177*, 109588. [CrossRef]

53. Song, H.-Y.; Sun, J.-J.; Li, M. Enhancement of monolayer HfSe$_2$ thermoelectric performance by strain engineering: A DFT calculation. *Chem. Phys. Lett.* **2021**, *784*, 139109. [CrossRef]
54. Zhou, W.X.; Cheng, Y.; Chen, K.Q.; Xie, G.; Wang, T.; Zhang, G. Thermal conductivity of amorphous materials. *Adv. Funct. Mater.* **2019**, *30*, 1903829. [CrossRef]
55. Cai, Y.; Zhang, G.; Zhang, Y.-W. Polarity-reversed robust carrier mobility in monolayer MoS$_2$ nanoribbons. *J. Am. Chem. Soc.* **2014**, *136*, 6269–6275. [CrossRef] [PubMed]
56. Rawat, A.; Jena, N.; Dimple; De Sarkar, A. A comprehensive study on carrier mobility and artificial photosynthetic properties in group VI B transition metal dichalcogenide monolayers. *J. Mater. Chem. A* **2018**, *6*, 8693–8704. [CrossRef]
57. Zhang, P.; Jiang, E.; Ouyang, T.; Tang, C.; He, C.; Li, J.; Zhang, C.; Zhong, J. Potential thermoelectric candidate monolayer silicon diphosphide (SiP$_2$) from a first-principles calculation. *Comput. Mater. Sci.* **2021**, *188*, 110154. [CrossRef]
58. Li, G.; Yao, K.; Gao, G. Strain-induced enhancement of thermoelectric performance of TiS$_2$ monolayer based on first-principles phonon and electron band structures. *Nanotechnology* **2018**, *29*, 015204. [CrossRef]

Article

Monolayer SnI₂: An Excellent p-Type Thermoelectric Material with Ultralow Lattice Thermal Conductivity

Qing-Yu Xie [1,2], Peng-Fei Liu [1,3], Jiang-Jiang Ma [1,3], Fang-Guang Kuang [4], Kai-Wang Zhang [2,*] and Bao-Tian Wang [1,3,5,*]

Citation: Xie, Q.-Y.; Liu, P.-F.; Ma, J.-J.; Kuang, F.-G.; Zhang, K.-W.; Wang, B.-T. Monolayer SnI₂: An Excellent p-Type Thermoelectric Material with Ultralow Lattice Thermal Conductivity. *Materials* 2022, 15, 3147. https://doi.org/10.3390/ma15093147

Academic Editor: Israel Felner

Received: 18 March 2022
Accepted: 24 April 2022
Published: 26 April 2022

Publisher's Note: MDPI stays neutral with regard to jurisdictional claims in published maps and institutional affiliations.

Copyright: © 2022 by the authors. Licensee MDPI, Basel, Switzerland. This article is an open access article distributed under the terms and conditions of the Creative Commons Attribution (CC BY) license (https://creativecommons.org/licenses/by/4.0/).

[1] Institute of High Energy Physics, Chinese Academy of Sciences (CAS), Beijing 100049, China; qyxie@ihep.ac.cn (Q.-Y.X.); pfliu@ihep.ac.cn (P.-F.L.); majj88@ihep.ac.cn (J.-J.M.)
[2] School of Physics and Optoelectronics, Xiangtan University, Xiangtan 411105, China
[3] Spallation Neutron Source Science Center (SNSSC), Dongguan 523803, China
[4] School of Physics and Electronic Information, Gannan Normal University, Ganzhou 341000, China; kuangfg1987@126.com
[5] Collaborative Innovation Center of Extreme Optics, Shanxi University, Taiyuan 030006, China
* Correspondence: kwzhang@xtu.edu.cn (K.-W.Z.); wangbt@ihep.ac.cn (B.-T.W.)

Abstract: Using density functional theory and semiclassical Boltzmann transport equation, the lattice thermal conductivity and electronic transport performance of monolayer SnI₂ were systematically investigated. The results show that its room temperature lattice thermal conductivities along the zigzag and armchair directions are as low as 0.33 and 0.19 W/mK, respectively. This is attributed to the strong anharmonicity, softened acoustic modes, and weak bonding interactions. Such values of the lattice thermal conductivity are lower than those of other famous two-dimensional thermoelectric materials such as MoO₃, SnSe, and KAgSe. The two quasi-degenerate band valleys for the valence band maximum make it a p-type thermoelectric material. Due to its ultralow lattice thermal conductivities, coupled with an ultrahigh Seebeck coefficient, monolayer SnI₂ possesses an ultrahigh figure of merits at 800 K, approaching 4.01 and 3.34 along the armchair and zigzag directions, respectively. The results indicate that monolayer SnI₂ is a promising low-dimensional thermoelectric system, and would stimulate further theoretical and experimental investigations of metal halides as thermoelectric materials.

Keywords: thermoelectrics; electronic transport; thermal transport

1. Introduction

With more than 60% of energy in the world lost in the form of waste heat, the thermoelectric system has attracted widespread attention, since it directly converts the waste heat to electric energy through the Seebeck effect. It has the advantages of small size, high reliability, no pollutants, and a feasibility in a wide temperature range. Such a system is widely used in aerospace exploration and industrial production, such as in space probes, thermoelectric generators, and precise temperature controls [1,2]. The converting efficiency of TE materials is ruled by the dimensionless figure of merit (ZT); $ZT = S^2 \sigma T/(\kappa_L + \kappa_e)$, where S, σ, T, κ_L, and κ_e are the Seebeck coefficient, electrical conductivity, absolute temperature, lattice thermal conductivity, and electronic thermal conductivity, respectively. The electronic transport properties S, σ, and κ_e, have a complex coupling relationship, and are difficult to decouple, even though they can be modified via carrier concentration [3–5]. Historically, two aspects were established to enhance ZT: one aspect is to optimize carrier concentration by the band structure, engineered to enhance the power factor (σS^2) [6–9], and the other aspect is to reduce the lattice thermal conductivity, via alloying and nanostructuring [10]. Alternatively, it is more attractive to seek materials with an intrinsic low lattice thermal conductivity (generally associated with complex crystal structures [11]), strong anharmonicity [12,13], lone pair electrons [14,15], and liquid-like behavior [16–18], etc.

The group IVA metal dihalides are candidates for semiconductor optical devices and perovskite solar cells, due to their excellent properties, such as the visible-range band gap and the thickness-dependent band structure [19–22]. However, the application of the Pb-based materials, such as the layered 2H-PbI$_2$, has been greatly limited by their toxicity and environmental unfriendliness [23–25]. In addition, the surface of the bulk SnI$_2$ is generally very rough and accompanied by many defects, which strongly scatters carriers [26]. Fortunately, its vdW monolayer has been experimentally realized, via molecular beam epitaxy [27]. The low dimensionality provides an effective conductive channel for carriers, and also suppresses phonon thermal transport [28]. Thus, monolayer SnI$_2$ is a preferable option over bulk SnI$_2$ and monolayer PbI$_2$ in use as a TE material, and deserves to be carefully explored in a systematic study.

In this work, combining first-principles calculations and the semiclassical Boltzmann transport equation, the TE properties of monolayer SnI$_2$ were systematically explored. Results show that the intrinsic ultralow lattice thermal conductivity originates from the strong anharmonicity, weak bonding, and softened acoustic modes. The Grüneisen parameter, phonons scattering phase space, and phonon relaxation time were calculated to understand the micro-mechanism of the phonon transports. The two quasi-degenerate band valleys for the valence band maximum (VBM) in its electronic band structure led to a p-type TE material. The maximum ZT value along the armchair and zigzag directions at 800 K reach 4.01 and 3.34, respectively, using optimal p-type doping. These results indicate that monolayer SnI$_2$ exhibits an extraordinary TE response, and is an ideal material for TE applications.

2. Computational Methods

The first-principles calculations were implemented in the Vienna ab initio simulation package (VASP, VASP.5.3, Wien, Austria) [29]. The generalized gradient approximation (GGA) [30,31] in the Perdew–Burke–Ernzerhof (PBE) [32] form was employed to deal with the exchange–correlation functional, with a cutoff of 300 eV on a $9 \times 9 \times 1$ Monkhorst–Pack k-mesh. To screen the interactions between adjacent images, the length of the unit cell of 20 Å was used along the z direction. The geometry structure was fully relaxed, with a criterion of convergence for residual forces of 0.001 eV/Å, and the total energy difference converged to within 10^{-8} eV/Å. To obtain an accurate band gap and electronic transport performance, the Heyd–Suseria–Ernzerhof (HSE06) [33] method was employed.

The electronic transport properties of the monolayer SnI$_2$ were calculated by solving the semiclassical Boltzmann transport equation, utilizing the BoltzTraP code (Georg K. H. Madsen, Århus, Denmark) [34] with a dense $35 \times 35 \times 1$ k-mesh. The constant relaxation time approach (CRTA) was used, since the relaxation time is not strongly dependent on the energy scale of k_BT, and has accurately predicted TE properties of multitudinous materials. In this work, the electrons relaxation time was calculated using the deformation potential (DP) theory [35], which considers the primarily acoustic phonon scatterings, but ignores effects of the optical phonons in the single parabolic band (SPB) model. Based on the rigid band approximation (RBA) [36] and CRTA, the transport coefficients S, σ, and κ_e can be obtained by:

$$S = \frac{ek_B}{\sigma} \int \Xi(\varepsilon)(-\frac{\partial f_0}{\partial \varepsilon})\frac{\varepsilon - \mu}{k_BT}d\varepsilon \quad (1)$$

$$\sigma = e^2 \int \Xi(\varepsilon)(-\frac{\partial f_0}{\partial \varepsilon})\frac{\varepsilon - \mu}{k_BT}d\varepsilon \quad (2)$$

$$\kappa_e = \kappa_0 - \sigma S^2 T = L\sigma T \quad (3)$$

$$\Xi = \sum v_k v_k \tau_k \quad (4)$$

where k_B, f_0, and $\Xi(\varepsilon)$ are the Boltzmann constant, the Fermi–Dirac distribution function, and the transport distribution function, respectively. The calculation of the relaxation time τ was extremely difficult, due to the various complex scattering mechanism in the crystals. The relaxation time was calculated based on the DP theory in the SPB model, which considers the predominant scatterings between carriers and acoustic phonons in the low-energy region. In fact, the scattering matrix element ($\left|M(\vec{k},\vec{k'})\right|^2$) of the acoustic phonons in the long-wavelength can be approximated as $k_B T E_1^2/C_{ii}$, where C_{ii} and E_1 are the elastic and DP constants, respectively. Thus, for the 2D material, the carrier mobility and electrons relaxation time can be approached as [37,38]

$$\mu_{2D} = \frac{eh^3 C_{ii}}{8\pi^3 k_B T m_d m * E_1^2} \quad (5)$$

$$\tau = \frac{u m^*}{e} \quad (6)$$

where h, m^*, and m_d are the Planck constant, the effective mass along the transport direction, and the averaged effective mass, respectively.

The harmonic second-order interaction force constants (2nd IFCs) and phonon spectrum were calculated using the Phonopy package [39], using a 2 × 2 × 1 supercell with a 5 × 5 × 1 k-mesh using the finite-difference method [40]. The anharmonic third-order interaction force constants (3rd IFCs), which consider the interactions between the sixth-nearest-neighbor atoms, were obtained using the ShengBTE package (ShengBTE version 1.0.2, Wu Li, Grenoble, France; Phonopy version 2.11.0, Atsushi Togo, Sakyo, Japan) [41] with the same supercell. The lattice thermal conductivity and phonon transport properties were obtained using the self-consistent iterative solution of the Boltzmann transport equation, with a dense 36 × 36 × 1 mesh, which had a good convergence. Based on the Boltzmann transport equation, with the Fourier's low of heat conduction, the matrix elements of the phonon thermal conductivity can be expressed by [42,43]

$$\kappa_{p.\alpha\beta} = \frac{1}{N_q} \sum_\lambda c_\lambda v_{\lambda,\alpha} v_{\lambda,\beta} \tau_\lambda \quad (7)$$

where α and β are the Cartesian indices, N_q is the total number of q-points sampled in the first Brillouin zone, and c_λ, v_λ, and τ_λ are the mode-specific heat capacities, phonon group velocity, and the relaxation time, respectively. Here, the phonon thermal properties were calculated based on the 3rd IFCs, ignoring the fourth- and higher-order terms. This strategy described the phonon behavior of the most anharmonic materials [44,45]. To define the effective thickness of two-dimensional (2D) materials, the summation of interlayer distance and the vdW radii of the outermost surface atoms was adopted [44].

3. Results and Discussion

3.1. Geometry and Electronic Structure

Monolayer SnI_2 crystallizes in a hexagonal lattice with space group P-3m1 (164), as shown in Figure 1. The optimized lattice parameter is 4.57 Å, which is in good agreement with the previous experimental data of 4.48 Å [27]. The structure is analogous to H-MoS_2 [46], consisting of three layers, with Sn atoms as the middle layer, and I atoms as the upper and lower atomic layers. The electron localization function (ELF) provides a deeper insight to characterize the nature of, and strengthen, the chemical band. Figure 1c shows the calculated three-dimensional (3D) ELF map (isosurface level of 0.97). The ELF around I is in the shape of a "mushroom", suggesting the existence of the lone pair electrons. Moreover, the electron sharing is better visualized by the 2D ELF map in Figure 1d. The interstitial electrons between Sn and I are close to I atoms, and the value of the area between them is 0.5, which shows the characteristics of a free electron gas and indicates a weak bonding between Sn and I atoms. The electronic repulsion between the lone pair electrons and the Sn–I bonding electrons results in strong anharmonicity, such as in $CuSbS_2$ [14]. The complex structure and lone pair electrons are also beneficial to its low lattice thermal conductivity.

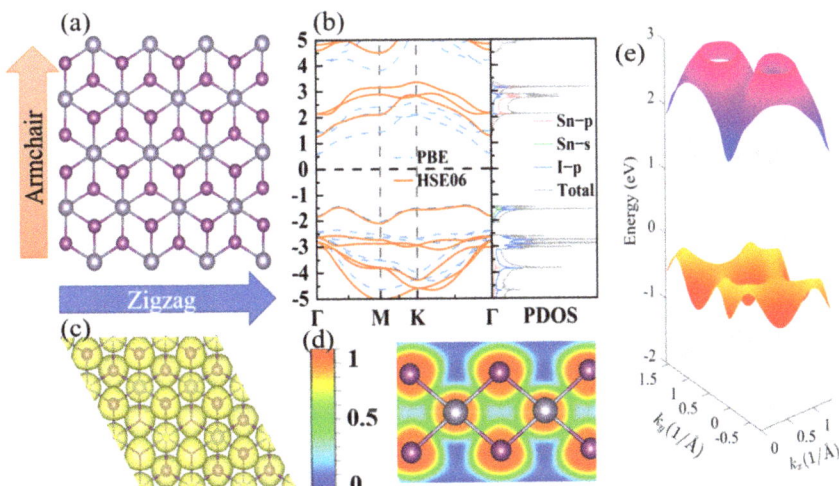

Figure 1. (a) Top view of the atomic structure of SnI$_2$ monolayer. (b) The electronic band structures calculated with PBE and HSE06 hybrid functional potentials. (c,d) the 3D and 2D ELF maps. (e) The 3D electronic band structure calculated with PBE.

The electronic structure plays a crucial role in characterizing the electronic transport properties. As shown in Figure 1b, the electronic band structures obtained from the PBE and HSE06 hybrid function potentials are presented. They exhibit analogous band structures, except for a more accurate bandgap (2.71 eV) that approaches the experiment value (2.9 eV) [27] for the latter structure. The conduction band mainly originates from the p orbitals of the Sn atom and the p orbitals of the I atom, whereas the valence band consists of the s orbital of the Sn atom and the p orbitals of the I atom. Below the VBM, there are two quasi-degenerate band valleys along the Γ–M and Γ–K directions with an energy difference of ~0.03 eV, which is far less than those in SnTe (~0.35 eV) [47] and PbTe (~0.15 eV) [1]. The multi-energy valley enhances the TE performance and is verified in many materials [48,49]. However, the behavior of band valleys degenerate do not exist for the conduction band minimum (CBM), which is located at the Γ point. Hence, it is expected that the TE performance of the p-type could be superior to that of the n-type.

3.2. Electronic Transport Properties

All the parameters for electronic transport, calculated according to the DP theory, are tabulated in Table 1. Based on the reasonable relaxation time, all the electronic transport coefficients, Seebeck coefficient S, electrical conductivity σ, and power factor (PF), as a function of temperature at the corresponding optimal carrier concentration for p-type SnI$_2$, range from 1×10^{13} to 1×10^{14} cm^{-2}, and are presented in Figure 2a–c. For comparison, those for n-type SnI$_2$ at the same condition are also plotted in Figure 2d–f. Here, the tiny anisotropy characteristics are ignored.

Table 1. The calculated DP constant E_1, elastic constant, effective mass, carrier mobility, and relaxation time along the zigzag and armchair directions in monolayer SnI$_2$ at 300 K.

Direction	Type	E_1 (eV)	C_{ii} (J/m^2)	m^* (m_0)	u (cm^{-2}V^{-1}S^{-1}))	τ (fs)
Zigzag	e	−4.44	17.92	0.59	42.52	17.41
	h	−4.43	17.92	0.73	40.16	16.72
Armchair	e	−4.25	17.42	0.84	30.48	15.17
	h	−4.49	17.42	0.68	35.14	13.60

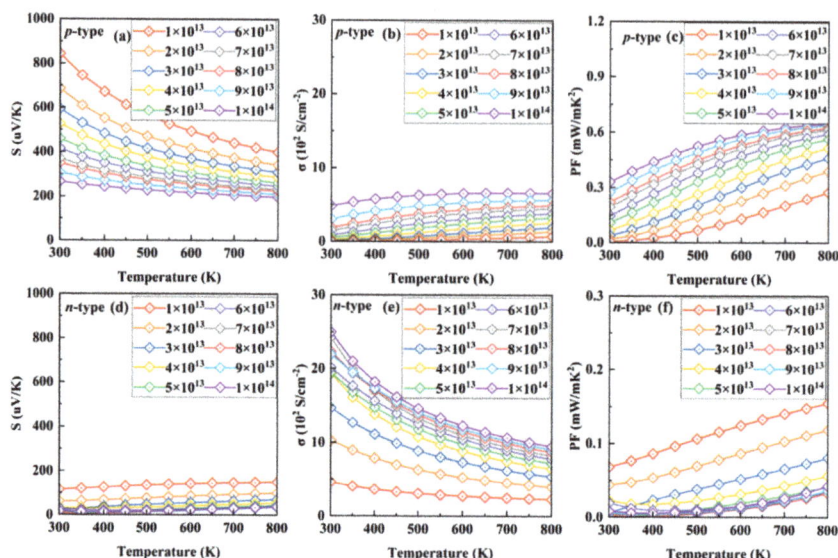

Figure 2. Seebeck coefficient, electronic conductivity, and PF as a function of temperature under various concentrations of p- (**a**–**c**) and n-type (**d**–**f**) doping.

Increasing the carrier concentration, the chemical potential enters in the deeper energy levels for both p- and n-types of SnI_2, resulting in a decreased Seebeck coefficient at the same temperature. It is clear that the values under p-type doping are always far larger than those under n-type at the same condition, which can be attributed to the multi-band character near the VBM, as shown in Figure 1b. For example, the absolute values of the S under p- and n-types of doping at 700 K are 439 and 146 uV/K, respectively, for a doping concentration of 1×10^{13} cm^{-2}. In addition, the trend of S under p-type doping is opposite to that under n-type. The former decreases while the latter increases upon heating.

Compared with S, the electronic conductivity σ exhibits a different behavior, due to the complete relationship between them, as Equations (1) and (2) show. Interestingly, in all cases, the electrical conductivities for p-type SnI_2 are always far less than those for n-type. For example, the values under p- and n-type doping are 0.46 and 2.43 S/cm^{-2}, respectively, at 700 K for a doping concentration of 1×10^{13} cm^{-2}. At 800 K, the electrical conductivity under p-type (n-type) increases from ~0.67 S/cm^{-2} (~2.24 S/cm^{-2}) for 1×10^{13} cm^{-2} to 6.35 S/cm^{-2} (9.43 S/cm^{-2}) for 1×10^{14} cm^{-2}.

Ultimately, the PF decouples the complete relationship between the Seebeck coefficient and electrical conductivity. It is clearly seen that the p-type SnI_2 possesses significantly higher PF values than those of the n-type system. This is in good agreement with the previous analysis for the electronic structure. At 800 K, the PF for p-type increases from ~0.27 mW/mK^2 for 1×10^{13} cm^{-2} to 0.62 mW/mK^2 for 1×10^{14} cm^{-2}. For the n-type system, however, the PF decreases from ~0.15 mW/mK^2 for 1×10^{13} cm^{-2} to 0.04 mW/mK^2 for 1×10^{14} cm^{-2}. Thus, it is expected that the p-type SnI_2 possesses a more excellent performance than that of the n-type. According to the Wiedemann–Franz law, $\kappa_e = L\sigma T$, where L is the Lorenz number, the classical value $L = (\pi k_B)^2/3e^2 \approx 2.44 \times 10^{-8}$ WΩK^{-2} is adopted. This value meets the result of $\kappa_0 - \sigma S^2 T$ [50]. Thus, the electronic conductivity is proportional to the electronic thermal conductivity. The results in the present work also obey this rule.

3.3. Phonon Transport Properties

As shown in Figure 3a, the calculated total κ_L is very low in the wide temperature of 300–800 K. Remarkably, the low room temperature phonon thermal conductivities of

0.33 and 0.19 W/mK along the armchair and zigzag directions, respectively, are fundamentally lower than those previously reported values for 2D MO$_3$ (1.57 W/mk) [51], SnSe (2.77 W/mK) [13], and CaP$_3$ (0.65 W/mK) [52]. Through analyzing the contributions of the acoustic phonon branches along the in-plane (TA and LA) or out-of-plane (ZA), as well as the optical phonon modes to the total κ_L, results show that the main contribution to the κ_L is from the ZA mode. In the following, we reveal the origins of such ultralow phonon conductivity, and present a comprehensive analysis to support this result.

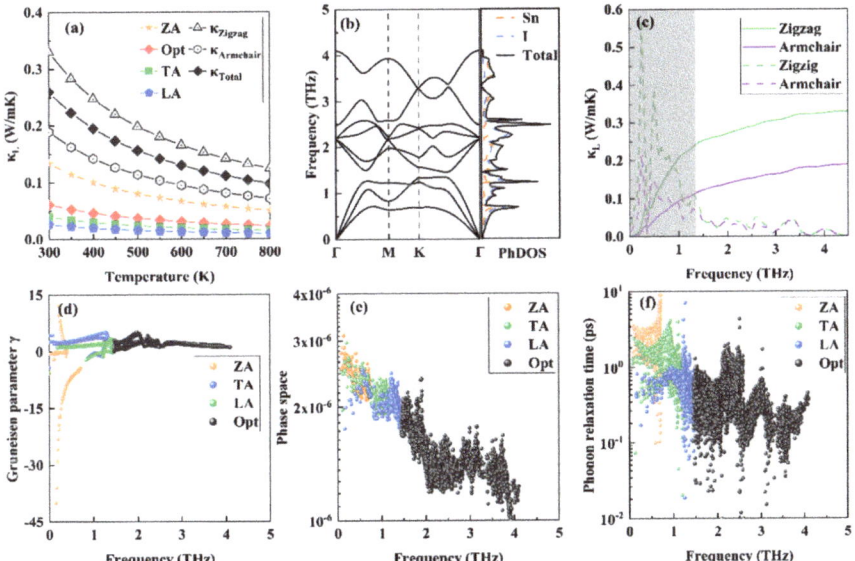

Figure 3. (a) Calculated lattice thermal conductivity. The contributions from the ZA, TA, LA, and optical modes to the total κ_L, as well as the lattice thermal conductivities along the armchair and the zigzag directions, are shown. (b) Phonon dispersion curves and corresponding PhDOS. (c) Cumulative thermal conductivity and the derivatives (dashed line) with respect to frequency. (d–f) Grüneisen parameter γ, phonon scattering phase space, and phonon relaxation time of the ZA, TA, LA, and optical modes.

The phonon dispersion curves are presented in Figure 3b. With three atoms in its primitive cell, there are three acoustic and six optical phonon modes for monolayer SnI$_2$. It is dynamically stable, since no imaginary frequency is observed. Near the long wave limit, the TA and LA branches are in linear trend, whereas the ZA branch exhibits a quadratic trend. These features are typical for 2D materials, and can be explained with the elastic theory of thin plate [53]. It is clearly seen that a narrow phonon gap of about 0.1 THz separates the phonon modes into the acoustic phonon part (0~1.35 THz) and the optical phonon part (1.45~4.10 THz). The cutoff acoustic phonon frequency, which is as low as 1.35 THz, is lower than those for SnSe (1.6 THz) and SnS (1.9 THz) [13]. The low-lying acoustic modes, as well as the soft mode for TA near the M point, imply the weak bonding between Sn and I atoms, consistent with the previous analysis. From the corresponding phonon density of states (PhDOS), the contributions from the I atomic vibrations are apparently larger than those from Sn, within the region of 2.8 to 4.10 THz. Both Sn and I evidently contribute in a wide energy range, implying the nature of covalent bonding [12]. In addition, it is recalled that the decoupling of the in-plane and out-of-plane phonon modes results in ultrahigh phonon thermal conductivity for graphene, due to its one atom plane nature [53,54]. However, such full decoupling behavior should not be observed for a finite thickness 2D system, such as the case of the present studied monolayer SnI$_2$.

To explain the anomalous thermal transport behavior of monolayer SnI$_2$, the cumulative lattice conductivity, and their derivatives, with respect to frequency at 300 K are calculated and presented in Figure 3c. Clearly, the κ_L is mainly caused by phonons of the cutoff acoustic part (the shadow region < 1.35 THz). Specifically, the contributions from ZA, TA, LA, and the optical branches are 51.21%, 15.14%, 10%, and 23.54%, respectively. For graphene, the low frequency (0–5 THz) also dominates the κ_L and the ZA mode, and contributes 75% to κ_L [54]. Comparing with graphene, the structure of the monolayer SnI$_2$ lacks mirror symmetry. Thus, the mirror symmetry does not reflect whether the ZA mode dominates the thermal transport [55].

Generally, there are two factors that dominate the κ_L: (i) anharmonic interaction matrix elements, and (ii) the inverse of phonon phase space volume. The Grüneisen parameter γ is generally employed to quantify the strength of anharmonicity [41]. Figure 3d shows the calculated γ of the ZA, TA, LA, and optical modes of the SnI$_2$ monolayer. As a large $|\gamma|$ implies strong anharmonicity, it can be seen that the ZA mode exhibits giant anharmonicity. The large $|\gamma|$ of this out-of-plane mode means that the anharmonicity of the bonding between Sn and I atoms in the vertical direction of the monolayer plate is strong. The phonon scattering phase space reveals all available scattering processes, which are ruled by the energy and (quasi)momentum conservation. As shown in Figure 3e, the total scatting phase space of the ZA, TA, and LA modes are 1.55×10^{-3}, 1.41×10^{-3}, and 1.35×10^{-3}, respectively, which confirms more abundant scattering channels of the ZA mode, than those of the TA and LA modes. The larger the phonon phase space, the greater the contribution to the κ_L, thus, validating the decreasing contributions to κ_L from the ZA, TA, and then LA modes. The phonon relaxation time provides a deeper microcosmic insight to understand the ultralow κ_L of SnI$_2$. Compared with the monolayer SnP$_3$ [38], monolayer SnI$_2$ possesses a shorter phonon relaxation time, implying an ultralow κ_L.

3.4. Thermoelectric Figure of Merit

By combining the phonon and electron transport coefficients, the ZT of SnI$_2$ under p- and n-types of doping as functions of temperature and carrier concentration are presented in Figure 4. Owing to the calculated thermal conductivity, along the zigzag direction is slightly larger than the armchair direction, and it is expected that the ZT along the armchair direction is higher than that along the zigzag direction. In addition, the p-type SnI$_2$ is obviously superior to the n-type, since the two quasi-degenerate band valleys in VBM leads to a larger Seebeck coefficient. Results show that monolayer SnI$_2$ is thermally stable up to 800 K, by performing ab initio molecular dynamic (AIMD) simulations (see Figure S1). The values of ZT are as high as 4.01 and 3.34 along the armchair and zigzag directions, respectively, around concentrations of 1.0×10^{13} and 1.2×10^{13} cm^{-2}, at 800 K. Such large values are better than the experimental value of 2.6 for the well-known TE material SnSe along the specific axis at 925 K [3]. In addition, the doping carrier concentrations at such levels have been realized experimentally in monolayer MoS$_2$ [56]. Therefore, such a high TE value for monolayer SnI$_2$ is possible, and indicates its excellent potential TE performance.

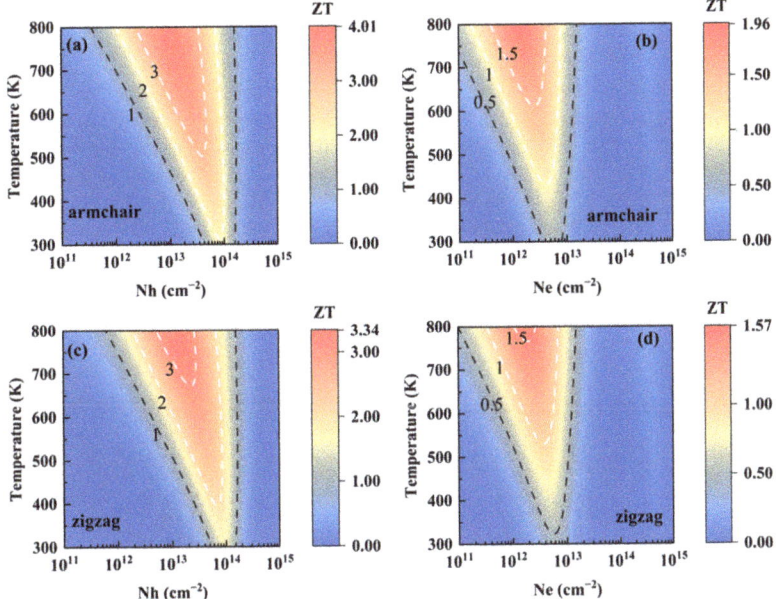

Figure 4. Contour maps of ZT as functions of both temperature and carrier concentration for monolayer SnI$_2$: (**a**) p-type and (**b**) n-type along the armchair direction, (**c**) p-type and (**d**) n-type along the zigzag direction.

To clearly compare the TE performances of the monolayer SnI$_2$ with some famous TE materials, the κ_L and the max ZT are listed in Table 2. According to these results, the monolayer SnI$_2$ is an emerging candidate for TE devices, due to its ultrahigh ZT. As seen, the κ_L of monolayer SnI$_2$ is very low, almost a twentieth of the monolayer SnP$_3$ [38], and is greatly lower than those of tin selenide and tin sulfide [13,48]. Such low value of the κ_L for monolayer SnI$_2$ make it a typical 2D TE system. After all, the electronic transport properties can be easily modulated in experiments, while the lattice thermal conductivity is very difficult to change. In addition, the κ_L values at the level of 0.1–0.5 W/mK in experiments are very few. To the best of our knowledge, the complex systems of CsAg$_5$Te$_3$ (0.18 W/mK) [57] and CsCu$_5$Se$_3$ (0.4~0.8 W/mK) [58], as well as the bulk superlattice material Bi$_4$O$_4$SeCl$_2$ (0.1 W/mK) [59], are typical low κ_L TE materials. Overall, the low thermal conductivity, as well as the high ZT, make monolayer SnI$_2$ a good TE material for low-dimensional devices.

Table 2. κ_L at 300 K and the max ZT at corresponding maximum thermodynamic temperature for monolayer SnI$_2$, as well as some typical TE materials.

Material	κ_L (W/mK)	ZT	
SnI$_2$	0.26	4.01 (800 K)	This work
SnS	1.5	1.00 (750 K)	Ref. [13]
SnSe	0.6	1.50 (750 K)	Ref. [13]
SnP3	4.97	3.46 (500 K)	Ref. [38]
SnSe	1.12	0.85 (900 K)	Ref. [48]
LaCuOSe	0.84	2.71 (900 K)	Ref. [50]

4. Conclusions

In summary, the TE properties of monolayer SnI$_2$ were studied using DFT and the semi-classical Boltzmann transport equation. The results indicate that this 2D material possesses intrinsically ultralow lattice thermal conductivity. Its strong anharmonicity, weak bonding, and softened acoustic branches greatly suppress the phonon transport, and result in an ultralow κ_L of 0.33 and 0.18 W/mK at 300 K along the zigzag and armchair directions, respectively. The p-type SnI$_2$ possesses a superior electric transport performance than the n-type one, due to the two quasi-degenerate band valleys in its VBM. The ZT at 800 K under p-type doping is as high as 4.01 along the armchair direction. Collectively, these results indicate the great advantages of monolayer SnI$_2$ for converting heat energy with high efficiency at high temperatures. Generally, when the ZT value of a material exceeds 1.0, it is considered as an ideal TE material. Therefore, the high ZT of monolayer SnI$_2$ demonstrates it as an emerging candidate for TE applications.

Supplementary Materials: The following supporting information can be downloaded at: https://www.mdpi.com/article/10.3390/ma15093147/s1, Figure S1: Free energy fluctuations with respect to time and equilibrium structures of SnI2 monolayer by AIMD simulations at 800 K.

Author Contributions: Conceptualization, Q.-Y.X., K.-W.Z. and B.-T.W.; methodology, Q.-Y.X., P.-F.L., J.-J.M. and B.-T.W.; software, B.-T.W.; validation, Q.-Y.X., K.-W.Z. and B.-T.W.; formal analysis, Q.-Y.X.; investigation, Q.-Y.X., P.-F.L. and F.-G.K.; resources, B.-T.W.; data curation, Q.-Y.X. and B.-T.W.; writing—original draft preparation, Q.-Y.X.; writing—review and editing, K.-W.Z. and B.-T.W.; supervision, K.-W.Z. and B.-T.W.; project administration, K.-W.Z. and B.-T.W.; funding acquisition, B.-T.W. All authors have read and agreed to the published version of the manuscript.

Funding: We gratefully acknowledge the financial support from the National Natural Science Foundation of China (Grants No. 12074381, No. 12005230, and No. 12104458), the Guangdong Basic and Applied Basic Research Foundation (Grant No. 2021A1515110587), and the Youth Science Foundation of Jiangxi Province (Grants 20171BAB211009).

Institutional Review Board Statement: Not applicable.

Informed Consent Statement: Not applicable.

Data Availability Statement: The data presented in this study are available on request from the corresponding authors. The data are not publicly available due to ongoing research in the project.

Conflicts of Interest: There are no conflicts to declare.

References

1. Pei, Y.; Shi, X.; LaLonde, A.; Wang, H.; Chen, L.; Snyder, G.J. Convergence of electronic bands for high performance bulk thermoelectrics. *Nature* **2011**, *473*, 66–69. [CrossRef] [PubMed]
2. Tan, G.; Zhao, L.D.; Kanatzidis, M.G. Rationally Designing High-Performance Bulk Thermoelectric Materials. *Chem. Rev.* **2016**, *116*, 12123–12149. [CrossRef] [PubMed]
3. Zhao, L.D.; Tan, G.; Hao, S.; He, J.; Pei, Y.; Chi, H.; Wang, H.; Gong, S.; Xu, H.; Dravid, V.P.; et al. Ultrahigh power factor and thermoelectric performance in hole-doped single-crystal SnSe. *Science* **2016**, *351*, 141–144. [CrossRef] [PubMed]
4. Zhang, X.; Zhao, L.-D. Thermoelectric materials: Energy conversion between heat and electricity. *J. Mater.* **2015**, *1*, 92–105. [CrossRef]
5. Zhao, L.D.; Chang, C.; Tan, G.J.; Kanatzidis, M.G. SnSe: A remarkable new thermoelectric material. *Energy Environ. Sci.* **2016**, *9*, 3044–3060. [CrossRef]
6. Zhao, L.D.; He, J.; Wu, C.I.; Hogan, T.P.; Zhou, X.; Uher, C.; Dravid, V.P.; Kanatzidis, M.G. Thermoelectrics with earth abundant elements: High performance p-type PbS nanostructured with SrS and CaS. *J. Am. Chem. Soc.* **2012**, *134*, 7902–7912. [CrossRef]
7. Zhao, L.D.; Wu, H.J.; Hao, S.Q.; Wu, C.I.; Zhou, X.Y.; Biswas, K.; He, J.Q.; Hogan, T.P.; Uher, C.; Wolverton, C.; et al. All-scale hierarchical thermoelectrics: MgTe in PbTe facilitates valence band convergence and suppresses bipolar thermal transport for high performance. *Energy Environ. Sci.* **2013**, *6*, 3346–3355. [CrossRef]
8. Pei, Y.Z.; May, A.F.; Snyder, G.J. Self-Tuning the Carrier Concentration of PbTe/Ag$_2$Te Composites with Excess Ag for High Thermoelectric Performance. *Adv. Energy Mater.* **2011**, *1*, 291–296. [CrossRef]
9. Zebarjadi, M.; Joshi, G.; Zhu, G.; Yu, B.; Minnich, A.; Lan, Y.; Wang, X.; Dresselhaus, M.; Ren, Z.; Chen, G. Power factor enhancement by modulation doping in bulk nanocomposites. *Nano Lett.* **2011**, *11*, 2225–2230. [CrossRef]
10. Shi, X.; Yang, J.; Salvador, J.R.; Chi, M.; Cho, J.Y.; Wang, H.; Bai, S.; Yang, J.; Zhang, W.; Chen, L. Multiple-filled skutterudites: High thermoelectric figure of merit through separately optimizing electrical and thermal transports. *J. Am. Chem. Soc.* **2011**, *133*, 7837–7846. [CrossRef]
11. Zhang, Y.; Skoug, E.; Cain, J.; Ozoliņš, V.; Morelli, D.; Wolverton, C. First-principles description of anomalously low lattice thermal conductivity in thermoelectric Cu-Sb-Se ternary semiconductors. *Phys. Rev. B* **2012**, *85*, 054306. [CrossRef]

12. Liu, P.-F.; Bo, T.; Xu, J.; Yin, W.; Zhang, J.; Wang, F.; Eriksson, O.; Wang, B.-T. First-principles calculations of the ultralow thermal conductivity in two-dimensional group-IV selenides. *Phys. Rev. B* **2018**, *98*, 235426. [CrossRef]
13. Guo, R.Q.; Wang, X.J.; Kuang, Y.D.; Huang, B.L. First-principles study of anisotropic thermoelectric transport properties of IV-VI semiconductor compounds SnSe and SnS. *Phys. Rev. B* **2015**, *92*, 115202. [CrossRef]
14. Feng, Z.Z.; Jia, T.T.; Zhang, J.H.; Wang, Y.X.; Zhang, Y.S. Dual effects of lone-pair electrons and rattling atoms in $CuBiS_2$ on its ultralow thermal conductivity. *Phys. Rev. B* **2017**, *96*, 235205. [CrossRef]
15. Nielsen, M.D.; Ozolins, V.; Heremans, J.P. Lone pair electrons minimize lattice thermal conductivity. *Energy Environ. Sci.* **2013**, *6*, 570–578. [CrossRef]
16. Liu, H.; Yuan, X.; Lu, P.; Shi, X.; Xu, F.; He, Y.; Tang, Y.; Bai, S.; Zhang, W.; Chen, L.; et al. Ultrahigh thermoelectric performance by electron and phonon critical scattering in $Cu_2Se_{1-x}I_x$. *Adv. Mater.* **2013**, *25*, 6607–6612. [CrossRef]
17. Wu, B.; Zhou, Y.; Hu, M. Two-Channel Thermal Transport in Ordered-Disordered Superionic Ag_2Te and Its Traditionally Contradictory Enhancement by Nanotwin Boundary. *J. Phys. Chem. Lett.* **2018**, *9*, 5704–5709. [CrossRef]
18. Luo, Y.; Yang, X.; Feng, T.; Wang, J.; Ruan, X. Vibrational hierarchy leads to dual-phonon transport in low thermal conductivity crystals. *Nat. Commun.* **2020**, *11*, 2554. [CrossRef]
19. Ravindran, P.; Delin, A.; Ahuja, R.; Johansson, B.; Auluck, S.; Wills, J.M.; Eriksson, O. Optical properties of monoclinic SnI_2 from relativistic first-principles theory. *Phys. Rev. B* **1997**, *56*, 6851–6861. [CrossRef]
20. Zhong, M.; Zhang, S.; Huang, L.; You, J.; Wei, Z.; Liu, X.; Li, J. Large-scale 2D PbI_2 monolayers: Experimental realization and their indirect band-gap related properties. *Nanoscale* **2017**, *9*, 3736–3741. [CrossRef]
21. Yagmurcukardes, M.; Peeters, F.M.; Sahin, H. Electronic and vibrational properties of PbI_2: From bulk to monolayer. *Phys. Rev. B* **2018**, *98*, 085431. [CrossRef]
22. Zhang, Y.M.; Hou, B.; Wu, Y.; Chen, Y.; Xia, Y.-J.; Mei, H.; Kong, M.R.; Peng, L.; Shao, H.Z.; Cao, J.; et al. Towards high-temperature electron-hole condensate phase in monolayer tetrels metal halides: Ultra-long excitonic lifetimes, phase diagram and exciton dynamics. *Mater. Today Phys.* **2022**, *22*, 100604. [CrossRef]
23. Sandoval, S.; Kepic, D.; Perez Del Pino, A.; Gyorgy, E.; Gomez, A.; Pfannmoeller, M.; Tendeloo, G.V.; Ballesteros, B.; Tobias, G. Selective Laser-Assisted Synthesis of Tubular van der Waals Heterostructures of Single-Layered PbI_2 within Carbon Nanotubes Exhibiting Carrier Photogeneration. *ACS Nano* **2018**, *12*, 6648–6656. [CrossRef]
24. Peng, B.; Mei, H.D.; Zhang, H.; Shao, H.Z.; Xu, K.; Ni, G.; Jin, Q.Y.; Soukoulis, C.M.; Zhu, H.Y. High thermoelectric efficiency in monolayer PbI_2 from 300 K to 900 K. *Inorg. Chem. Front.* **2019**, *6*, 920–928. [CrossRef]
25. Fan, Q.; Huang, J.W.; Dong, N.N.; Hong, S.; Yan, C.; Liu, Y.C.; Qiu, J.S.; Wang, J.; Sun, Z.Y. Liquid Exfoliation of Two-Dimensional PbI_2 Nanosheets for Ultrafast Photonics. *ACS Photonics* **2019**, *6*, 1051–1057. [CrossRef]
26. Howie, R.A.; Moser, W.; Trevena, I.C. The crystal structure of tin(II) iodide. *Acta Crystallogr. B Struct. Sci. Cryst. Eng. Mater.* **1972**, *28*, 2965–2971. [CrossRef]
27. Yuan, Q.Q.; Zheng, F.; Shi, Z.Q.; Li, Q.Y.; Lv, Y.Y.; Chen, Y.; Zhang, P.; Li, S.C. Direct Growth of van der Waals Tin Diiodide Monolayers. *Adv. Sci.* **2021**, *8*, 2100009. [CrossRef]
28. Arab, A.; Li, Q. Anisotropic thermoelectric behavior in armchair and zigzag mono- and fewlayer MoS_2 in thermoelectric generator applications. *Sci. Rep.* **2015**, *5*, 13706. [CrossRef]
29. Kresse, G.; Furthmuller, J. Efficient iterative schemes for ab initio total-energy calculations using a plane-wave basis set. *Phys. Rev. B Condens. Matter.* **1996**, *54*, 11169–11186. [CrossRef]
30. Blochl, P.E. Projector augmented-wave method. *Phys. Rev. B Condens. Matter* **1994**, *50*, 17953–17979. [CrossRef]
31. Blochl, P.E.; Jepsen, O.; Andersen, O.K. Improved tetrahedron method for Brillouin-zone integrations. *Phys. Rev. B Condens. Matter.* **1994**, *49*, 16223–16233. [CrossRef] [PubMed]
32. Kresse, G.; Joubert, D. From ultrasoft pseudopotentials to the projector augmented-wave method. *Phys. Rev. B* **1999**, *59*, 1758–1775. [CrossRef]
33. Heyd, J.; Scuseria, G.E. Assessment and validation of a screened Coulomb hybrid density functional. *J. Chem. Phys.* **2004**, *120*, 7274–7280. [CrossRef] [PubMed]
34. Madsen, G.K.H.; Singh, D.J. BoltzTraP. A code for calculating band-structure dependent quantities. *Comput. Phys. Commun.* **2006**, *175*, 67–71. [CrossRef]
35. Bardeen, J.; Shockley, W. Deformation Potentials and Mobilities in Non-Polar Crystals. *Phys. Rev.* **1950**, *80*, 72–80. [CrossRef]
36. Lee, M.S.; Mahanti, S.D. Validity of the rigid band approximation in the study of the thermopower of narrow band gap semiconductors. *Phys. Rev. B* **2012**, *85*, 165149. [CrossRef]
37. Cai, Y.; Zhang, G.; Zhang, Y.W. Polarity-reversed robust carrier mobility in monolayer MoS_2 nanoribbons. *J. Am. Chem. Soc.* **2014**, *136*, 6269–6275. [CrossRef]
38. Zhu, X.L.; Liu, P.F.; Zhang, J.; Zhang, P.; Zhou, W.X.; Xie, G.; Wang, B.-T. Monolayer SnP_3: An excellent p-type thermoelectric material. *Nanoscale* **2019**, *11*, 19923–19932. [CrossRef]
39. Baroni, S.; de Gironcoli, S.; Dal Corso, A.; Giannozzi, P. Phonons and related crystal properties from density-functional perturbation theory. *Rev. Mod. Phys.* **2001**, *73*, 515–562. [CrossRef]
40. Togo, A.; Oba, F.; Tanaka, I. First-principles calculations of the ferroelastic transition between rutile-type and $CaCl_2$-type SiO_2 at high pressures. *Phys. Rev. B* **2008**, *78*, 134106. [CrossRef]

41. Li, W.; Carrete, J.; Katcho, N.A.; Mingo, N. ShengBTE: A solver of the Boltzmann transport equation for phonons. *Comput. Phys. Commun.* **2014**, *185*, 1747–1758. [CrossRef]
42. Tong, Z.; Li, S.; Ruan, X.; Bao, H. Comprehensive first-principles analysis of phonon thermal conductivity and electron-phonon coupling in different metals. *Phys. Rev. B* **2019**, *100*, 144306. [CrossRef]
43. Li, S.; Wang, A.; Hu, Y.; Gu, X.; Tong, Z.; Bao, H. Anomalous thermal transport in metallic transition-metal nitrides originated from strong electron–phonon interactions. *Mater. Today Phys.* **2020**, *15*, 100256. [CrossRef]
44. Wang, Y.X.; Xu, N.; Li, D.Y.; Zhu, J. Thermal Properties of Two Dimensional Layered Materials. *Adv. Funct. Mater.* **2017**, *27*, 1604134. [CrossRef]
45. Peng, B.; Zhang, H.; Shao, H.; Xu, Y.; Zhang, R.; Lu, H.; Zhang, D.W.; Zhu, H. First-Principles Prediction of Ultralow Lattice Thermal Conductivity of Dumbbell Silicene: A Comparison with Low-Buckled Silicene. *ACS Appl. Mater. Interfaces* **2016**, *8*, 20977–20985. [CrossRef]
46. Onodera, T.; Morita, Y.; Nagumo, R.; Miura, R.; Suzuki, A.; Tsuboi, H.; Hatakeyama, N.; Endou, A.; Takaba, H.; Dassenoy, F.; et al. A computational chemistry study on friction of h-MoS(2). Part II. Friction anisotropy. *J. Phys. Chem. B* **2010**, *114*, 15832–15838. [CrossRef]
47. Zhao, L.D.; Zhang, X.; Wu, H.; Tan, G.; Pei, Y.; Xiao, Y.; Chang, C.; Wu, D.; Chi, H.; Zheng, L.; et al. Enhanced Thermoelectric Properties in the Counter-Doped SnTe System with Strained Endotaxial SrTe. *J. Am. Chem. Soc.* **2016**, *138*, 2366–2373. [CrossRef]
48. Wang, D.Y.; He, W.K.; Chang, C.; Wang, G.T.; Wang, J.F.; Zhao, L.D. Thermoelectric transport properties of rock-salt SnSe: First-principles investigation. *J. Mater. Chem. C* **2018**, *6*, 12016–12022. [CrossRef]
49. Huang, S.; Wang, Z.Y.; Xiong, R.; Yu, H.Y.; Shi, J. Significant enhancement in thermoelectric performance of Mg_3Sb_2 from bulk to two-dimensional mono layer. *Nano Energy* **2019**, *62*, 212–219. [CrossRef]
50. Wang, N.; Li, M.; Xiao, H.; Zu, X.; Qiao, L. Layered LaCuOSe: A Promising Anisotropic Thermoelectric Material. *Phys. Rev. Appl.* **2020**, *13*, 024038. [CrossRef]
51. Tong, Z.; Dumitrica, T.; Frauenheim, T. Ultralow Thermal Conductivity in Two-Dimensional MoO_3. *Nano Lett.* **2021**, *21*, 4351–4356. [CrossRef] [PubMed]
52. Zhu, X.-L.; Liu, P.-F.; Wu, Y.-Y.; Zhang, P.; Xie, G.; Wang, B.-T. Significant enhancement of the thermoelectric properties of CaP_3 through reducing the dimensionality. *Adv. Mater.* **2020**, *1*, 3322–3332. [CrossRef]
53. Liu, D.; Every, A.G.; Tománek, D. Continuum approach for long-wavelength acoustic phonons in quasi-two-dimensional structures. *Phys. Rev. B* **2016**, *94*, 165432. [CrossRef]
54. Lindsay, L.; Li, W.; Carrete, J.; Mingo, N.; Broido, D.A.; Reinecke, T.L. Phonon thermal transport in strained and unstrained graphene from first principles. *Phys. Rev. B* **2014**, *89*, 155426. [CrossRef]
55. Wang, H.; Zhou, E.; Duan, F.; Wei, D.; Zheng, X.; Tang, C.; Ouyang, T.; Yao, Y.; Qin, G.; Zhong, J. Unique Arrangement of Atoms Leads to Low Thermal Conductivity: A Comparative Study of Monolayer Mg_2C. *J. Phys. Chem. Lett.* **2021**, *12*, 10353–10358. [CrossRef]
56. Hippalgaonkar, K.; Wang, Y.; Ye, Y.; Qiu, D.Y.; Zhu, H.; Wang, Y.; Moore, J.; Louie, S.G.; Zhang, X. High thermoelectric power factor in two-dimensional crystals of MoS_2. *Phys. Rev. B* **2017**, *95*, 115407. [CrossRef]
57. Lin, H.; Tan, G.; Shen, J.N.; Hao, S.; Wu, L.M.; Calta, N.; Malliakas, C.; Wang, S.; Uher, C.; Wolverton, C.; et al. Concerted Rattling in $CsAg_5Te_3$ Leading to Ultralow Thermal Conductivity and High Thermoelectric Performance. *Angew. Chem. Int. Ed. Engl.* **2016**, *55*, 11431–11436. [CrossRef]
58. Ma, N.; Li, Y.Y.; Chen, L.; Wu, L.M. alpha-$CsCu_5Se_3$: Discovery of a Low-Cost Bulk Selenide with High Thermoelectric Performance. *J. Am. Chem. Soc.* **2020**, *142*, 5293–5303. [CrossRef]
59. Gibson, Q.D.; Zhao, T.; Daniels, L.M.; Walker, H.C.; Daou, R.; Hebert, S.; Zanella, M.; Dyer, M.S.; Claridge, J.B.; Slater, B.; et al. Low thermal conductivity in a modular inorganic material with bonding anisotropy and mismatch. *Science* **2021**, *373*, 1017–1022. [CrossRef]

MDPI AG
Grosspeteranlage 5
4052 Basel
Switzerland
Tel.: +41 61 683 77 34

Materials Editorial Office
E-mail: materials@mdpi.com
www.mdpi.com/journal/materials

Disclaimer/Publisher's Note: The title and front matter of this reprint are at the discretion of the . The publisher is not responsible for their content or any associated concerns. The statements, opinions and data contained in all individual articles are solely those of the individual Editors and contributors and not of MDPI. MDPI disclaims responsibility for any injury to people or property resulting from any ideas, methods, instructions or products referred to in the content.

www.ingramcontent.com/pod-product-compliance
Lightning Source LLC
LaVergne TN
LVHW070643100526
838202LV00013B/868